Composites are used extensively in engineering applications. A constant concern is the effect of foreign object impacts on composite structures because significant damage can occur and yet be undetectable by visual inspection. Such impacts can range from the most ordinary at low velocity – a tool dropped on a product – to the hypervelocity impact of space debris on a spacecraft.

This book brings together the latest developments in this important new research area. It explains how damage develops during impact, the effect of impact-induced damage on the mechanical behavior of structures, and methods of damage prediction and detection. Numerous examples are included to illustrate these topics.

Written for graduate students, as well as for researchers and practicing engineers working with composite materials, this book presents state-of-the-art knowledge on impact dynamics, yet requires only basic understanding of the mechanics of composite materials.

IMPACT ON COMPOSITE STRUCTURES

IMPACT ON COMPOSITE STRUCTURES

SERGE ABRATE
Southern Illinois University at Carbondale

CAMBRIDGE
UNIVERSITY PRESS

CAMBRIDGE UNIVERSITY PRESS
Cambridge, New York, Melbourne, Madrid, Cape Town, Singapore, São Paulo

Cambridge University Press
The Edinburgh Building, Cambridge CB2 2RU, UK

Published in the United States of America by Cambridge University Press, New York

www.cambridge.org
Information on this title: www.cambridge.org/9780521473897

First published 1998
This digitally printed first paperback version 2005

A catalogue record for this publication is available from the British Library

Library of Congress Cataloguing in Publication data
Abrate, Serge, 1953–
Impact on composite structures / Serge Abrate.
p. cm.
Includes bibliographical references and index.
ISBN 0-521-47389-6 (hc)
1. Composite materials – Impact testing. 2. Composite
materials – Mechanical properties. I. Title.
TA418.9.C6A25 1998
624.1′89 – dc21 97-16551
CIP

ISBN-13 978-0-521-47389-7 hardback
ISBN-10 0-521-47389-6 hardback

ISBN-13 978-0-521-01832-6 paperback
ISBN-10 0-521-01832-3 paperback

Contents

Preface

As composite materials are used more extensively, a constant source of concern is the effect of foreign objects impacts. Such impacts can reasonably be expected during the life of the structure and can result in internal damage that is often difficult to detect and can cause severe reductions in the strength and stability of the structure. This concern provided the motivation for intense research resulting in hundreds of journal and conference articles. Important advances have been made, and many aspects of the problem have been investigated. One would need to study a voluminous literature in order to get an appreciation for this new research area. After writing three comprehensive literature reviews on the topic of impact on composite materials, I felt that there was a need to present this material in book form. The study of impact on composite structures involves many different topics, including contact mechanics, structural dynamics, strength, stability, fatigue, damage mechanics, and micromechanics. Impacts are simple events with many complicated effects, and what appears as a logical conclusion in one situation seems to be completely reversed in another. This variety and complexity have kept me interested in this area for several years, and I hope to communicate that to the readers.

This book attempts to present this new body of knowledge in a unified, detailed, and comprehensive manner. It assumes only a basic knowledge of solid mechanics and the mechanics of composite materials and can be used as a text for a graduate-level course or for self-study. Exercise problems are proposed in order to challenge the user's understanding of the material. It is hoped that, after each section, the reader will gain an appreciation of the problems being addressed and the methods available and will be able to implement some of these methods. For example, after Chapter 3, "Impact Dynamics," the reader should understand what is involved in the selection of a mathematical model for predicting the contact force history and should be able to implement what appears to be the logical choice for the particular problem being considered.

Researchers and practitioners in industry would appreciate this comprehensive presentation of the state of the art in this area and use this book as a reference. A long list of references is included for those interested in more details.

Current knowledge in certain areas has reached a certain level of maturity. For example, there is a general agreement on how to model the local indentation of a composite plate by smooth indentors. In other areas, current methods are not completely satisfactory, particularly in the areas of damage prediction and the prediction of the residual properties of impact-damaged composites. The book covers both low-velocity and ballistic impacts. Hypervelocity impacts encountered when spacecraft are hit by meteorites or space debris are in a different class, where impact velocities are measured in km/s. Too few studies dealing with this type of impacts on composite structures can be found, so this aspect of the problem was omitted. Recently, a new area of research dealing with smart materials and smart structures has developed, and one of the objectives has been to use distributed sensors and actuators to monitor the health of the structure. Several attempts were made to determine if and where an impact occurred or to determine the location and size of impact damage. This type of study is beyond the scope of this book, because it would require discussing the modeling of such sensors and actuators, control methods, and identification procedures. However, the methods for modeling the impact dynamics and for modeling the behavior of damaged structures discussed in this book will be very useful for modeling such smart structures.

Upon reading this book, it will be obvious that I am indebted to the numerous investigators who have contributed to the development of this body of knowledge. I have made an effort to include every significant development and to acknowledge the source. A book is usually a reflection of the author's choices regarding what needed to be included and what could be omitted. A balance must be struck between the desire to be clear and comprehensive and the risk of overburdening the reader. In this case, almost every chapter could have been expanded to full book length, making it sometimes difficult to choose. However, I hope readers will gain a sufficient understanding of each aspect of the problem that they will be able to pursue its study on their own. The extensive list of references and the several available literature surveys should be helpful in this regard.

Finally, I would like to thank my editor, Florence Padgett, for being so supportive of me in this endeavor. My wife Jayne and my sons Denis and Marc deserve thanks for allowing me to devote much of my free time to this project. Special thanks to my parents, who encouraged me to pursue my dreams and helped me get the best education. I wish my father were still with us to see this book completed. He would have been very proud.

1
Introduction

During the life of a structure, impacts by foreign objects can be expected to occur during manufacturing, service, and maintenance operations. An example of in-service impact occurs during aircraft takeoffs and landings, when stones and other small debris from the runway are propelled at high velocities by the tires. During the manufacturing process or during maintenance, tools can be dropped on the structure. In this case, impact velocities are small but the mass of the projectile is larger. Laminated composite structures are more susceptible to impact damage than a similar metallic structure. In composite structures, impacts create internal damage that often cannot be detected by visual inspection. This internal damage can cause severe reductions in strength and can grow under load. Therefore, the effects of foreign object impacts on composite structures must be understood, and proper measures should be taken in the design process to account for these expected events. Concerns about the effect of impacts on the performance of composite structures have been a factor in limiting the use of composite materials. For these reasons, the problem of impact has received considerable attention in the literature. The objective with this book is to present a comprehensive view of current knowledge on this very important topic.

A first step in gaining some understanding of the problem is to develop mathematical models for predicting the force applied by the projectile on the structure during impact. In order to predict this contact force history, the model should account for the motion of the structure, the motion of the projectile, and the local deformations in the contact zone. A detailed description of the contact between the impactor and the structure during impact would be difficult to obtain and is not required as part of the impact dynamics analysis. Instead, what is needed is a contact law relating the contact force to the indentation, which is defined as the motion of the projectile relative to that of the target. With the material systems commonly used, strain rate effects are negligible such that static and dynamic contact laws are identical, and so statically determined contact laws can be used in the dynamics analysis. Many investigators analyzed the loading phase of the indentation process and predicted the contact law for beams and plates with various boundary conditions and several indentor shapes. The indentation process introduces damage and permanent deformations in the contact zone; as a result, the unloading curve differs from the loading curve. In many cases, multiple impacts occur and the reloading curve is again different. While some simple analyses of the unloading and reloading phases have been presented, a complete and accurate characterization of the contact law for a particular target–indentor combination can only be obtained experimentally. A set of equations are commonly used to describe the contact behavior between solid laminates and smooth indentors, with a small number of constants to be determined from experiments. Contact problems are discussed in Chapter 2.

An important aspect of a model for predicting the impact dynamics is to accurately describe the dynamic behavior of the target. For some simple cases, the motion of the structure can be modeled by a simple spring–mass system or by assuming a quasi-static behavior and considering the balance of energy in the system. But in general, more sophisticated beam, plate, or shell theories are required to model the structure. Two- and three-dimensional elasticity models are also employed. For each particular case, a choice must be made between a number of theories developed to account for complicating factors such as transverse shear deformation and rotary inertia. In addition, one must select to use variational approximation methods or finite element methods if exact solutions are not available for the case at hand. In Chapter 3, several structural theories are presented and their accuracies are discussed, particularly in regard to the impact problem. Several examples of impact dynamics analyses are presented for impacts on beams, plates, and shells. Often tests cannot be performed on full scale prototypes, so small specimens must be used. In other cases, tests or analyses have been performed on two specimens of different sizes. Methods developed to scale results obtained from a small-scale model to a full-size prototype are discussed in Chapter 3.

Different test apparatus are used to simulate various types of impact. Drop-weight testers are used to simulate low-velocity impacts typical of the tool-drop problem, in which a large object falls onto the structure with low velocity. Air-gun systems, in which a small projectile is propelled at high speeds, are used to simulate the type of impacts encountered during aircraft takeoffs and landings. Testing conditions should replicate the actual impacts the structure is designed to withstand, and so the choice of the proper test apparatus is important. Descriptions of several common types of impact testers are given in Chapter 4, where damage development, the morphology of impact damage, and experimental techniques for damage detection are also described. Impact damage consists of delaminations, matrix cracking, and fiber failure. Extensive experimental investigations have revealed definite patterns for the shape of the damage and its initiation and growth. It is generally accepted that during low-velocity impacts, damage is initiated by matrix cracks which create delaminations at interfaces between plies with different fiber orientations. For stiff structures, matrix cracks start on the impacted face of the specimen due to high contact stresses. Damage propagates downward by a succession of intra-ply cracks and interface delaminations to give what is called a pine tree pattern. For thin specimens, bending stresses cause matrix cracking in the lowest ply, and damage progresses from the nonimpacted face up toward the projectile giving a reverse pine tree appearance. Many studies demonstrated that no delaminations are induced at the interface between plies with the same fiber orientation. The delaminated area at an interface depends on the fiber orientations in the plies adjacent to that interface. Of the large number of experimental methods that have been developed for damage assessment, some result in the complete destruction of the specimen and others are nondestructive, but most methods are used to determine the state of damage after impact. A few methods can be used to monitor the development of damage during impact. Once the failure modes involved are known and methods to assess damage are available, the next task is to determine what the governing parameters are. The properties of the matrix, the reinforcing fibers, and the fiber–matrix interfaces affect impact resistance in several ways, along with effects of layup, thickness, size, boundary conditions, and the shape, mass, and velocity of the projectile. Techniques for improving both impact resistance and impact tolerance will also be discussed.

The prediction of impact damage is a difficult task for which complete success may not be possible and is probably not necessary. A large number of matrix cracks and delaminations are created in the impact zone, and the introduction of each new crack creates a redistribution of stresses. Thus it is not realistic to attempt to track every detail of damage development during the dynamic analysis of the impact event. Some purely qualitative models explain the orientation and size of delaminations at the various interfaces through the thickness of the laminate. Other models predict the onset of delaminations by predicting the appearance of the first matrix crack. Models for estimating the final state of delaminations are available. Chapter 5 deals with models for predicting impact damage.

Impact-induced delaminations result in drastic reductions in strength, stiffness, and stability of the laminate, which explains why impact is of such concern to the designers and users of composite structures and why such effort has been directed at understanding the problem. Reductions are particularly significant in compression, where the effect of delaminations is more readily visible. The extent of these reductions has been characterized experimentally, and several mathematical models are used to predict the residual strength, as discussed in Chapter 6.

Chapter 7 deals with high-velocity impacts. Sometimes the expression "high-velocity impact" refers to impacts resulting in complete perforation of the target. However, it must be recognized that complete perforation can be achieved both under low-velocity impacts when overall deflections of the target are taking place and under high-velocity impacts when the deformations of the target are localized in a small region near the point of impact. Here, high-velocity impacts are defined so that the ratio between impact velocity and the velocity of compressive waves propagating through the thickness is larger than the maximum strain to failure in that direction. This implies that damage is introduced during the first few travels of the compressive wave through the thickness when overall plate motion is not yet established. The ballistic limit is defined as the initial velocity of the projectile that results in complete penetration of the target with zero residual velocity. For impact velocities above the ballistic limit, the residual velocity of the projectile is also of interest. General trends for ballistic impacts (including the effect of several important parameters) are known from experimental results, and models for predicting the ballistic limit are available. Another area to be discussed in this chapter is the design of composite armor, which usually involves using one layer of ceramic material to blunt or fragment the projectile and another layer of composite material to absorb the residual kinetic energy.

Many repair procedures for composite structures with impact damage have been reported. In selecting a particular repair technique, several factors must be considered, including the severity of the damage, the loads to be sustained, and the equipment available at the location where the repair procedure will be performed. In some cases, a successful repair cannot be made and the part must be replaced. In other cases, only cosmetic repairs are needed. In yet other cases, either bolted repairs or bonded repairs are needed. Proper design of the repair patch, machining, surface preparation, prefabrication of patches, and bonding are important phases of a successful repair procedure. The basic problems and most promising approaches are reported in Chapter 8.

In many applications, sandwich structures with laminated facings are used because of the well-known advantages of this type of construction. The facings are loaded primarily in tension or compression to resist bending while the core resists the shear stresses. As a result, sandwich structures achieve a great bending rigidity, but in the transverse direction the rigidity remains low. A low rigidity in the transverse direction results in a low contact stiffness and lower contact forces during impact. Contact laws for sandwich structures are completely different from those of monolithic laminates and are dominated by the deformation of the core. Usually cores are made out of honeycomb materials, foam, or balsa wood. All these materials are considered to be part of the family of cellular materials and have a similar stress–strain behavior when loaded in compression in the transverse direction. The choice of a particular material is based on considerations of performance, weight, and cost. Models for predicting the behavior of a sandwich panel under a smooth indentor are discussed in Chapter 9. Impact creates damage to the facings, the core, and the core-facing interfaces. The evolution of damage with impact energy and its effects on the residual properties of sandwich panels are also discussed in that chapter.

2
Contact Laws

2.1 Introduction

Local deformations in the contact region must be accounted for in the analysis in order to accurately predict the contact force history. The indentation, defined as the difference between the displacement of the projectile and that of the back face of the laminate, can be of the same order as or larger than the overall displacement of the laminate. One could consider the projectile and the structure as two solids in contact and then analyze the impact problem as a dynamic contact problem. However, this approach is computationally expensive and cannot describe the effect of permanent deformation and local damage on the unloading process. The unloading part of the indentation process can be modeled only using experimentally determined contact laws. To predict the contact force history and the overall deformation of the target, a detailed model of the contact region is not necessary. A simple relationship between the contact force and the indentation, called the *contact law*, has been used by Timoshenko (1913) to study the impact of a beam by a steel sphere. This approach has been used extensively since then and is commonly used for the analysis of impact on composite materials.

Although the impact event is a highly dynamic event in which many vibration modes of the target are excited, statically determined contact laws can be used in the impact dynamics analysis of low-velocity impacts because strain rate and wave propagation effects are negligible with commonly used material systems. Hunter (1957) calculated the energy lost by elastic waves during the impact of a sphere on an elastic half-space. The ratio between the energy lost in the half-space and the initial kinetic energy of the sphere is given by

$$1.04\left(\frac{V}{(E/\rho)^{\frac{1}{2}}}\right)^{\frac{3}{5}}. \tag{2.1}$$

Therefore, the energy loss is negligible as long as the initial velocity of the sphere, V, is small compared with $c = (E/\rho)^{1/2}$, the phase velocity of compressive elastic waves in the solid. In many cases, the contact area is small and the behavior is similar to that observed during the indentation of a half-space. The indentation is independent of the overall deflection of the structure. When the contact area becomes large, the overall deflection of the structure affects the pressure distribution under the indentor; this interaction must be accounted for accurately.

This chapter presents the basic results from contact mechanics needed to model the indentation of composite materials during the impact by a foreign object. The objective of such models is the accurate prediction of the contact force history and the overall response of the structure.

2.2 Contact between Two Isotropic Elastic Solids

Detailed accounts of the study of contact between two smooth elastic solid pioneered by Hertz are given in several books (Barber 1992, Gladwell 1980, Johnson 1985, Timoshenko and Goodier 1970). Essential results from Hertz theory of contact as given here without derivation. For two isotropic bodies of revolution (Fig. 2.1), contact occurs in a circular zone of radius a in which the normal pressure p varies as

$$p = p_0 \left[1 - \left(\frac{r}{a}\right)^2\right]^{\frac{1}{2}} \tag{2.2}$$

where p_0 is the maximum contact pressure at the center of the contact zone and r is the radial position of an arbitrary point in the contact zone (Fig. 2.2). Defining the parameters R and E as

$$\frac{1}{R} = \frac{1}{R_1} + \frac{1}{R_2}, \tag{2.3}$$

$$\frac{1}{E} = \frac{1 - \nu_1^2}{E_1} + \frac{1^2 - \nu_2^2}{E_2} \tag{2.4}$$

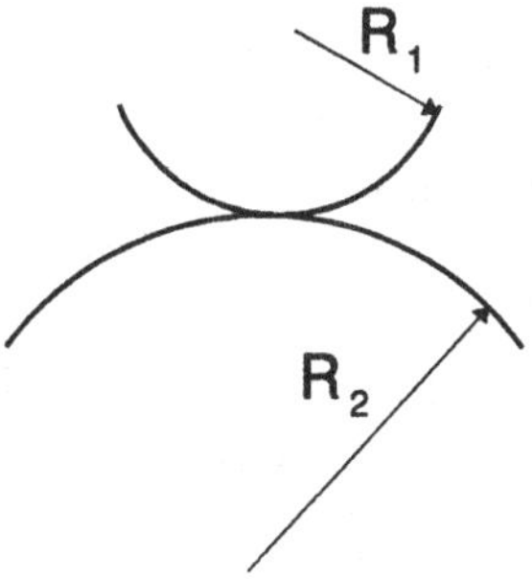

Figure 2.1. Two bodies of revolution for Hertzian analysis of contact.

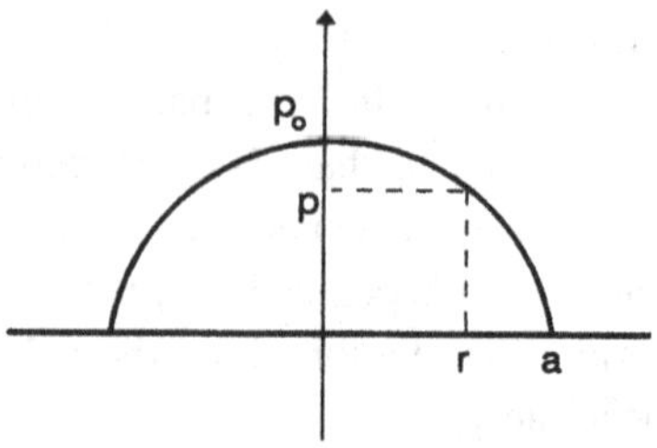

Figure 2.2. Pressure distribution in the contact zone.

where R_1 and R_2 are the radii of curvature of the two bodies. The Young moduli and Poisson's ratios of the two bodies are E_1, ν_1, and E_2, ν_2, respectively. Without loss of generality, the subscript 1 is taken to denote properties of the indentor and subscript 2 identifies properties of the target. The radius of the contact zone a, the relative displacement α, and the maximum contact pressure are given by

$$a = \left(3\frac{PR}{4E}\right)^{\frac{1}{3}} \tag{2.5}$$

$$\alpha = \frac{a^3}{R} = \left(\frac{9P^2}{16RE^2}\right)^{\frac{1}{3}} \tag{2.6}$$

$$p_0 = \frac{3P}{2\pi a^2} = \left(\frac{6PE^2}{\pi^3 R^2}\right)^{\frac{1}{3}}. \tag{2.7}$$

The force indentation law is expressed as

$$P = k\alpha^{\frac{3}{2}} \tag{2.8}$$

where P is the contact force, α is the indentation, and the contact stiffness k is given by

$$k = \frac{4}{3}ER^{\frac{1}{2}}. \tag{2.9}$$

Equation (2.8) is usually referred to as *Hertz contact law* or the *Hertzian law of contact* and is found to apply for a wide range of cases, even if all the assumptions made in the derivation of the theory are not satisfied. For example, (2.8) also applies for laminated composites, even though these are not homogeneous and isotropic materials. With composite materials, permanent deformations are introduced in the contact zone for relatively low force levels, but the material is assumed to remain linear elastic in the analysis. Matrix cracks, fiber fracture, and interply delaminations can be introduced in the contact zone, but overall the contact law follows (2.8).

The experimental determination of the contact law usually involves the use of a displacement sensor which directly measures the displacement of the indentor relative to the back face of the specimen. The plot of the applied force versus the indentation measured in this way follows the general form given by (2.8). Least-squares fits of the experimental data usually show that the exponent of α in that equation is close to $\frac{3}{2}$, and k is close to the value given by (2.9). Some investigators present plots of the contact force versus the displacement of the indentor that show a significant discontinuity when delaminations are introduced. The displacement of the indentor is the sum of the indentation and the deflection of the structure. The observed discontinuity corresponds to the reduction in the apparent stiffness of the structure caused by the introduction of delaminations.

2.3 Indentation of Beams

During the indentation of beams, local deformations are superposed onto the overall deflection of the beam. When the beam deflections are small, the interaction between the local indentation and the overall deflection of the beam is negligible. Therefore, the indentation law and the pressure distribution in the contact zone are the same as for the indentation of a half-space. However, for flexible beams under large indentors, the beam may wrap around the indentor, leading to significant changes in the normal pressure distribution.

First, consider the bending of a Bernoulli-Euler beam and neglect local deformations. This simple analysis will provide some understanding of the physical behavior of a beam loaded by a large indentor. From the free body diagram of an infinitesimal beam element (Fig. 2.3), the equilibrium equations for a planar beam are

$$\frac{dV}{dx} + p = 0, \qquad \frac{dM}{dx} + V = 0, \tag{2.10}$$

where V is the shear force, M is the bending moment, and p is the applied force per unit length. With the Bernoulli-Euler beam theory, plane sections are assumed to remain planar and perpendicular to the neutral axis of the beam.

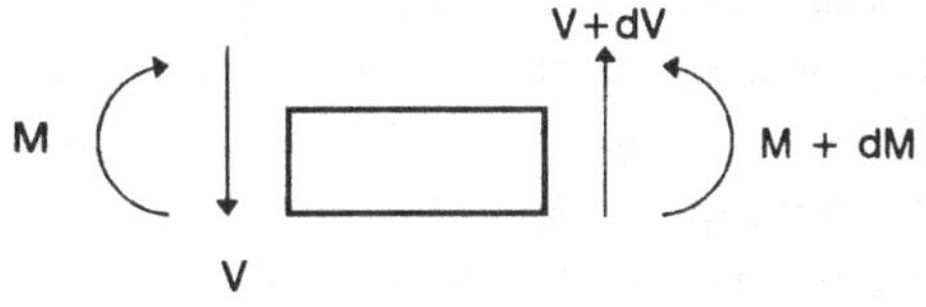

Figure 2.3. Free body diagram of beam element.

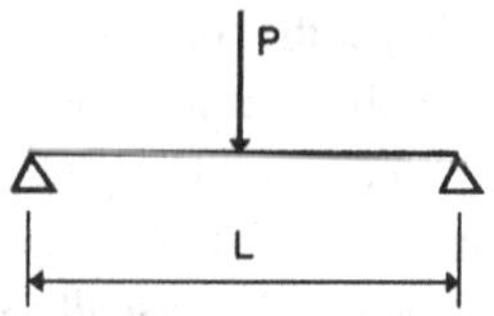

Figure 2.4. Simply supported beam with concentrated force applied in center.

The axial displacements are then related to the transverse displacement by

$$u(x,z) = -z\frac{dw(x)}{dx}, \tag{2.11}$$

and the axial strain is

$$\epsilon_{xx} = -z\frac{d^2w}{dx^2}. \tag{2.12}$$

The bending moment is related to the transverse displacement by

$$\begin{aligned} M &= -\int_A \sigma_{xx} z dA = -\int_A E\epsilon_{xx} z dA \\ &= E\int_A z^2 dA \frac{d^2w}{dx^2} = EI\frac{d^2w}{dx^2}. \end{aligned} \tag{2.13}$$

The bending moment distribution in a simply supported beam of length L subjected to a concentrated force applied in the center (Fig. 2.4) is given by

$$M = \frac{P}{2}\left(\frac{L}{2} - x\right) \tag{2.14}$$

for $x > 0$. The curvature of the beam is given by the second derivative of the deflection. Using (2.13) and (2.14), the load required to bend the beam to a radius R equal to that of the indentor is found to be

$$P = 4\frac{EI}{RL}, \tag{2.15}$$

which indicates that wrapping is more likely to occur for long flexible beams under a large impactor.

As the load increases, the beam comes in contact with the indentor over a distance a (Fig. 2.5). In that region, the radius of curvature is constant and equal to the radius of the impactor. Equation (2.13) implies that the bending moment is also constant, and (2.10) indicates that the shear force is uniformly equal to zero and that the pressure under the impactor is also zero. The beam is then loaded by two concentrated forces of magnitude $P/2$ located at $+a$ and $-a$. The distance a is such that the curvature of the beam at $x = a$ is equal to the

Table 2.1. *Analytical studies of contact between a beam and a rigid indentor*

Boundary conditions	Material	Indentor shape	Reference
Simply supported	Isotropic	Cylindrical	Keer and Miller (1983b) Sankar and Sun (1983) Shankar (1987)
	Orthotropic	Cylindrical	Sankar (1989a,b) Keer and Ballarini (1983a) Sun and Sankar (1985)
Clamped-clamped	Isotropic	Cylindrical	Keer and Miller (1983b) Sankar and Sun (1983)
	Orthotropic	Cylindrical	Keer and Ballarini (1983a)
Cantilever	Isotropic	Cylindrical	Keer and Silva (1970) Keer and Schonberg (1986a)
		Flat	Keer and Schonberg (1986a)
	Orthotropic	Cylindrical	Keer and Schonberg (1986b) Sankar (1989a)
		Flat	Keer and Schonberg (1986b)

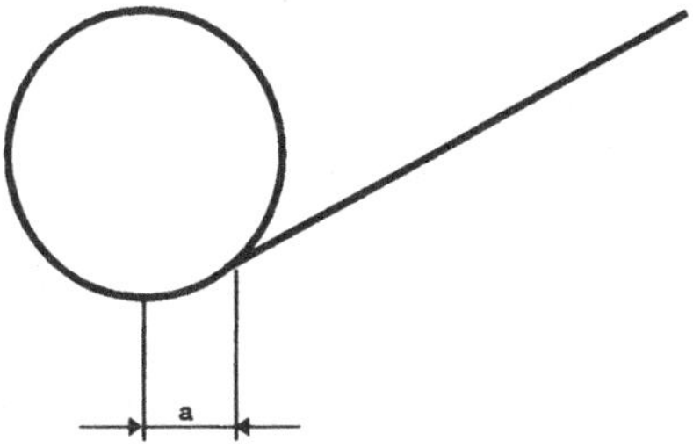

Figure 2.5. Beam wrapping around the indentor.

curvature of the indentor. For $a < x < L/2$, the bending moment is given by (2.14), and using (2.13) we find

$$a = \frac{L}{2} - 2\frac{EI}{PR} \tag{2.16}$$

for values of P larger than that given by (2.15). The conclusion from this simple analysis is that, neglecting local deformations, as a concentrated contact force would produce radius of curvature smaller than the radius of the indentor, the beam is loaded by two forces separated by a distance $2a$. When local deformations are included, the pressure distribution is expected to be such that the maximum pressure is reached close to the edge of the contact zone.

Beam theory does not predict those local deformations, and various approaches have been used to address the problem (Table 2.1). These investigators used various methods to study the loading portion of the contact laws for either

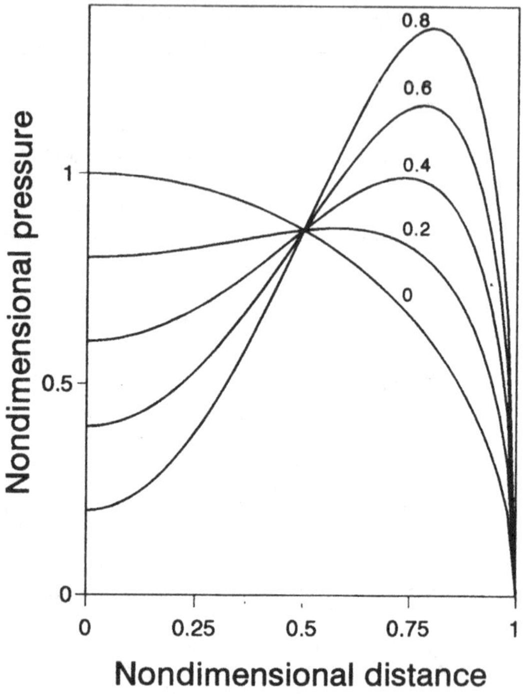

Figure 2.6. Pressure distribution under a cylindrical indentor in contact with a simply supported beam ($\beta = 0, 0.2, 0.4, 0.6, 0.8$).

isotropic or orthotropic beams. Of particular interest in these studies is the pressure distribution in the contact zone. Sankar (1989a) gave results for simply supported beams of length l centrally loaded by a rigid cylindrical indentor of radius R. The nondimensional contact pressure is given by

$$\bar{p} = (1 - \bar{x}^2)^{\frac{1}{2}}[1 - \beta(1 - 4\bar{x}^2)] \tag{2.17}$$

where $\bar{x} = x/c$, with c being the radius of the contact zone. The nondimensional contact pressure is related to the actual pressure by

$$\bar{p} = \frac{\pi bc}{2P} p(x), \tag{2.18}$$

with $\bar{c} = c/l$, $\beta = 8.75B\bar{c}^4$, $B = \frac{\pi}{32}\frac{D_2}{D_1}(l/h)^3$ and $D_1 = E_1/(1 - \nu_{12}^2)$, $D_2 = 2E_2/(\lambda_1 + \lambda_2)$. λ_1 and λ_2 are the roots of the bi-quadratic equation

$$S_{11}\lambda^4 - (2S_{12} + S_{66})\lambda^2 + S_{22} = 0. \tag{2.19}$$

The components of the compliance matrix of the material are defined as

$$S_{11} = \frac{1}{E_1}, \quad S_{22} = \frac{1}{E_2}, \quad S_{12} = -\frac{\nu_{12}}{E_1}, \quad S_{66} = \frac{1}{G_{12}}. \tag{2.20}$$

For small values of the parameter β, the contact pressure follows a Hertzian distribution (Eq. 2.2), but as β increases, the pressure is distributed more towards the edge of the contact zone as predicted by the beam analysis as shown in Fig. 2.6.

This discussion indicates that as the beam becomes more flexible, the pressure distribution under the indentor becomes significantly different than that between the same indentor and a half-space. While this phenomenon may not have a major effect on the contact law, it will significantly affect the stress distribution under the impactor and should be considered when attempting to predict impact damage. Many investigators have studied the contact between smooth indentors and beams with various support conditions (Table 2.1); the reader is referred to these studies for further details.

2.4 Indentation of Plates

Similarly, several analytical studies of the contact between a smooth indentor and laminated plates have been published (see Table 2.2). Many times the plate was modeled as an homogeneous orthotropic plate, but in some of the latest studies, the laminate nature of the composite is considered to determine the effect of the stacking sequence and the material properties on the contact law.

An analysis of the indentation of simply supported laminated plates under rigid spheres using three-dimensional elasticity theory (Wu and Yen 1994) indicates that the stacking sequence has little effect on the contact law confirming previous experimental results (Tan and Sun 1985, Yang and Sun 1982). The contact force is found to be roughly proportional to the transverse modulus of

Table 2.2. *Analytical studies of contact between a plate and a rigid indentor*

Material	Indentor shape	Reference
Isotropic	Sphere	Essenburg (1962) Keer and Miller (1983c) Chen and Frederick (1993)
Transversely isotropic	Sphere	Sankar (1985a) Cairns and Lagace (1987)
Laminate	Sphere	Wu et al. (1993)
Laminate	Sphere	Wu and Yen (1994)
Laminate	Sphere	Suemasu et al. (1994)

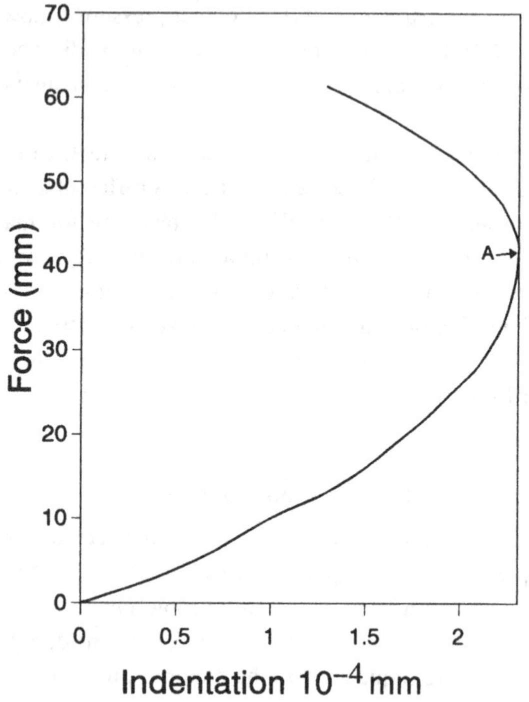

Figure 2.7. Contact law predicted by Wu et al. 1993.

the composite and independent of the longitudinal modulus, which had been suggested earlier by Yang and Sun (1982) based on experimental observations. When the contact force is small, plate thickness has essentially no effect on the indentation law. The contact behavior is then essentially the same as that for the indentation of a half-space. For larger indentations, the contact area increases, significant deviations from Hertzian contact law are observed, and changes in pressure distribution are similar to those shown in Fig. 2.6. In that case, laminate thickness has a significant effect on the force–indentation behavior.

Wu et al. (1993) studied the contact between a composite laminate and a cylindrical indentor and showed that as the contact force increases, the contact behavior changes from a small contact area stage to a large contact area stage. During the small contact area stage, the size of the contact zone increases slowly and almost linearly with the contact force. As the contact force increases above a critical value, the size of the contact zone increases dramatically, and the contact pressure distribution changes dramatically from an Hertzian-type distribution to a saddle-type distribution of the type shown in Fig. 2.6. This characterizes the beginning of the large contact area. In Fig. 2.7, the small contact area phase corresponds to the curve between the origin and point A. Beyond point A, the indentation actually decreases as the force keeps increasing because of this redistribution of contact pressure over a larger area.

Wu and Shyu (1993) showed that the contact force for the onset of delamination is independent of indentor size and that indentation consists of two stages. In the small indentation stage, the plate is still intact and the stacking sequence has insignificant effect on the force–indentation relationship. For large loads, damage becomes important, and the force–indentor movement curves differ for different layups. The contact behavior during low-velocity impacts is the same as in static tests.

2.5 Indentation of a Laminate

A more practical approach consists of determining the relationship between the contact force and the indentation experimentally. Sun and coworkers (Dan-Jumbo et al. 1989, Gu and Sun 1987, Sun et al. 1993, Tan and Sun 1985, Yang and Sun 1982) studied the static indentation of laminated composites. Typically, during impact the contact force increases to maximum value and then decreases back to zero. In some cases, multiple impacts and reloading occurs. Therefore, contact laws should include the unloading and reloading phases.

During the first loading phase, the contact law closely follows Hertz's law of contact (2.8). The parameter n in that equation is usually determined experimentally, but it can also be estimated if the radius of the indentor and the elastic properties of the impactor and the target are known using (2.9), with R given by (2.3). E in (2.9) is given by (2.4), E_2 being the transverse modulus of the composite. Poisson's ratio of the composite is often taken to be zero (Yang and Sun 1982) in (2.4).

Equation (2.8) describes the contact law during the loading phase of the indentation process. However, permanent indentation occurs even at relatively low loading levels, and the unloading phase of the process is significantly different from the loading phase. This phenomenon was observed by Crook (1952) for the indentation of steel plates by spherical indentors. During the unloading phase, the contact law suggested by Crook is

$$P = P_m[(\alpha - \alpha_o)/(\alpha_m - \alpha_o)]^{2.5}, \tag{2.21}$$

where P_m is the maximum force reached before unloading, α_m is the maximum indentation, and α_o is the permanent indentation. α_o is zero when the maximum indentation remains below a critical value α_{cr}. When $\alpha_m > \alpha_{cr}$,

$$\alpha_o = \alpha_m\left[1 - (\alpha_{cr}/\alpha_m)^{\frac{2}{5}}\right]. \tag{2.22}$$

During subsequent reloading, the reloading curve is distinct from the unloading curve but always returns to the point where unloading began (Yang and Sun 1982). The unloading curve is modeled by

$$P = P_m[(\alpha - \alpha_o)/(\alpha_m - \alpha_o)]^{\frac{3}{2}}. \tag{2.23}$$

The parameter α_o does not necessarily correspond to the permanent indentation of the laminate, even though (2.23) indicates that $P = 0$ when $\alpha = \alpha_o$. α_o is selected so that (2.23) fits the experimental unloading curve using a least squares fit procedure. The parameter α_o is related to the actual permanent indentation α_p and the maximum indentation α_m during the loading phase by

$$\alpha_o = \beta(\alpha_m - \alpha_p) \tag{2.24}$$

when $\alpha_m > \alpha_{cr}$; $\alpha_o = 0$ otherwise. The permanent indentation α_p and the parameter β are determined from experiments. However, many experiments need to be performed in order to determine these parameters, whereas a single unloading curve is necessary to determine α_{cr} when (2.22) is used.

Yang and Sun (1982) presented experimental results for glass-epoxy and graphite-epoxy laminates. The elastic properties of the glass-epoxy material system were

$$E_1 = 39.3 \text{ GPa}, \quad E_2 = 8.27 \text{ GPa}, \quad G_{12} = 4.14 \text{ GPa}, \quad \nu_{12} = 0.26. \tag{2.25}$$

For the loading phase, the contact law follows (2.8); for a 6.35-mm-radius indentor, the experimentally determined contact law is

$$P_{\text{exp}} = 1.60 \times 10^4 \alpha^{\frac{3}{2}}, \tag{2.26}$$

and the predicted contact law is

$$P_{\text{pred}} = 1.90 \times 10^4 \alpha^{\frac{3}{2}}. \tag{2.27}$$

The critical indentation is 0.1016 mm (.004 in.). Figure 2.8 shows the loading curve given by (2.26) up to a maximum indentation of 0.3 mm and the corresponding unloading curve obtained from (2.21) and (2.22). If the indentation remains under 0.1016 mm, the unloading curve is identical to the loading curve, and as the maximum indentation becomes larger, the difference between loading and unloading paths is more pronounced. The area under the loading curve is the energy used to indent the laminate and is given by

$$U = \frac{2}{5} k \alpha_{\max}^{\frac{5}{2}}. \tag{2.28}$$

The area under the unloading curve is the energy recovered during the unloading

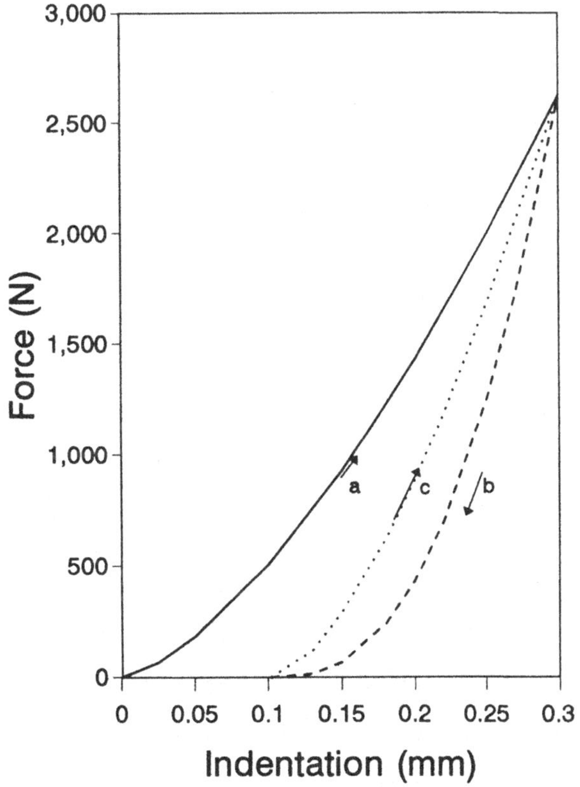

Figure 2.8. Indentation of glass-epoxy laminate from Yang and Sum (1982). (a) Loading; (b) unloading; (c) reloading.

process and is calculated using

$$U' = \frac{2}{7} P_{\max}(\alpha_m - \alpha_o) \tag{2.29}$$

For the case shown in Fig. 2.8, the ratio $U'/U = 0.463$ indicates that only 46.3% of the energy stored during the indentation process is recovered during unloading. This explains why, during the impact of a rigid structure where most of the impact energy is used for local indentation, the loading and unloading parts of the contact force history are significantly different. The reloading curve given by (2.23) is also shown in Fig. 2.8. After a first loading and unloading cycle, the target appears to be stiffer.

Tests conducted on graphite-epoxy specimens with elastic properties

$$E_1 = 120.7 \text{ GPa}, \quad E_2 = 7.93 \text{ GPa}, \quad G_{12} = 5.52 \text{ GPa}, \quad \nu_{12} = 0.30 \tag{2.30}$$

showed that the critical indentation was 0.0803 mm (3.16×10^{-3} in.).

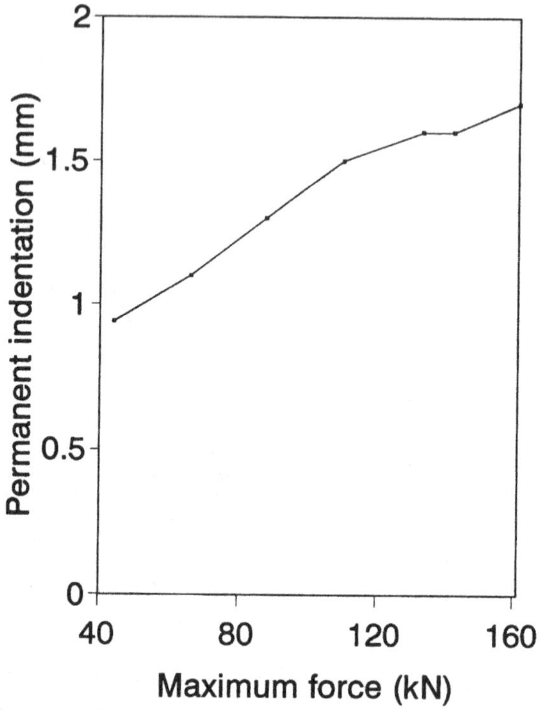

Figure 2.9. Permanent indentation versus maximum contact force for the Kevlar-epoxy laminate tested by Meyer (1988).

Meyer (1988) tested Kevlar-epoxy laminates with elastic properties

$$E_1 = 35.4 \text{ GPa}, \quad E_2 = 22.0 \text{ GPa}, \quad \nu_{12} = 0.175, \quad G_{12} = 7.36 \text{ GPa}. \tag{2.31}$$

A 15 cm × 15 cm × 2 cm piece cut from a cylindrical missile motor case with a 92.2 cm radius was continuously supported by a 92.2 cm steel platen and indented by a 25.4-mm-radius spherical indentor. The loading portion of the contact law is modelled by the equation

$$P = 26 \times 10^9 \alpha^{2.25}. \tag{2.32}$$

During the unloading phase the behavior is modeled by

$$P = P_m[(\alpha - \alpha_o)/(\alpha_m - \alpha_o)]^q \tag{2.33}$$

where the permanent indentation α_o was shown to increase with the maximum load P_m (Fig. 2.9), and the exponent q has an average value of 4.1. It must be noted that the exponent in (2.32) is 2.25, compared to 1.5 for Hertzian contact,

Table 2.3. *Experimental studies of contact*

Sun (1977)
Lin and Lee (1990a)
Lee et al. (1989)
Meyer (1988)
Sun et al. (1993)
Cairns (1991)
Swanson and Rezaee (1990)
Wu and Shyu (1993)
Yang and Sun (1982)
Tan and Sun (1985)
Dan-Jumbo et al. (1989)

and that $q = 4.1$ for this Kevlar-epoxy laminate whereas for glass-epoxy and graphite epoxy the usual value is 2.5.

The contact laws presented in this section are used extensively in the dynamic analysis of the impact event. Further experimental studies of contact can be found in the references listed in Table 2.3.

2.6 Elastoplastic Models

With composites, permanent deformations are introduced when the indentation exceeds a small critical value. The material behavior in the transverse direction is governed by the properties of the matrix, and plastic behavior can be expected. Several investigators presented simple analyses for the indentation of a thin layer bonded to a rigid substrate (Cairns 1991, Christoforou 1993, Christoforou and Yigit 1994, Conway et al. 1970). First, the material is assumed to behave elastically until a critical indentation is reached. Then, the contact area is divided into a plastic zone and an elastic zone as the loading increases and a new contact law is obtained.

Consider a laminate of thickness h, supported by a rigid substrate and indented by a rigid sphere of radius R_i (Fig. 2.10). The indentation is assumed to occur as if the material immediately under the indentor was subjected to uniaxial compression in the z-direction and the deformation takes place as shown in Fig. 2.10a. From that figure, the indentation α can be written in terms of the radius of the sphere, R_i, and the contact radius, R_c, as

$$\alpha = R_i - \left(R_i^2 - R_c^2\right)^{\frac{1}{2}}. \qquad (2.34)$$

The contact radius can be expressed as

$$R_c = (2\alpha R_i - \alpha^2)^{\frac{1}{2}}. \qquad (2.35)$$

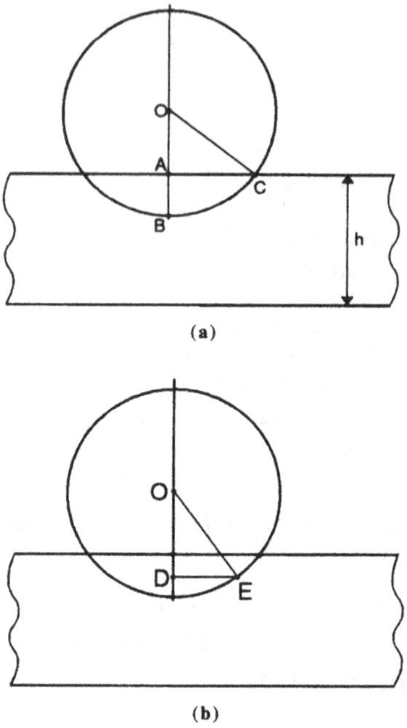

Figure 2.10. Indentation of thin layer supported by a rigid substrate.

In general, the indentation is much smaller than the radius of the indentor; considering the simplification made in this analysis, the second term on the right-hand side of (2.35) is neglected, yielding

$$R_c = (2\alpha R_i)^{\frac{1}{2}}. \tag{2.36}$$

The displacements under the indentor can be written directly from Fig. 2.10b as

$$\delta(r) = \alpha - R_i \left[1 - \left(1 - \left(\frac{r}{R_i} \right)^2 \right)^{\frac{1}{2}} \right]. \tag{2.37}$$

For small indentations, the contact radius is much smaller than the radius of the indentor, and the bracketed quantity can be simplified so that

$$\delta(r) = \alpha - \frac{r^2}{2R_i}. \tag{2.38}$$

The transverse normal strain is assumed to as uniform through the thickness and is given by $\epsilon_{zz} = \delta(r)/h$. Assuming that the material behaves elastically, the contact force is given by

$$P = \frac{2\pi E}{h} \int_0^{R_c} \delta(r) r \, dr, \tag{2.39}$$

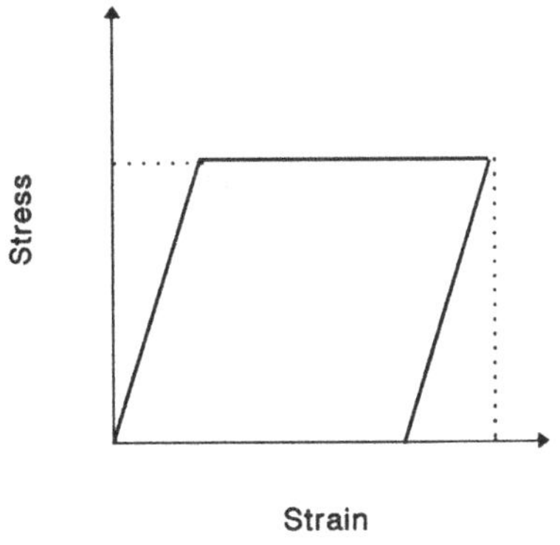

Figure 2.11. Stress-strain relations for elastoplastic analysis.

where E is the modulus of elasticity in the transverse direction. Using (2.36) and (2.38), Eq. (2.39) gives the contact law

$$P = \frac{\pi E R_i}{h}\alpha^2 \tag{2.40}$$

for elastic contact. The model indicates that the contact force increases with α^2 instead of $\alpha^{3/2}$, as predicted by Hertz law of contact.

Permanent indentations are observed even for low values of the contact force. To account for these effects, assume that in the transverse direction the stress-strain behavior of the composite material is elastoplastic (Fig. 2.11). Because deformations are larger at the center of the contact zone, stresses will reach the yield strength first in the center, and the contact area can be divided into a plastic zone of radius R_p and an elastic zone between R_p and R_c. At the junction between the two zones, the stress is equal to the strength of material Z_c so that

$$Z_c = \frac{E}{h}\delta(R_p). \tag{2.41}$$

Using (2.38), we find the radius of the plastic zone as

$$R_p = \left[2R_i\left(\alpha - \frac{Z_c h}{E}\right)\right]^{\frac{1}{2}} \tag{2.42}$$

as long as the indentation α is larger than the critical indentation

$$\alpha_{cr} = \frac{Z_c h}{E}. \tag{2.43}$$

Yang and Sun (1982) assumed that the critical indentation is a material property, but (2.43) indicates that it depends on the laminate thickness in addition to the compressive strength and the elastic modulus. The contact force is calculated

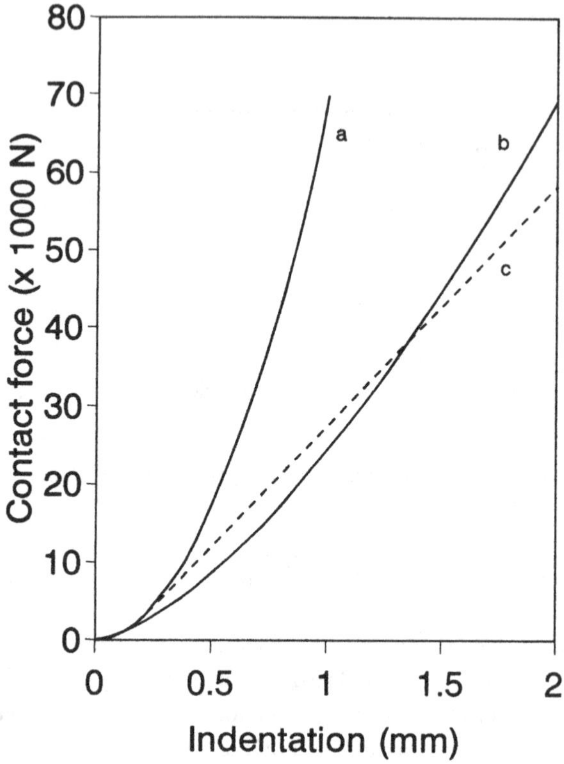

Figure 2.12. Comparison of results from (a) elastic analysis (Eq. 2.40), (b) Hertzian contact law (Eq. 2.8), and (c) elastoplastic analysis (Eqs. 2.40, 2.45).

using

$$P = \pi R_p^2 Z_c + \frac{2\pi E}{h} \int_{R_p}^{R_c} \delta(r) r \, dr, \tag{2.44}$$

and using (2.36), (2.37), (2.42), and (2.44), we find the contact law

$$P = \pi R_i Z_c (2\alpha - \alpha_{cr}) \tag{2.45}$$

for the loading phase when $\alpha > \alpha_{cr}$.

During the unloading phase, it is assumed that for $r < R_p$, the material will follow the elastic unloading path shown in Fig. 2.11 and that the stress-strain relation is given by

$$\sigma = Z_c + E(\epsilon - \epsilon_m). \tag{2.46}$$

The contact force during the unloading phase is given by

$$P = \pi R_p^2 Z_c - \frac{2\pi E}{h} \int_0^{R_p} \left(\alpha_m - \frac{r^2}{2R_i}\right) r \, dr + \frac{2\pi E}{h} \int_0^{R_c} \delta(r) r \, dr \tag{2.47}$$

with $R_p = [2R_i(\alpha_m - Z_c h/E)]^{1/2}$. After integration, (2.47) gives the contact

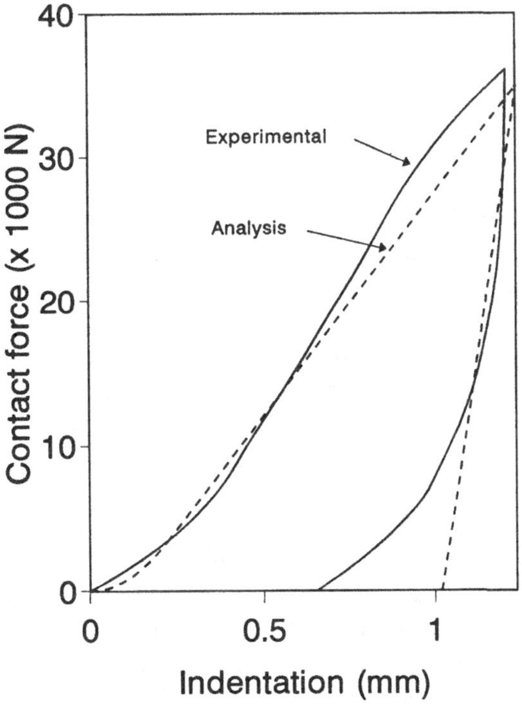

Figure 2.13. Comparison of experimental results with predictions of elastoplastic analysis (Eqs. 2.40, 2.45, 2.48).

law for unloading

$$P = \frac{\pi R_i E}{h}\left(\alpha^2 - \alpha_o^2\right) \tag{2.48}$$

for $\alpha > \alpha_o$, with the permanent indentation α_o defined as

$$\alpha_o = \alpha_m - \alpha_{cr}. \tag{2.49}$$

Swanson and Rezaee (1990) conducted indentation tests on carbon-epoxy $[0,\pm 68,0]_S$ laminates supported by rigid platen. The elastic properties of the material are

$$E_1 = 169 \text{ GPa}, \quad E_2 = 6.72 \text{ GPa}, \quad \nu_{12} = 0.266, \quad G_{12} = 5.26 \text{ GPa}, \tag{2.50}$$

the thickness of the laminate is 2.4 mm, and the diameter of the indentor is 15.88 mm. The Hertz law of contact can adequately model the loading phase but for large loads (above 36,000 N), contact laws deviate significantly from Hertz law during the loading part of the indentation process. That smooth deviation is attributed to severe damage introduced during indentation of specimens that are back-supported by a hardened steel block. Figure 2.12 indicates that for these load levels the indentor has penetrated more than half the thickness. The elastoplastic analysis also provides a reasonable fit to the experimental data as shown in Fig. 2.13.

2.7 Conclusion

It is generally accepted that for most current material systems such as graphite-epoxy or glass-epoxy, the indentation of a composite structure can be modeled using a statically determined contact law. The loading, unloading, and reloading portions of the force–indentation curves are modeled using (2.8), (2.21)–(2.23) with parameters usually obtained from experiments. Various aspects of the contact problem have been examined in details both experimentally then analytically, but these equations are now used by most investigators in mathematical models for the analysis of the impact dynamics.

2.8 Exercise Problems

2.1 Plot the loading phase of the contact laws for glass-epoxy, graphite-epoxy, and Kevlar-epoxy laminates under a 10-mm-radius indentor as the contact force increases from 0 to 1000 N. The material properties for these materials are given by (2.25), (2.30), and (2.31), respectively.

2.2 For the three materials in problem 2.1, plot the size of the contact zone and the maximum contact pressure as a function of the contact force. Discuss implications for stress distribution in the contact zone.

2.3 Starting with the Hertzian contact law, derive an expression for the energy required to produce a maximum indentation α_{max}. What would be the energy needed if the contact force increases linearly to the same maximum level?

2.4 Consider a simply supported rectangular steel beam indented by a steel sphere. The indentation is expected to follow the Hertz law of contact. Find the overall deflection of the beam and the local indentation, and discuss the relative magnitude of these two deformations. When is one negligible compare to the other?

2.5 When unloading is governed by (2.21), derive an expression for the energy recovered during the unloading process.

2.6 Many investigators neglect local indentation altogether, and others assume that both the loading and the unloading phases are governed by Hertz law and obtain excellent agreement with experimental results. Discuss when local indentation effects can be neglected and when they must be considered, and to what extent.

2.7 Thermoplastic composites are known to be more damage resistant. Explain how a lower transverse modulus would affect the contact stiffness, the size of the contact area, the pressure distribution under the indentor, and therefore impact resistance.

3
Impact Dynamics

3.1 Introduction

A first step in gaining some understanding of the effect of impact by a foreign object on a structure is to predict the structure's dynamic response to such an impact. Predictions are made using a mathematical model that appropriately accounts for the motion of the projectile, the overall motion of the target, and the local deformations in the area surrounding the impact point. In general, details of the local interaction between the projectile and the target are not needed to predict the contact force history. A particular beam, plate, or shell theory is selected, and the local deformation in the through-the-thickness direction, which is not accounted for in such theories, is included through the use of an appropriate contact law. Contact laws relate the contact force to the indentation, which is the difference between the displacement of the projectile and the displacement of the target at the point of impact. With most low-velocity impacts, small amounts of damage are introduced in a small zone surrounding the impact point, and the dynamic properties of the structures usually are not affected by the presence of damage. Therefore, impact dynamic analyses generally do not attempt to model damage as it develops during the impact event.

The choice of a particular structural theory must be based on careful consideration of the effect of complicating factors such as transverse shear deformation and rotary inertia. For example, for beams, one can select to use the Bernoulli-Euler beam theory, the Timoshenko beam theory, or one of several other available beam theories. Similar choices are available for plates and shells, and the structure also can be modeled as a two- or three-dimensional solid. In this chapter, we will first present brief derivations of several beam, plate, and shell theories outlining the basic assumptions. Results from these different models are compared using static loading examples, free vibration examples, and wave propagation examples.

Exact solutions to the equations of motion are available for some special cases of beams, plates, and shells subjected to dynamic loading. These solutions can be used effectively, but generally, approximate solutions must be sought. The finite element method is used often because it can be readily applied to a wide range of geometries and boundary conditions. Variational approximation methods are also used since in some cases they can provide more efficient solutions. The implementation of such methods, particularly with reference to the impact problem, is also discussed at length.

To reduce cost, often tests are performed on small specimens rather than on full-scale prototypes. Whether such a scaling is possible and how to relate the results obtained on a small-scale model to the expected results on a full-scale model are also discussed in this chapter.

3.2 Beam Theories

In this section we present the basic assumptions and the development of three beam theories: a higher-order theory, the Timoshenko beam theory, and the Bernoulli-Euler beam theory. Results from these theories for static loading, free vibration, and wave propagation examples will be compared in order to determine the applicability of each theory.

3.2.1 Bending of a Beam by a Uniform Load

The following example, treated as a two-dimensional elasticity problem without any assumptions for a particular beam theory, is discussed to bring some understanding about the general behavior of beams, insight into why some of the assumptions in the various beam theories are made, and an understanding of each theory's limitations.

Consider an isotropic beam of narrow rectangular cross section supported at the ends and subjected to a uniformly distributed load q (see Fig. 3.1). For this two-dimensional problem, the theory of elasticity (Timoshenko and Goodier 1970, pp. 46–48) indicates that the three nonzero stress components are

$$\sigma_{xx} = -\frac{q}{2I}(L^2 - x^2)z + \frac{q}{2I}\left(\frac{2}{3}z^3 - \frac{h^2}{10}z\right) - \frac{q}{2I}\left(x^2 z - \frac{2}{3}z^3\right) \tag{3.1a}$$

$$\sigma_{zz} = -\frac{q}{2I}\left(\frac{z^3}{3} - z\frac{h^2}{4} + \frac{h^3}{12}\right) \tag{3.1b}$$

$$\sigma_{xz} = -\frac{q}{2I}\left(\frac{h^2}{4} - z^2\right)x \tag{3.1c}$$

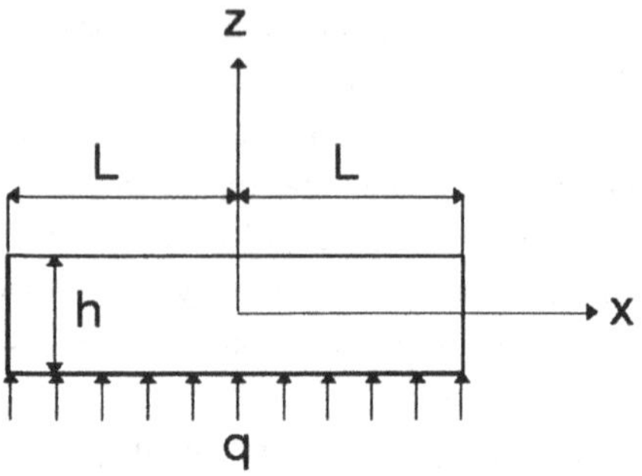

Figure 3.1. Simply supported beam subjected to uniform loading.

where I is the moment of inertia of the beam. Equation (3.1c) indicates that the transverse shear stress σ_{xz} follows a parabolic distribution through-the-thickness of the laminate. The first term in (3.1a) represents the linear distribution through-the-thickness of the normal stress σ_{xx} that is predicted by the elementary beam theory. The second term in (3.1a) is the correction due the effect of the transverse normal stress σ_{zz}, given by (3.1b). The normal stress σ_{xx} reaches a maximum at $x = 0$ and $z = h/2$, at that point the second term in (3.1a) is less than 1% of the first term when $L/h > 2.58$. Therefore, for most beams this correction term can be neglected, and the elementary beam theory is sufficiently accurate for most applications. In addition (3.1b) indicates that the transverse normal stress σ_{zz} is independent of x and has a maximum value of q at the bottom of the beam. The maximum value of σ_{zz} is less than 1% of the maximum value of σ_{xx} when $L/h > 5.8$. Therefore, for most beams this stress component is very small, its contribution to the total strain energy is small, and its effects are neglected in most beam theories.

From two-dimensional elasticity, the displacements at an arbitrary location are

$$u = \frac{q}{2EI}\left[\left(L^2 x - \frac{x^3}{3}\right) z + x\left(\frac{2}{3}z^2 - \frac{1}{10}h^2 z\right) + \nu x\left(\frac{y^3}{3}z^3 - \frac{h^2}{4}z + \frac{h^3}{12}\right)\right] \tag{3.2a}$$

$$v = -\frac{q}{2EI}\left\{\frac{z^4}{12} - \frac{h^2 z^2}{8} + \frac{h^3 z}{12} + \nu\left[(L^2 - x^2)\frac{z^2}{8} + \frac{z^4}{96} - \frac{h^2 z^2}{20}\right]\right\} - \frac{q}{2EI}\left[L^2\frac{x^2}{2} - \frac{x^4}{12} - \frac{h^2 x^2}{20} + \left(1 + \frac{\nu}{2}\right)\frac{h^2 x^2}{4}\right] + \delta \tag{3.2b}$$

where u is the axial displacement; v is the transverse displacement; and δ is the transverse displacement of the center of the beam ($x = z = 0$), which is given by

$$\delta = \frac{5}{24}\frac{qL^4}{EI}\left[1 + \frac{3}{5}\frac{h^2}{L^2}\left(\frac{4}{5} + \frac{\nu}{2}\right)\right], \tag{3.3}$$

where ν is the Poisson's ratio for the material. The factor before the brackets in (3.3) is the deflection predicted by the elementary beam theory, and the second term inside the bracket represents the effect of shear deformation. As the ratio L/h becomes large, the effect of shear deformation becomes small. For a typical value of $\nu = 0.3$, this correction factor is less than 0.01 when $L/h > 15$. Therefore, for long beams the effect of shear deformations can be neglected as in the elementary beam theory.

In the beam theories to be discussed here, the transverse normal strain is neglected, and the three theories differ in the way shear deformation is accounted for. With the higher-order theory, the transverse shear strain is assumed to follow a parabolic distribution and vanish on the top and bottom surfaces, as suggested by (3.1c). With the Timoshenko beam theory, the transverse shear strain is assumed to be constant through-the-thickness; with the Bernoulli-Euler beam theory (also called the elementary beam theory), the effect of transverse shear deformation is ignored altogether.

3.2.2 Higher-Order Beam Theory

The development of the present higher-order beam theory is based on two hypotheses:

(H1) *The transverse shear strain vanishes on the top and bottom surfaces and follows a parabolic distribution through-the-thickness.*

(H2) *The transverse normal strain can be neglected.*

These two hypotheses can be expressed as

$$\epsilon_{xz} = \frac{\partial u}{\partial z} + \frac{\partial w}{\partial x} = \gamma\left(1 - \frac{4z^2}{h^2}\right) \tag{3.4a}$$

$$\epsilon_{zz} = \frac{\partial w}{\partial z} = 0 \tag{3.4b}$$

where γ is the magnitude of the shear stress along the x-axis. Equation (3.4b) yields

$$w = w_o(x), \tag{3.5}$$

and, after substitution into (3.4a), the axial displacement can be expressed as

$$u = u_o(x) + z(\gamma - w_{o,x}) - \frac{4z^3}{3h^2}\gamma. \tag{3.6}$$

Considering only bending deformations ($u_o(x) = 0$), and introducing the new variable $\psi = \gamma - w_{o,x}$, we obtain

$$u = z\psi - \frac{4z^3}{3h^2}(\psi + w_{o,x}). \tag{3.7}$$

From this equation, a physical interpretation of the function Ψ is that it represents the rotation of a small line segment initially perpendicular to the x-axis at $z = 0$. Equations (3.5) and (3.7) are the kinematic assumptions of the theory from which we get the two nonzero strain components

$$\begin{aligned} \epsilon_{xx} &= \frac{\partial u}{\partial x} = z\psi_{,x} - \frac{4z^3}{3h^2}(\psi + w_{o,x})_{,x} \\ \epsilon_{xz} &= (\psi + w_{o,x})\left(1 - 4\frac{z^2}{h^2}\right). \end{aligned} \tag{3.8}$$

The strain energy, kinetic energy, and potential energy of the external forces are given by

$$\begin{aligned} U &= \frac{1}{2}\int_V (\sigma_{xx}\epsilon_{xx} + \sigma_{xx}\epsilon_{xz})dV, \quad T = \frac{1}{2}\int_V \rho(\dot{u}^2 + \dot{w}^2)\,dV, \\ V &= -\int_0^L p(x)w\,dx. \end{aligned} \tag{3.9}$$

Hamilton's principle (Reddy 1984)

$$\delta\left(\int_{t_1}^{t_2}(U + V - T)dt\right) = 0 \tag{3.10}$$

is used to derive the equations of motion:

$$\begin{aligned} &-\frac{8}{15}GA(\psi + w_{o,x})_{,x} + \frac{EI}{21}w_{o,xxxx} - \frac{16}{105}EI\psi_{,xxx} \\ &\quad = q - \rho A w_{o,tt} + \frac{\rho I}{21}w_{o,xxtt} - \frac{16}{105}\rho I\psi_{,xtt} \\ &\frac{8}{15}GA(\psi + w_{o,x}) + \frac{16}{105}EIw_{o,xxx} - \frac{68}{105}EI\psi_{,xx} \\ &\quad = \frac{16}{105}\rho I w_{o,xtt} - \frac{68}{105}\rho I\psi_{,tt}. \end{aligned} \tag{3.11a,b}$$

The essential boundary conditions are that either Ψ, $w_{o,x}$, or w_o are prescribed, and the corresponding natural boundary conditions are that either

$$\frac{68}{105}EI\psi_{,x} - \frac{16}{105}EIw_{o,xx}, \tag{3.12a}$$

$$-\frac{1}{21}EIw_{o,xx}+\frac{16}{105}EI\psi_{,xx}+GA(\psi+w_{o,x})$$

$$+\frac{16}{105}\rho I\psi_{,tt}-\frac{1}{21}\rho I w_{o,xtt}, \text{ or} \tag{3.12b}$$

$$\frac{1}{21}EIw_{o,xx}-\frac{16}{105}EIw_{o,x} \tag{3.12c}$$

is prescribed at each end.

3.2.3 Timoshenko Beam Theory

With the Timoshenko beam theory, also called the first-order shear deformation theory, the transverse normal strain is neglected (H2), and assumption (H1) is replaced by:

(H1′) *The transverse shear strain is assumed to be uniformly distributed over the cross section.*

This is expressed by

$$\epsilon_{xz}=\theta(x). \tag{3.13}$$

Using (3.17a) and (3.18), the axial displacement becomes

$$u=z(\theta-w_{o,x})=z\psi, \tag{3.14}$$

and the strain components are

$$\epsilon_{xx}=z\psi_{,x},\quad \epsilon_{xz}=\psi+w_{o,x}. \tag{3.15}$$

Equation (3.14) indicates that cross sections remain planar and that Ψ represents the rotation of the cross section. The Timoshenko beam theory often is derived by assuming that the cross section remains planar but rotates independently of the transverse displacement w. Then, (3.14) is taken as a direct assumption on the kinematics of the deformation. With the strains given by (3.15), substituting into (3.9) and applying Hamilton's principle (3.10) yields the equations of motion of a Timoshenko beam:

$$\begin{aligned}\kappa GA(w_{,x}+\psi)_{,x}+q&=\rho Aw_{,tt}\\ EI\psi_{,xx}-\kappa GA(w_{,x}+\psi)&=\rho I\psi_{,tt}\end{aligned} \tag{3.16}$$

with boundary conditions where either

$$\begin{aligned}EI\psi_{,x}\quad&\text{or}\quad\psi\\ \kappa GA(\psi+w_{o,x})\quad&\text{or}\quad w_o\end{aligned} \tag{3.17}$$

are prescribed. The shear correction factor κ is introduced because the shear stress is not uniform through-the-thickness, as assumed (H1′). The use of this correction factor has been investigated extensively and is known to depend on the shape of the cross section. Cowper (1966) and Kaneko (1975) give values for many cross-sectional shapes.

3.2.4 Bernoulli-Euler Beam Theory

In the Bernoulli-Euler beam theory, assumption (H1) is replaced by:

(H1″) *The transverse normal shear strain is assumed to be negligible.*

That is,

$$\epsilon_{xz} = 0, \tag{3.18}$$

which implies that

$$u = -z w_{o,x}, \quad \epsilon_{xx} = -z w_{o,xx}. \tag{3.19}$$

Equation (3.19a) expresses what is often taken as the point of departure for the development of the elementary beam theory. That is, planar sections initially normal to the neutral axis remain planar and perpendicular to the neutral axis during deformation. Using Hamilton's principle (3.10), the well-known equation of motion

$$EI w_{o,xxxx} + \rho A w_{o,tt} = p \tag{3.20}$$

is obtained, and the boundary equations are that either

$$\begin{aligned} EI w_{o,xx} \quad &\text{or} \quad w_{o,x} \\ -EI W_{o,xxx} \quad &\text{or} \quad w \end{aligned} \tag{3.21}$$

are prescribed at the ends.

3.2.5 Static Loading on Beams

One way to assess the performance of the various beam theories is to compare the results from these various theories with those obtained from the elasticity solution for a simply supported beam subjected to a uniformly distributed static load. With the Bernoulli-Euler beam theory, after integrating (3.20) four times, the application of boundary conditions leads to

$$w = \frac{q}{EI}\left(\frac{x^4}{24} - \frac{x^2}{4}L^2 + \frac{5}{24}L^4\right). \tag{3.22}$$

With the Timoshenko beam theory, the transverse displacements and the rotation of the cross section are

$$w = \frac{q}{EI}\left(\frac{x^4}{24} - \frac{L^2}{4}x^2 + \frac{5L^4}{24}\right) + \frac{q}{2\kappa GA}(L^2 - x^2) \tag{3.23a}$$

$$\psi = \frac{q}{EI}\left(-\frac{x^3}{6} + \frac{x}{2}L^2\right). \tag{3.23b}$$

The first term in (3.23a) is the identical to the right-hand side of (3.22), and the second term represents the effect of shear deformation on the transverse displacements.

With the higher-order theory,

$$w = \frac{q}{EI}\left[\frac{x^4}{24} - \frac{x^2L^2}{4} + \frac{5}{24}L^4\right] + \frac{q}{2(\frac{14}{17})GA}(L^2 - x^2) \tag{3.24}$$

$$\psi = \frac{q}{EI}\left(\frac{xL^2}{2} - \frac{x^3}{6}\right) - \frac{2}{7}\frac{q}{GA}x. \tag{3.25}$$

The transverse displacement w in (3.24) is identical to that given by the Timoshenko beam theory (3.23a) except that the shear correction factor κ is now replaced by the factor $\frac{14}{17} = .8235$. Typically, κ is taken to be $\frac{5}{6} = 0.8333$, or $\frac{\pi^2}{12} = .8225$, for beams with rectangular cross sections. Thorough discussions of shear correction factors are given by Kaneko (1975) and Cowper (1966). Therefore, for this example the deflections predicted by the higher-order theory will be very close to those predicted using the Timoshenko beam theory.

From (3.24), the deflection at the center of the beam ($x = 0$) can be written as

$$w = \frac{5}{24}\frac{qL^2}{EI}\left[1 + \frac{17}{70}\frac{E}{G}\frac{h^2}{L^2}\right]. \tag{3.26}$$

The second term in the bracket is the ratio between the deflection due to shear deformation and that due to bending. As discussed before, the effect of shear deformation becomes small as the ratio h/L become small. Equation (3.26) also shows that the effect of shear deformation depends on the ratio between Young's modulus E and the shear modulus G. For isotropic materials, $E/G = 2(1+\nu)$, which for a typical value of Poisson's ratio of 0.3 gives $E/G = 2.6$. For a typical graphite-epoxy composite, the modulus in the fiber direction is $E_1 = 181$ GPa, and the shear modulus is $G_{12} = 7.17$ GPa. For a unidirectional laminate with fibers oriented in the x-direction, $E_1/G_{12} = 25.2$, and therefore the effect of shear deformation will be much more significant. Figure 3.2 shows a comparison of the effect of shear deformation on the deflection of a simply supported beam as predicted by the higher-order beam theory and by the

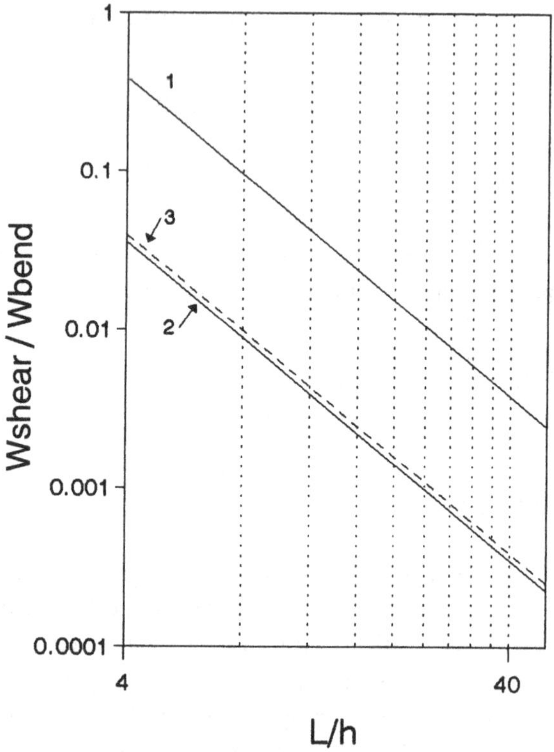

Figure 3.2. Effect of shear deformation on the deflection of a simply supported beam. 1: graphite-epoxy beam, HOT; 2: isotropic beam, elasticity solution; 3: isotropic beam, HOT.

elasticity solution for an isotropic beam with $\nu = 0.3$. As the length-to-height ratio becomes large, the effect of shear deformation becomes negligible. Excellent agreement is observed between the elasticity solution and the results from beam theory. For the graphite-epoxy beam, the effects of shear deformation are more significant (Fig. 3.2). However, it must be noted that in this case the shear deflection is less than 1% of the bending deflection when $L/h > 25$. Therefore, the elementary beam theory can be used when $L/h > 25$.

So far, this example indicates that shear deformations have a significant effect on the deflections for short beams and that this effect is more pronounced for composite beams. Now let us compare the stress distributions predicted by the beam theories with the elasticity solution. With the Bernoulli-Euler model, substituting (3.22) into (3.19b) and using Hooke's law, the axial stress is

$$\sigma_{xx} = \frac{q}{2I}(L^2 - x^2)z, \tag{3.27}$$

which is just the first term in (3.1a). The Timoshenko beam theory yields the same expression for the axial stress (3.27) and the following expression for the transverse shear stress:

$$\sigma_{xz} = -\frac{q}{\kappa A}x. \tag{3.28}$$

With the higher-order theory, the transverse shear stress is

$$\sigma_{xz} = -\frac{3}{2}\frac{q}{A}\left(1 - 4\frac{z^2}{h^2}\right)x \tag{3.29}$$

which is, for a rectangular beam, the exact result from the theory of elasticity. This indicates that the higher-order beam theory gives a much better prediction of the transverse shear stresses. The drawback is that the corresponding equations of motion and the boundary conditions are more complex than with the other theories.

3.2.6 Recovery of Accurate Stresses from Beam Models

If an accurate determination of the transverse shear stresses is needed, it is possible to use the simpler elementary beam theory to determine the axial stress and then to use this result to determine the shear stresses from the equations of equilibrium of two-dimensional elasticity. For example, with the simply supported beam under uniform loading, the axial stress is accurately predicted (3.27) by the Bernoulli-Euler beam theory.

Considering the beam as a two-dimensional solid, in the absence of body forces, equilibrium in the x-direction is governed by

$$\frac{\partial \sigma_{xx}}{\partial x} + \frac{\partial \sigma_{zx}}{\partial z} = 0. \tag{3.30}$$

From (3.30), the shear stress at location z is given by

$$\sigma_{zx}(x, z) = \int_{-\frac{h}{2}}^{z} \frac{\partial \sigma_{xx}(x, z^*)}{\partial x} dz^*. \tag{3.31}$$

Substituting the expression for σ_{xx} obtained from the Bernoulli-Euler beam theory (3.27), we find

$$\sigma_{zx} = -\frac{q}{2I}\left(z^2 - \frac{h^2}{4}\right) \tag{3.32}$$

which is precisely the exact solution for a rectangular beam obtained from the theory of elasticity (3.1c). Similarly, from two-dimensional elasticity, the equation of equilibrium in the z-direction is

$$\frac{\partial \sigma_{xz}}{\partial x} + \frac{\partial \sigma_{zz}}{\partial z} = 0. \tag{3.33}$$

Integrating through-the-thickness from the bottom of the beam to an arbitrary location z, we obtain

$$\sigma_{zz}(x,z) + q = -\int_{-\frac{h}{2}}^{z} \frac{\partial \sigma_{xz}(x,z^*)}{\partial x} dz^*. \tag{3.34}$$

Substituting the value of the shear stress given by (3.32), the exact value of the transverse normal stress given by (3.1c) is recovered from (3.34). Therefore, it is possible to accurately determine both the transverse normal stress and the transverse shear stress distributions from the results of the elementary beam theory.

3.2.7 Harmonic Wave Propagation

In order to compare the dynamic behavior of beams described by the three theories previously discussed, consider the propagation of free harmonic waves of the form

$$w = We^{i(kx-\omega t)}, \quad \psi = \Psi e^{i(kx-\omega t)} \tag{3.35a,b}$$

where k is the wavenumber and ω is the frequency. The wavenumber is related to the wavelength of the disturbance by $k = 2\pi/\lambda$, and the phase velocity is $c = \omega/k$. Studying the propagation of free harmonic waves in an infinite beam allows us to gain some understanding about the dynamic behavior of the beam.

Substituting w from (3.35a) into the equation of motion for a Bernoulli-Euler beam (3.20) gives $\omega^2 = \frac{EI}{\rho A}k^4$, or in terms of the phase velocity, $c = (\frac{EI}{\rho A})^{1/2}k$. This indicates that beams are dispersive because the phase velocity depends on the wavelength, and that waves with shorter wavelengths propagate faster than waves with longer wavelengths. With this beam theory, however, short wavelengths can propagate with infinite speeds, which is not physically possible. This problem is the direct result of the simplifying assumptions made in the development of that theory.

Substituting (3.35) into the equations of motion of a Timoshenko beam (3.16) gives the system of homogeneous equations:

$$\left(\begin{bmatrix} K_{11} & K_{12} \\ K_{21} & K_{22} \end{bmatrix} - \omega^2 \begin{bmatrix} M_{11} & M_{12} \\ M_{21} & M_{22} \end{bmatrix} \right) \begin{Bmatrix} W \\ \Psi \end{Bmatrix} = 0 \tag{3.36}$$

where

$$\begin{aligned} &K_{11} = \kappa GAk^2, \quad K_{22} = \kappa GA + EIk^2, \quad K_{12} = -K_{21} = i\kappa GAk, \\ &M_{11} = \rho A, \quad M_{22} = \rho I, \quad M_{12} = M_{21} = 0. \end{aligned} \tag{3.37}$$

For nontrivial solutions, the determinant of the matrix of the coefficients in

(3.36) must vanish, which gives the dispersion relation

$$\rho A \rho I \omega^4 - \omega^2[\rho A(EIk^2 + \kappa GA) + \kappa GA\rho Ik^2] + \kappa GAEIk^4 = 0. \quad (3.38)$$

This equation shows that for a given value of k there will be two solutions for the frequency or phase velocity. Plots of phase velocity versus wavenumbers are called *dispersion curves*, and in this case we will have two branches. For very long wavelengths, k approaches zero and (3.38) gives the two cutoff frequencies $\omega_1 = 0, \omega_2 = (GA/\rho I)^{1/2}$. The first branch corresponds to bending motion, and the second branch to shear deformation. It can be shown that a first approximation to the first branch is $\omega^2 = \frac{EI}{\rho A}k^4$, which means that for long wavelengths the Timoshenko beam theory will predict the same behavior as the elementary beam theory. For short wavelengths (large values of k), Eq. (3.38) implies that the phase velocities tend to two limiting values: $c_1 = (\kappa GA/\rho A)^{1/2}, c_2 = (EI/\rho I)^{1/2}$. Therefore, the phase velocities are bounded for short wavelengths.

The equations of motion of the Timoshenko beam can be written as uncoupled:

$$EIw_{,xxxx} + \rho Aw_{,tt} - \left[\rho I + \rho A\frac{EI}{\kappa GA}\right] w_{,xxtt} + \frac{\rho A\rho I}{\kappa GA} w_{,tttt} = 0 \quad (3.39)$$

$$EI\psi_{,xxxx} + \rho A\psi_{,tt} - \left[\rho I + \rho A\frac{EI}{\kappa GA}\right] \psi_{,xxtt} + \frac{\rho A\rho I}{\kappa GA} \psi_{,tttt} = 0. \quad (3.40)$$

Abramovich and Elishakoff (1990) neglected the $w_{,tttt}$ terms in these equations and obtained solutions to free and forced vibration problems. For free wave propagation, the effect of this simplification is to remove the $\rho A\ \rho I\ \omega^4$ term from the dispersion equation (3.38). In that case, there is only one branch of the dispersion curve, and the phase velocity tends to the limit $\frac{1}{c} = (\frac{\rho I}{EI} + \frac{\rho A}{GA})^{1/2}$ for short wavelengths.

After substituting (3.35) into the equations of motion for the higher-order beam theory (3.11), the dispersion relations are given by (3.36) with

$$\begin{aligned} K_{11} &= \frac{8}{15}GAk^2 + \frac{EI}{21}k^4, \quad K_{22} = \frac{8}{15}GA + \frac{68}{105}EIk^2. \\ K_{12} &= -i\left(\frac{8}{15}GAk - \frac{16}{105}EIk^3\right), \quad K_{21} = -K_{12}, \\ M_{11} &= \rho A + \frac{\rho I}{21}k^2, \quad M_{22} = \frac{68}{105}\rho I, \\ M_{12} &= \frac{16}{105}\rho Iki, \quad M_{21} = -M_{12}. \end{aligned} \quad (3.41)$$

The two cutoff frequencies for long wavelengths are $\omega_1 = 0, \omega_2 = (\frac{14}{17}\frac{GA}{\rho I})^{1/2}$. The shear correction factor κ used in the Timoshenko beam theory is now

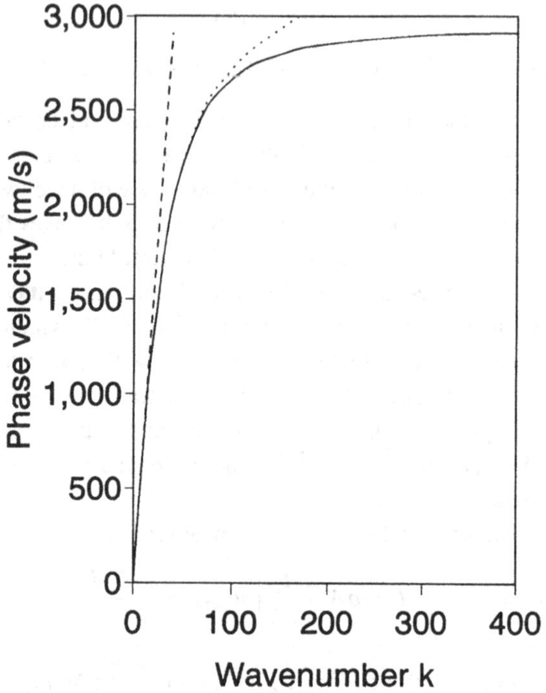

Figure 3.3. Dispersion curves for a steel beam with rectangular cross section. Solid line: Timoshenko's beam theory; dashed line: Bernoulli-Euler beam theory; dotted line: higher-order theory.

replaced by $\frac{14}{17} = 0.8235$ in the expression for the second cutoff frequency. In many cases, κ is taken to be $\frac{5}{6} = 0.8225$ for beams with rectangular cross sections. For short wavelengths, the phase velocity tends to limit $c = (EI/\rho I)^{1/2}$ for both modes.

Example 3.1: Wave Propagation in Isotropic Beam To compare the three beam theories, consider the example of a steel beam with $E = 210$ GPa, $\nu = 0.3$, $h = b = .05$ m, $\rho = 7850$ kg/m^3. For $k < 100$, the higher-order theory and the Timoshenko beam theory give very similar results (Fig. 3.3). The higher-order theory will not bring about drastic changes in the predictions of the dynamic behavior of the beam. For the first mode, the phase velocity predicted by the Timoshenko beam has nearly reached its asymptotic value of 2928 m/s when $k = 400$, while with the higher-order theory the phase velocity keeps increasing to eventually reach a much higher asymptotic value of 5172 m/s. Figure 3.3 also shows the results obtained from the Bernoulli-Euler beam theory which indicate that this theory is applicable only for very long wavelengths λ or small values of $k(k = 2\pi/\lambda)$.

3.2.8 Free Vibrations of Beams

Another way to study the dynamics of the various beam theories is to study the free vibrations of finite beams. The natural frequencies of a simply supported beam of length L are identical to the frequencies of free harmonic waves with wavenumbers $k = i\pi/L$. The study of harmonic wave propagation indicated that differences between the Timoshenko beam model and those from the higher-order theory are small. Therefore, the higher-order theory will not be considered in this section. Exact solutions for simple beams with any combination of boundary conditions at the ends will be developed for both the Bernoulli-Euler and the Timoshenko beam theories. An example of a clamped-clamped beam will be used to illustrate the effect of the slenderness ratio, shear deformations and rotary inertia on the natural frequencies.

For free vibrations of a Bernoulli-Euler beam, the displacements are taken in the form

$$w(x,t) = W(x)\sin(\omega t). \tag{3.42}$$

Substituting (3.42) into the equation of motion (3.20) gives the ordinary differential equation

$$EI\frac{d^4W}{dx^4} - \rho A\omega^2 W = 0. \tag{3.43}$$

Seeking a solution of the form $W = e^{\beta^* x}$, we find by direct substitution that (3.43) is satisfied for four distinct values of β^*

$$\beta_1^* = i\beta, \quad \beta_2^* = -i\beta, \quad \beta_3^* = \beta, \quad \beta_4^* = -\beta \tag{3.44}$$

where $\beta = (\omega^2\rho A/EI)^{1/4}$. This indicates that the general solution is of the form

$$W(x) = A\sin\beta x + B\cos\beta x + C\sinh\beta x + D\cosh\beta x \tag{3.45}$$

where the constants A, B, C, and D have to be determined from the boundary conditions. Since there are two boundary conditions at each end of the beam, using (3.45), a system of four equations of the form

$$[K] = \begin{Bmatrix} A \\ B \\ C \\ D \end{Bmatrix} = 0 \tag{3.46}$$

will be obtained. For each of the natural frequencies of the beam, the determinant of the matrix K vanishes so that a nontrivial solution can be obtained for the four constants A–D. Therefore, the natural frequencies for beams with any combinations of natural frequencies at the ends can be determined by finding the zeroes of a 4×4 determinant.

Exact solutions for the free vibration of Timoshenko beams are obtained by substituting

$$w = W e^{\beta x} \sin \omega t, \quad \psi = X e^{\beta x} \sin \omega t \tag{3.47}$$

into the equations of motion (3.16), which leads to

$$\begin{bmatrix} (GA\beta^2 + \rho A \omega^2) & GA\beta \\ GA\beta & (-EI\beta^2 - \rho I \omega^2 + GA) \end{bmatrix} \begin{Bmatrix} W \\ X \end{Bmatrix} = 0. \tag{3.48}$$

Nontrivial solutions are obtained when the matrix of the coefficients vanishes, which implies that β satisfies the bi-quadratic equation

$$\beta^4 + \beta^2 \omega^2 \left(\frac{\rho A}{GA} + \frac{\rho I}{EI} \right) + \frac{\rho A \cdot \rho I}{GA \cdot EI} \omega^4 - \frac{\rho A}{EI} \omega^2 = 0. \tag{3.49}$$

This equation will have two solutions with the corresponding mode shapes

$$\phi_1 = \begin{Bmatrix} \varphi_{11} \\ \varphi_{12} \end{Bmatrix}, \quad \phi_2 = \begin{Bmatrix} \varphi_{21} \\ \varphi_{22} \end{Bmatrix} \tag{3.50}$$

where

$$\begin{aligned} \phi_{11} &= -GA\beta_1, \quad \phi_{12} = GA\beta_1^2 + \rho A \omega^2 \\ \phi_{21} &= -GA\beta_2, \quad \phi_{22} = GA\beta_2^2 + \rho A \omega^2. \end{aligned} \tag{3.51}$$

The general solution is

$$\begin{aligned} W &= A\varphi_{11} \sin \beta_1 x + B\varphi_{11} \cos \beta_1 x \\ &\quad + C\varphi_{21} \sinh \beta_2 x + C\varphi_{21} \cosh \beta_2 x \\ X &= A\varphi_{12} \cos \beta_1 x + B\varphi_{12} \sin \beta_1 x \\ &\quad + C\varphi_{22} \cosh \beta_2 x + C\varphi_{22} \sinh \beta_2 x. \end{aligned} \tag{3.52}$$

Applying the boundary conditions leads again to a system of four equations of the form given by (3.46) and the natural frequencies are obtain by setting the determinant of the matrix K to zero.

Example 3.2: Free Vibrations of Clamped-Clamped Rectangular Beam Levinson (1981b) studied the example of a clamped-clamped steel beam with $E =$ 210 GPa, $G = 80.77$ GPa, $\rho = 7850$ kg/m^3, $L = 0.5$ m, and slenderness

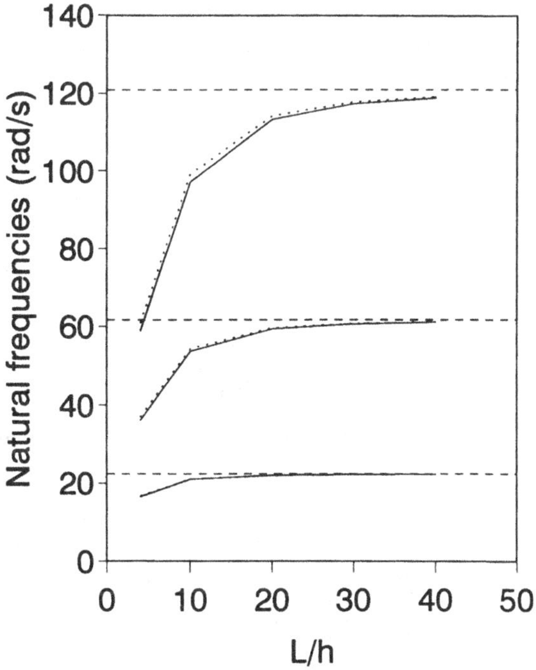

Figure 3.4. Effect of length-to-height ratio on the natural frequencies of clamped-clamped rectangular beams. Solid lines: Timoshenko beam theory; dotted lines: Timoshenko beam theory with $\rho I = 0$; dashed lines: Bernoulli-Euler beam theory.

ratios $L/h = 4, 10$, where L is the total length and h is the depth of the beam. With the present approach, all the results given by Levinson were duplicated. In addition, Fig. 3.4 shows the effect of increasing the ratio L/h from 4 to 50 on the first three natural frequencies. As in the static case, the effect of shear deformation is significant for shorter beams, and as the ratio L/h increases, the difference between the results from the Timoshenko and Bernoulli-Euler beam models becomes negligible. The effects of shear deformation and rotatory inertia are more pronounced for higher modes. The effect of rotary inertia, while present, is small (Fig. 3.4).

For more complex beams systems, finite element models are often used; many finite element models that include the effect of shear deformation and rotary inertia have been proposed. Finite element models are expected to converge to the exact solution as the number of elements increases. The finite element model presented by Friedman and Kosmatka (1993) can be used to analyze Bernoulli-Euler beams and Timoshenko beams. Many other finite element models for Timoshenko beams are available in the literature. This one has the advantage of being free of shear locking problems and therefore can be used to study thin beams. Also, the mass and stiffness matrices are given explicitly. The stiffness matrix is

$$[K] = \frac{EI}{(1+\phi)L}\begin{bmatrix} 12 & 6L & -12 & 6L \\ & (4+\phi)L^2 & -6L & (2-\phi)L^2 \\ & & 12 & -6L \\ \text{symmetric} & & & (4+\phi)L^2 \end{bmatrix} \tag{3.53}$$

where

$$\phi = \frac{12}{L^2}\left(\frac{EI}{\kappa GA}\right). \tag{3.54}$$

The mass matrix is the sum of two matrices: the matrix associated with translational inertia

$$[M_{\rho A}] = \frac{\rho AL}{210(1+\phi)^2}\begin{bmatrix} m_{11} & m_{12} & m_{13} & m_{14} \\ & m_{22} & m_{23} & m_{24} \\ & & m_{33} & m_{34} \\ \text{symmetric} & & & m_{44} \end{bmatrix}, \tag{3.55}$$

with

$$\begin{aligned} m_{11} &= m_{33} = 70\phi^2 + 147\phi + 78, \quad m_{12} = (35\phi^2 + 77\phi + 44)\frac{L}{4}, \\ m_{13} &= 35\phi^2 + 63\phi + 27, \quad m_{14} = -(35\phi^2 + 63\phi + 26)\frac{L}{4}, \\ m_{22} &= m_{44} = (7\phi^2 + 14\phi + 8)\frac{L^2}{4}, \quad m_{23} = (35\phi^2 + 63\phi + 26)\frac{L}{4} \\ m_{24} &= -(7\phi^2 + 14\phi + 6)\frac{L^2}{4}, \quad m_{34} = -(35\phi^2 + 77\phi + 44)\frac{L}{4}, \end{aligned} \tag{3.56}$$

and the matrix associated with rotatory inertia

$$[M_{\rho I}] = \frac{\rho I}{30(1+\phi)^2 L}\begin{bmatrix} \bar{m}_{11} & \bar{m}_{12} & \bar{m}_{13} & \bar{m}_{14} \\ & \bar{m}_{22} & \bar{m}_{23} & \bar{m}_{24} \\ & & \bar{m}_{33} & \bar{m}_{34} \\ \text{symmetric} & & & \bar{m}_{44} \end{bmatrix}, \tag{3.57}$$

$$\begin{aligned} &\bar{m}_{11} = \bar{m}_{13} = \bar{m}_{33} = 36, \quad \bar{m}_{12} = \bar{m}_{14} = -\bar{m}_{34} = -(15\phi - 3)L, \\ &\bar{m}_{22} = \bar{m}_{44} = (10\phi^2 + 5\phi + 4)L^2, \quad \bar{m}_{23} = -(15\phi - 3)L, \\ &\bar{m}_{24} = (5\phi^2 - 5\phi - 1)L^2. \end{aligned} \tag{3.58}$$

To ascertain the accuracy of that finite element, consider the error in the eigenvalues obtained using uniform meshes with increasing numbers of elements.

The error is defined as

$$E_i = \frac{\lambda_i}{\lambda_i^E} - 1 = \left(\frac{\omega_i}{\omega_i^E}\right)^2 - 1, \tag{3.59}$$

where the subscript i denotes the ith natural frequency and the superscript E indicates the exact value of the natural frequency or eigenvalue.

Example 3.3: Convergence of Finite Element Analysis of Beam Vibrations The example treated by Friedman and Kosmatka (1993) is a simply supported beam with

$$EI = 1, \quad \rho A = 1, \quad h/L = 0.2, \quad \nu = 0.3 \tag{3.60}$$

with the shear correction factor

$$\kappa = \frac{10(1+\nu)}{12+11\nu}. \tag{3.61}$$

Because of symmetry, only half of the beam needs to be modeled. Figure 3.5 shows that the error varies with h^2, h being the length of one element – or

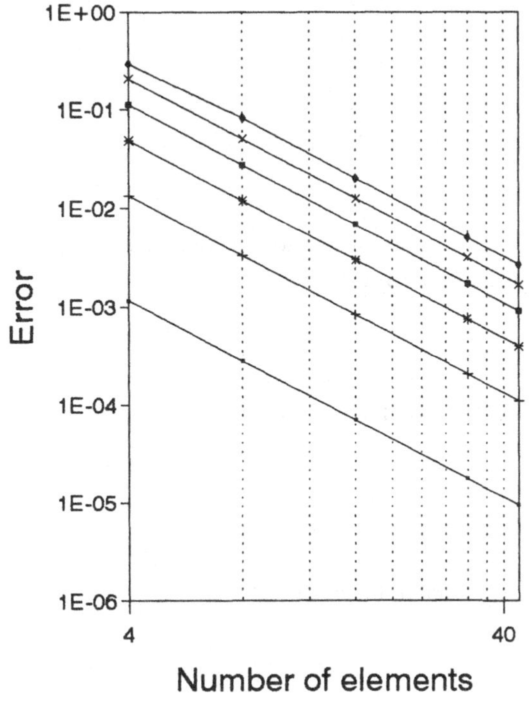

Figure 3.5. Modelization error in finite element analysis of free vibration of beams.

Table 3.1. *Natural frequencies of simply supported beam*

Mode	32 Elts.	Exact	Error
1	9.28200	9.281920	1.767e−5
2	32.24894	32.24561	2.066e−4
3	61.72164	61.69867	7.447e−4
4	93.81159	93.73128	1.714e−3
5	126.88143	126.6825	3.143e−3
6	160.26995	159.8679	5.036e−3

in other words, E varies with N^{-2}, N being the number of elements in the discretization. As expected, the error increases for higher modes. Table 3.1 shows the accuracy of the first six natural frequencies obtained using a uniform mesh with 32 elements.

3.2.9 Effect of Geometrical Nonlinearities

Since many composite structures are thin and flexible, moderately large deformations can be expected during impact. The effect of membrane stiffening on the impact dynamics and the contact force history should be investigated. Large displacements induce a coupling between axial and transverse displacements. For a Bernoulli-Euler beam, the kinematic assumption is written as

$$u = u_o - z w_{,x} \tag{3.62}$$

where u_o is the axial displacement and w is the transverse displacement. For moderately large displacements the axial strain is given by

$$\epsilon_{xx} = u_{,x} + \frac{1}{2} w_{,x}^2 = u_{o,x} - z w_{,xx} + \frac{1}{2} w_{,x}^2. \tag{3.63}$$

When the material follows Hooke's law, the strain energy in the beam is given by

$$U = \frac{E}{2} \int_0^L \epsilon_{xx}^2 dV = \frac{1}{2} \int_0^L \left\{ EA \left(u_{o,x} + \frac{1}{2} w_{,x}^2 \right)^2 + EI w_{,xx}^2 \right\} dx. \tag{3.64}$$

The kinetic energy stored in the beam is given by

$$T = \frac{\rho}{2} \int_0^L \left[(\dot{u}_o - z \dot{w}_{,x})^2 + \dot{w}_{,x}^2 \right] dV = \frac{1}{2} \int_0^L \left[\rho A (\dot{u}_o^2 + \dot{w}^2) + \rho I \dot{w}_{,x}^2 \right] dx. \tag{3.65}$$

Hamilton's principle (3.10) is used to derive the equations of motion

$$EA\left(u_{o,x}+\frac{1}{2}w_{,x}^2\right)_{,x}=\rho A\,\ddot{u}_o$$
$$EA\left[\left(u_{o,x}+\frac{1}{2}w_{,x}^2\right)w_{,x}\right]_{,x}-EIw_{,xxxx}-\rho A\ddot{w}+\rho I\ddot{w}_{,xx}=0 \tag{3.66a,b}$$

and the boundary conditions where either

$$u \quad\text{or}\quad -EA\left(u_{o,x}+\frac{1}{2}w_{,x}^2\right)$$
$$w_{,x} \quad\text{or}\quad -EIw_{,xx} \tag{3.67}$$
$$w \quad\text{or}\quad EIw_{,xxx}-EA\left(u_{o,x}+\frac{1}{2}w_{,x}^2\right)-\rho I\ddot{w}_{,x}$$

are prescribed at the ends. Since the effects of inplane inertia and rotary inertia are usually negligible, the right-hand side of (3.66a) and the ρI term in (3.66b) are neglected, and the equations of motion now read

$$\left(u_{o,x}+\frac{1}{2}w_{,x}^2\right)_{,x}=0$$
$$EIw_{,xxxx}+\rho A\ddot{w}=EA\left[\left(u_{o,x}+\frac{1}{2}w_{,x}^2\right)w_{,x}\right]_{,x} \tag{3.68a,b}$$

Integrating (3.68a) gives

$$u_{o,x}+\frac{1}{2}w_{,x}^2=C(t). \tag{3.69}$$

Integrating again gives

$$C(t)=\frac{1}{2L}\int_0^L w_{,x}^2dx. \tag{3.70}$$

After substitution into (3.68b),

$$EIw_{,xxxx}+\rho A\ddot{w}=\frac{EA}{2L}\left(\int_0^L w_{,x}^2dx\right)w_{,xx}. \tag{3.71}$$

For simply supported beams, a one term Rayleigh-Ritz approximation can be obtained by taking

$$w=\alpha\sin\frac{\pi x}{L}. \tag{3.72}$$

Substituting into (3.71), multiplying by $\sin\frac{\pi x}{L}$, integrating over the length of the beam, and using the change of variables $\tau=\omega t$, where $\omega^2=(\frac{\pi}{L})^4\frac{EI}{\rho A}$, yields the well-known Duffing equation:

$$\ddot{\alpha}+\alpha+\epsilon\alpha^3=0 \tag{3.73}$$

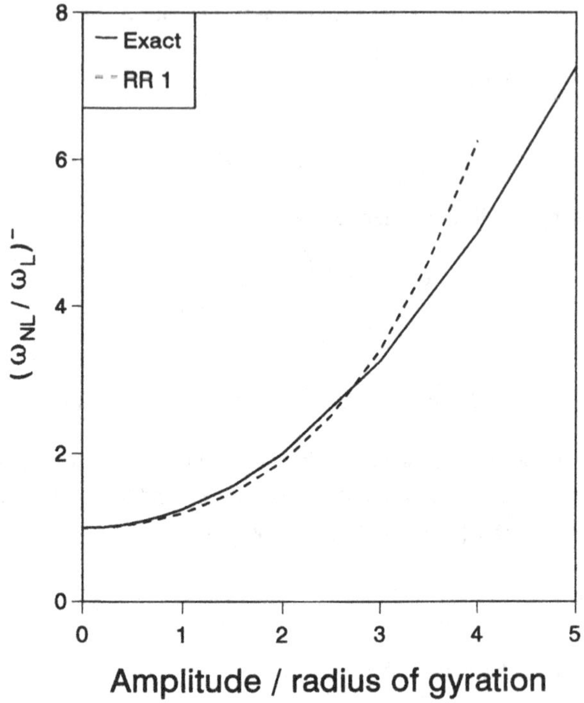

Figure 3.6. Effect of geometrical nonlinearities on the free vibrations of beams.

where

$$\epsilon = \frac{A}{4I} = \frac{1}{4r^2} \tag{3.74}$$

and r is the radius of gyration of the cross section. Using the Linstedt-Poincare perturbation approach, the ratio of the nonlinear and linear fundamental frequencies is obtained as

$$\frac{\omega_{NL}}{\omega_L} = 1 + \frac{3}{8}\epsilon\bar{\alpha}^2 = 1 + \frac{3}{32}\left(\frac{\bar{\alpha}}{r}\right)^2 \tag{3.75}$$

where $\bar{\alpha}$ is the vibration amplitude. Figure 3.6 shows that when the amplitude remains small compared with the radius of gyration of the beam cross section, the nonlinear effects remain small and a linear model can be used.

3.2.10 Summary

Any beam theory is based on a set of assumptions aimed at reducing the three-dimensional elasticity problem to a one-dimensional problem. In this section, we reviewed the basic assumptions made in the development of three basic theories: the Bernoulli-Euler beam theory, the Timoshenko beam theory, and a higher-order theory. The equations of motions and boundary conditions were derived from Hamilton's principle in a consistent way for each theory. The example of a simply supported beam subjected to a uniformly distributed load was used to compare the stresses and deflections predicted by the various beam theory to elasticity results. The objective was to evaluate the validity of the assumptions made in the development of the theories and to develop some general guidelines concerning the applicability of these theories. This example showed that for slender beams, the Bernoulli-Euler beam theory, which is the simplest of the three, provides very accurate results, but for shorter beams, the Timoshenko beam, which accounts for transverse shear deformations, is required. Shear deformation effects depend on the ratio of the elastic and shear moduli and therefore are more pronounced for composite beams, since the ratio is usually much higher for composites than for steel or aluminum.

An analysis of harmonic wave propagation in beams shows that for most practical cases, the higher-order theory gives results that are very close to those predicted by the Timoshenko beam theory. Therefore, the increased complexity of the theory is not justified by any improvement in the results. Free vibration analyses show the importance of transverse shear deformation and rotary inertia on the dynamic behavior of beams. The effect of large deformations is investigated by considering the large-amplitude vibration of simply supported beams. Results show that as long as the amplitude remains small relative to the radius of gyration of the cross section, nonlinear effects are negligible.

In many cases, impact tests are performed on beam specimens, and the beam theories discussed in this section will be used in the development of mathematical models for studying the impact dynamics. In addition, the knowledge about the various assumptions and the effects of shear deformations and rotary inertia gained through this study of beam theories will be directly applicable to the development of plate and shells theories examined in the following sections.

3.2.11 Exercise Problems

3.1 Using the Timoshenko beam theory, derive an expression for the static deflection of a cantilever subjected to a concentrated force P at the end. Show the effect of the slenderness h/L ratio on the magnitude of the shear deflection relative to the bending deflection.

3.2 Following the approach shown Section 3.28, formulate the free vibration problem for a cantilever Bernoulli-Euler beam. Write down the four boundary conditions in the form given by (3.46), and expand the 4×4 determinant to obtain the characteristic equation.

3.3 Derive a beam theory neglecting the transverse normal strain and assuming that the transverse shear strain varies as a cosine function of z that vanishes on the top and bottom surfaces and reaches a maximum for $z = 0$.

3.4 Determine the axial stresses and the transverse shear stresses in a simply supported, rectangular, Timoshenko beam subjected to a sinusoidally varying distributed load that vanishes at both ends. Compare with the results obtained using the Bernoulli-Euler beam theory and determine the effect of shear deformation. Use the approach shown in Section 3.2.6 to determine the transverse shear distribution from the results of the Bernoulli-Euler beam theory.

3.5 Write a computer program to calculate the natural frequencies of Timoshenko beams using the exact approach discussed in Section 3.2.7.

3.3 Plate Theories

From the discussion on the effect of shear deformation and rotary inertia on the static and dynamic behavior of beams, it is expected that these two factors will have similar effects on the behavior of plates, particularly for plates made out of composite materials. The first-order shear deformation theory (FSDT), also known as the Mindlin or Mindlin-Reissner plate theory, which is an extension of the Timoshenko beam theory, will be presented first. For thin plates, the effect of shear deformation and rotatory inertia are small, and the theory can be reduced to the classical plate theory (CPT), which is the equivalent of the Bernoulli-Euler theory for beams. Higher-order plate theories are also available and will be discussed.

3.3.1 First-Order Shear Deformation Theory

Following the discussion on beam theory, the development of a plate theory following the lines of the Timoshenko beam theory proceeds by assuming that the transverse normal strain ϵ_{zz} is negligible and that the two transverse shear strains are constant through-the-thickness. These two assumptions, which can be written as

$$\epsilon_{zz} = \frac{\partial w}{\partial z} = 0 \tag{3.76a}$$

$$\epsilon_{xz} = \frac{\partial u}{\partial z} + \frac{\partial w}{\partial x} = \gamma_1 \tag{3.76b}$$

$$\epsilon_{yz} = \frac{\partial v}{\partial z} + \frac{\partial w}{\partial y} = \gamma_2, \tag{3.76c}$$

imply that the displacements are of the form

$$\begin{aligned} u(x,y,z) &= u_o(x,y) + z(\gamma_1 - w_{o,x}) \\ v(x,y,z) &= v_o(x,y) + z(\gamma_2 - w_{o,x}) \\ w(x,y,z) &= w_o(x,y). \end{aligned} \tag{3.77}$$

The change of variables

$$\psi_x = \gamma_1 - w_{o,x}, \quad \psi_y = \gamma_2 - w_{o,y} \tag{3.78}$$

gives the displacement field of the first-order shear deformation theory in its usual form:

$$\begin{aligned} u(x,y,z) &= u_o(x,y) + z\psi_x(x,y) \\ v(x,y,z) &= v_o(x,y) + z\psi_y(x,y) \\ w(x,y,z) &= w_o(x,y) \end{aligned} \tag{3.79}$$

where u_o, v_o, w_o are the displacement components of a point on the midplane. ψ_x and ψ_y are the rotations of the normal to the midplane about the y- and x-axes respectively. Equations (3.79) are often postulated as an assumption on the kinematics of the deformation in which a line segment initially normal to the midplane of the plate is assumed to remain straight but not necessarily normal to the midsurface during deformation. The infinitesimal strain components, defined as

$$\begin{aligned} &\epsilon_{xx} = \frac{\partial u}{\partial x}, \quad \epsilon_{yy} = \frac{\partial v}{\partial y}, \quad \epsilon_{zz} = \frac{\partial w}{\partial z}, \\ &\epsilon_{xy} = \frac{\partial u}{\partial y} + \frac{\partial v}{\partial x}, \quad \epsilon_{xz} = \frac{\partial u}{\partial z} + \frac{\partial w}{\partial x}, \quad \epsilon_{yz} = \frac{\partial v}{\partial z} + \frac{\partial w}{\partial y}, \end{aligned} \tag{3.80}$$

can be evaluated from (3.79):

$$\begin{aligned} &\epsilon_{xx} = \frac{\partial u_o}{\partial x} + z\frac{\partial \psi_x}{\partial x}, \quad \epsilon_{yy} = \frac{\partial v_o}{\partial y} + z\frac{\partial \psi_y}{\partial y}, \\ &\epsilon_{zz} = 0, \quad \epsilon_{xy} = \frac{\partial u_0}{\partial y} + \frac{\partial v_o}{\partial x} + z\left(\frac{\partial \psi_x}{\partial y} + \frac{\partial \psi_y}{\partial x}\right), \\ &\epsilon_{xz} = \frac{\partial w_o}{\partial x} + \psi_x, \quad \epsilon_{yz} = \frac{\partial w_o}{\partial y} + \psi_y. \end{aligned} \tag{3.81}$$

The plate is subjected to a distributed load q over the surface, normal inplane force resultants $\hat{N}_n$ over a part of the boundary Γ_1, $\hat{N}_s$ over Γ_2, moment resultants $\hat{M}_n$ and $\hat{M}_s$ over Γ_3 and Γ_4 respectively, and transverse shear $\hat{Q}_n$ forces over Γ_5. The inplane force resultants are defined as

$$(N_x, N_y, N_{xy}) = \int_{-\frac{h}{2}}^{\frac{h}{2}} (\sigma_{xx}, \sigma_{yy}, \sigma_{xy}) dz; \tag{3.82}$$

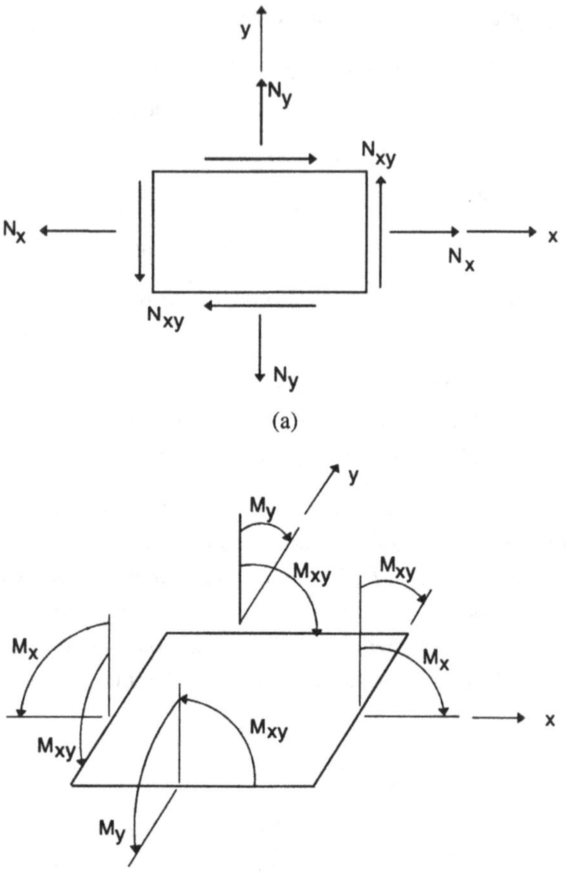

Figure 3.7. Force and moment resultants on a plate.

the bending moments as

$$(M_x, M_y, M_{xy}) = \int_{-\frac{h}{2}}^{\frac{h}{2}} (\sigma_{xx}, \sigma_{yy}, \sigma_{xy}) z \, dz; \tag{3.83}$$

and the shear force resultants as

$$(Q_x, Q_y) = \int_{-\frac{h}{2}}^{\frac{h}{2}} (\sigma_{xz}, \sigma_{yz}) \, dz. \tag{3.84}$$

The force and moment resultants acting on a small plate element are shown in Fig. 3.7a,b. Along a line oriented by the unit normal vector $\bar{n}$ (Fig. 3.7c), the inplane force resultants and the moment resultants are related to the x- and

y-components as

$$\begin{aligned} N_n &= N_x n_x^2 + N_y n_y^2 + 2n_x n_y N_{xy} \\ N_s &= (N_y - N_x)n_x n_y + N_{xy}\left(n_x^2 - n_y^2\right) \\ M_n &= M_x n_x^2 + M_y n_y^2 + 2n_x n_y M_{xy} \\ M_s &= (M_y - M_x)n_x n_y + M_{xy}\left(n_x^2 - n_y^2\right). \end{aligned} \tag{3.85}$$

The strain energy in a solid is defined as

$$U = \frac{1}{2}\int_V (\sigma_{xx}\epsilon_{xx} + \sigma_{yy}\epsilon_{yy} + \sigma_{zz}\epsilon_{zz} + \sigma_{xy}\epsilon_{xy} + \sigma_{xz}\epsilon_{xz} + \sigma_{yz}\epsilon_{yz})\, dV. \tag{3.86}$$

For a plate, using the strain-displacement relations (3.81),

$$\begin{aligned} U = \frac{1}{2}\int_\Omega \int_{-\frac{h}{2}}^{\frac{h}{2}} &\left\{ \sigma_{xx}\frac{\partial u_o}{\partial x} + \sigma_{yy}\frac{\partial v_o}{\partial y} + \sigma_{xy}\left(\frac{\partial u_o}{\partial y} + \frac{\partial v_o}{\partial x}\right) \right. \\ &+ z\left[\sigma_{xx}\frac{\partial \psi_x}{\partial x} + \sigma_{yy}\frac{\partial \psi_y}{\partial y} + \sigma_{xy}\left(\frac{\partial \psi_x}{\partial y} + \frac{\partial \psi_y}{\partial x}\right)\right] \\ &\left. + \sigma_{xz}\left(\frac{\partial w_o}{\partial x} + \psi_x\right) + \sigma_{yz}\left(\frac{\partial w_o}{\partial y} + \psi_y\right)\right\} dz\, d\Omega. \end{aligned} \tag{3.87}$$

Integrating over the thickness and using definitions for the force and moment resultants (3.82)–(3.84), we obtain the strain energy as

$$\begin{aligned} U = \frac{1}{2}\int_\Omega &\left\{ N_x\frac{\partial u_o}{\partial x} + N_y\frac{\partial v_o}{\partial y} + N_{xy}\left(\frac{\partial u_o}{\partial y} + \frac{\partial v_o}{\partial x}\right) \right. \\ &+ M_x\frac{\partial \psi_x}{\partial x} + M_y\frac{\partial \psi_y}{\partial x} + M_{xy}\left(\frac{\partial \psi_x}{\partial y} + \frac{\partial \psi_y}{\partial x}\right) \\ &\left. + Q_x\left(\frac{\partial w_o}{\partial x} + \psi_x\right) + Q_y\left(\frac{\partial w_o}{\partial y} + \psi_y\right)\right\} d\Omega. \end{aligned} \tag{3.88}$$

With the notation

$$\begin{aligned} \{N\} &= [N_x, N_y, N_{xy}]^{\mathrm{T}} \\ \{M\} &= [M_x, M_y, M_{xy}]^{\mathrm{T}} \\ \{\epsilon^o\} &= \left[\epsilon_{xx}^o, \epsilon_{yy}^o, \epsilon_{xz}^o\right]^{\mathrm{T}} = \left[\frac{\partial u_o}{\partial x}, \frac{\partial v_o}{\partial y}, \frac{\partial u_o}{\partial y} + \frac{\partial v_o}{\partial x}\right]^{\mathrm{T}} \\ \{K\} &= [K_x, K_y, K_{xy}]^{\mathrm{T}} = \left[\frac{\partial \psi_x}{\partial x}, \frac{\partial \psi_y}{\partial y}, \frac{\partial \psi_x}{\partial y} + \frac{\partial \psi_y}{\partial x}\right]^{\mathrm{T}} \end{aligned} \tag{3.89}$$

where the ϵ^o are called the midplane strains and $\{K\}$ the plate curvatures, the strain energy can be rewritten as

$$U = \frac{1}{2}\int_{\Omega}\left[\{N\}^{\mathrm{T}}\{\epsilon^o\} + \{M\}^{\mathrm{T}}\{\kappa\} + [Q_x, Q_y]\begin{Bmatrix}\epsilon_{xz}\\ \epsilon_{yz}\end{Bmatrix}\right] dV. \tag{3.90}$$

The potential energy of the external forces is given by

$$\begin{aligned} V = &-\int_{\Omega} q w_o d\Omega - \int_{\Gamma} \bar{N}_n u_{no} d\Gamma - \int_{\Gamma_2} \hat{N}_s u_{so}\, d\Gamma \\ &- \int_{\Gamma_3} \hat{M}_n \psi_n d\Gamma - \int_{\Gamma_4} \hat{M}_s \psi_n d\Gamma - \int_{\Gamma_5} \hat{Q}_n w_o\, d\Gamma. \end{aligned} \tag{3.91}$$

The kinetic energy of the plate is defined as

$$T = \frac{1}{2}\int_V \rho[(\dot{u})^2 + (\dot{v})^2 + (\dot{w})^2] dV. \tag{3.92}$$

Using (3.79) we find

$$T = \frac{1}{2}\int_{\Omega}\int_{-\frac{h}{2}}^{\frac{h}{2}} \rho\left[(\dot{u}_o + z\dot{\psi}_x)^2 + (\dot{v}_o + z\dot{\psi}_y) + (\dot{w}_o)^2\right] dz\, d\Omega \tag{3.93}$$

which can be written as

$$\begin{aligned} T = \frac{1}{2}\int_{\Omega} &\{I_1(\dot{u}_o^2 + \dot{v}_o^2 + \dot{w}_o^2) + 2I_2(\dot{u}_o\dot{\psi}_x + \dot{v}_o\dot{\psi}_y) \\ &+ I_3(\dot{\psi}_x^2 + \dot{\psi}_y^2)\} d\Omega \end{aligned} \tag{3.94}$$

where the inertia terms are defined as

$$(I_1, I_2, I_3) = \int_{-\frac{h}{2}}^{\frac{h}{2}} \rho(1, z, z^2) dz. \tag{3.95}$$

I_1 is the mass per unit length, I_3 is usually called the rotary inertia, and I_2 is a coupling term that vanishes if the plate is symmetric about the xy-plane.

The stress-strain relations for a single layer of orthotropic material are (Vinson and Sierakowski 1987)

$$\begin{Bmatrix}\sigma_{xx}\\ \sigma_{yy}\\ \sigma_{xy}\end{Bmatrix} = \begin{bmatrix}\bar{Q}_{11} & \bar{Q}_{12} & \bar{Q}_{16}\\ \bar{Q}_{12} & \bar{Q}_{22} & \bar{Q}_{26}\\ \bar{Q}_{16} & \bar{Q}_{26} & \bar{Q}_{66}\end{bmatrix}\begin{Bmatrix}\epsilon_{xx}\\ \epsilon_{yy}\\ \epsilon_{xy}\end{Bmatrix}. \tag{3.96}$$

The strain-displacement relations (3.81) can be written as

$$\begin{Bmatrix}\epsilon_{xx}\\ \epsilon_{yy}\\ \epsilon_{xy}\end{Bmatrix} = \begin{Bmatrix}\frac{\partial u_o}{\partial x}\\ \frac{\partial v_o}{\partial y}\\ \frac{\partial u_o}{\partial y} + \frac{\partial v_o}{\partial x}\end{Bmatrix} + z\begin{Bmatrix}\frac{\partial \psi_x}{\partial x}\\ \frac{\partial \psi_y}{\partial y}\\ \frac{\partial \psi_x}{\partial y} + \frac{\partial \psi_y}{\partial x}\end{Bmatrix} \tag{3.97}$$

or

$$\{\epsilon\} = \{\epsilon_o\} + z\{\kappa\} \tag{3.98}$$

where $\{\epsilon_o\}$ is the vector of midplane strains and $\{\kappa\}$ denotes the plate curvatures. Using the definitions (3.82)–(3.84) we find

$$\begin{Bmatrix} N_x \\ N_y \\ N_{xy} \\ M_x \\ M_y \\ M_{xy} \end{Bmatrix} = \begin{bmatrix} A_{11} & A_{12} & A_{16} & B_{11} & B_{12} & B_{16} \\ A_{12} & A_{22} & A_{26} & B_{12} & B_{22} & B_{26} \\ A_{16} & A_{26} & A_{66} & B_{16} & B_{26} & B_{66} \\ B_{11} & B_{12} & B_{16} & D_{11} & D_{12} & D_{16} \\ B_{12} & B_{22} & B_{26} & D_{12} & D_{22} & D_{26} \\ B_{16} & B_{26} & B_{66} & D_{16} & D_{26} & D_{66} \end{bmatrix} \begin{Bmatrix} \epsilon_{xx} \\ \epsilon_{yy} \\ \epsilon_{xy} \\ \kappa_x \\ \kappa_y \\ \kappa_{xy} \end{Bmatrix} \tag{3.99}$$

where

$$A_{ij} = \sum \bar{Q}_{ij}|_k (z_k - z_{k-1}) \tag{3.100}$$

$$B_{ij} = \frac{1}{2} \sum \bar{Q}_{ij}|_k \left(z_k^2 - z_{k-1}^2\right) \tag{3.101}$$

$$D_{ij} = \frac{1}{3} \sum \bar{Q}_{ij}|_k \left(z_k^3 - z_{k-1}^3\right). \tag{3.102}$$

For a general laminate, (3.99) shows extension-bending coupling. That is, the application of only inplane forces will produce both inplane and out-of-plane deformations. This coupling disappears when a symmetric layup is used. In that case, the B matrix vanishes.

$$\begin{Bmatrix} Q_y \\ Q_x \end{Bmatrix} = \begin{bmatrix} A_{44} & A_{45} \\ A_{45} & A_{55} \end{bmatrix} \begin{Bmatrix} \psi_y + \frac{\partial w_o}{\partial y} \\ \psi_x + \frac{\partial w_o}{\partial x} \end{Bmatrix} \tag{3.103}$$

$$A_{ij} = k_i k_j \int_{-\frac{h}{2}}^{\frac{h}{2}} Q_{ij}\, dz, \quad i, j = 4, 5. \tag{3.104}$$

k_i and k_j are two shear correction factors that are used to account for the nonuniformity and parabolic shape of the shear stress distribution through-the-thickness, as indicated by (3.81). k_i and k_j are usually taken as 5/6.

Applying Hamilton's principle (3.10) gives the equations of motion in terms of the force and moment resultants:

$$\frac{\partial N_x}{\partial x} + \frac{\partial N_{xy}}{\partial y} = I_1 \ddot{u}_o + I_2 \ddot{\psi}_x \tag{3.105}$$

$$\frac{\partial N_{xy}}{\partial x} + \frac{\partial N_y}{\partial y} = I_1 \ddot{v}_o + I_2 \ddot{\psi}_y \tag{3.106}$$

$$\frac{\partial Q_x}{\partial x} + \frac{\partial Q_y}{\partial y} + q = I_1 \ddot{w}_o \tag{3.107}$$

$$\frac{\partial M_x}{\partial x} + \frac{\partial M_{xy}}{\partial y} - Q_x = I_3\ddot{\psi}_x + I_2\ddot{u}_o \tag{3.108}$$

$$\frac{\partial M_{xy}}{\partial x} + \frac{\partial M_y}{\partial y} - Q_y = I_3\ddot{\psi}_y + I_2\ddot{v}_o. \tag{3.109}$$

Substituting (3.99) into (3.105)–(3.109) gives the equations of motions for a general laminate in terms of the displacements. These equations are very lengthy and will not be reproduced here. Instead we present below the equations of motion for a symmetrically laminated plate ($[B] = 0$). This does not present a serious restriction since symmetric laminates are used in most applications.

$$\frac{\partial}{\partial x}\left[A_{45}\left(\psi_y + \frac{\partial w_o}{\partial y}\right) + A_{55}\left(\psi_x + \frac{\partial w_o}{\partial x}\right)\right] + \frac{\partial}{\partial y}\left[A_{44}\left(\psi_y + \frac{\partial w_o}{\partial y}\right) + A_{45}\left(\psi_x + \frac{\partial w_o}{\partial x}\right)\right] + q = I_1\dot{w}_o \tag{3.110}$$

$$\frac{\partial}{\partial x}\left[D_{11}\frac{\partial \psi_x}{\partial x} + D_{12}\frac{\partial \psi_y}{\partial y} + D_{16}\left(\frac{\partial \psi_x}{\partial y} + \frac{\partial \psi_y}{\partial x}\right)\right] + \frac{\partial}{\partial y}\left[D_{16}\frac{\partial \psi_x}{\partial x} + D_{26}\frac{\partial \psi_y}{\partial y} + D_{66}\left(\frac{\partial \psi_x}{\partial y} + \frac{\partial \psi_y}{\partial x}\right)\right] - A_{45}\left(\psi_y + \frac{\partial w_o}{\partial y}\right) - A_{55}\left(\psi_x + \frac{\partial w_o}{\partial x}\right) = I_3\ddot{\psi}_x \tag{3.111}$$

$$\frac{\partial}{\partial x}\left[D_{16}\frac{\partial \psi_x}{\partial x} + D_{26}\frac{\partial \psi_y}{\partial y} + D_{66}\left(\frac{\partial \psi_x}{\partial y} + \frac{\partial \psi_y}{\partial x}\right)\right] + \frac{\partial}{\partial y}\left[D_{12}\frac{\partial \psi_x}{\partial x} + D_{22}\frac{\partial \psi_y}{\partial y} + D_{26}\left(\frac{\partial \psi_x}{\partial y} + \frac{\partial \psi_y}{\partial x}\right)\right] - A_{44}\left(\psi_y + \frac{\partial w_o}{\partial y}\right) - A_{45}\left(\psi_x + \frac{\partial w_o}{\partial y}\right) = I_3\ddot{\psi}_y \tag{3.112}$$

3.3.2 Classical Plate Theory (CPT)

As with the elementary beam theory, for the classical plate theory the effect of shear deformation and rotary inertia are neglected in addition to the transverse normal strain. Therefore, the functions γ_1 and γ_2 in (3.76b,c) are set to zero; according to (3.78),

$$\frac{\partial w_o}{\partial x} + \psi_x = 0, \qquad \frac{\partial w_o}{\partial y} + \psi_y = 0. \tag{3.113}$$

Functions ψ_x and ψ_y can be eliminated from the formulation, and the displacements are now in the form

$$
\begin{aligned}
u(x, y, z) &= u_o(x, y) - z\frac{\partial w_o}{\partial x}(x, y) \\
v(x, y, z) &= v_o(x, y) - z\frac{\partial w_o}{\partial x}(x, y) \\
w(x, y, z) &= w_o(x, y).
\end{aligned} \tag{3.114}
$$

This amounts to assuming that line segments normal to the midplane before deformation remain straight and remain perpendicular to the midsurface during deformation. The midplane strains remain unchanged, and the plate curvatures become

$$
\{K\} = \left[-\frac{\partial^2 w}{\partial x^2}, \quad -\frac{\partial^2 w}{\partial y^2}, \quad -2\frac{\partial^2 w}{\partial x \partial y}\right]^{\mathrm{T}}. \tag{3.115}
$$

Substituting into (3.108) and (3.109) and combining these two equations gives the following equation in terms of the moment resultants:

$$
\begin{aligned}
&\frac{\partial^2 M_x}{\partial x^2} + 2\frac{\partial^2 M_{xy}}{\partial x \partial y} + \frac{\partial^2 M_y}{\partial y^2} + q - I_1 \ddot{w}_o \\
&= I_3\left\{-\frac{\partial^2 \ddot{w}_o}{\partial x^2} - \frac{\partial^2 \ddot{w}_o}{\partial y^2}\right\} + I_2\left\{\frac{\partial \ddot{u}_o}{\partial x} + \frac{\partial \ddot{v}_o}{\partial y}\right\}
\end{aligned} \tag{3.116}
$$

Since the effect of shear deformation has already been neglected, and since as has been shown for beams the effect of rotary inertia is of the same order of magnitude or smaller, it is reasonable also to neglect rotatory inertia ($I_3 = 0$). For a symmetric layup, $I_2 = 0$, $B_{ij} = 0$, and the transverse motion is decoupled from the inplane motion. Therefore, the transverse displacements must satisfy a single equation of motion:

$$
\begin{aligned}
q = {} & D_{11}\frac{\partial^4 w_o}{\partial x^4} + 2(D_{12} + 2D_{66})\frac{\partial^4 w_o}{\partial x^2 \partial y^2} + D_{22}\frac{\partial^4 w_o}{\partial x^2 \partial y^2} \\
& + 4D_{16}\frac{\partial^4 w_o}{\partial x^3 \partial y} + 4D_{26}\frac{\partial^4 w_o}{\partial x \partial y^3} + I_1 \ddot{w}_o.
\end{aligned} \tag{3.117}
$$

Equation (3.117) is the equation of motion for symmetrically laminated plates according to the classical plate theory and is expressed in terms of the transverse displacement w.

3.3.3 Higher-Order Plate Theories (HOPT)

Following the same approach as for the higher-order beam theory (Section 3.2.2), a higher-order plate theory is derived by assuming that:

(H1) The transverse shear strains ϵ_{xz} and ϵ_{yz} have a parabolic variation through-the-thickness and vanish on the top and bottom surfaces.
(H2) The transverse normal strain ϵ_{zz} varies linearly through-the-thickness.

These two assumptions can be expressed as

$$\epsilon_{xz} = \frac{\partial u}{\partial z} + \frac{\partial w}{\partial x} = \epsilon_1\left(1 - 4\frac{z^2}{h^2}\right), \tag{3.118a}$$

$$\epsilon_{yz} = \frac{\partial v}{\partial z} + \frac{\partial w}{\partial y} = \epsilon_2\left(1 - 4\frac{z^2}{h^2}\right), \tag{3.118b}$$

$$\epsilon_{zz} = \frac{\partial w}{\partial z} = \epsilon_3\, z. \tag{3.118c}$$

Equation (3.118c) implies that

$$w(x, y, z, t) = w_o(x, y, t) + \frac{z^2}{2}\epsilon_3. \tag{3.119}$$

After substitution into (3.118a,b), we find

$$\begin{aligned} u(x, y, z, t) &= u_o(x, y, t) + z(\epsilon_1 - w_{o,x}) - \frac{4z^3}{3h^2}\epsilon_1 - \frac{z^3}{6}\epsilon_{3,x}, \\ v(x, y, z, t) &= v_o(x, y, t) + z(\epsilon_2 - w_{o,y}) - \frac{4z^3}{3h^2}\epsilon_2 - \frac{z^3}{6}\epsilon_{3,y}. \end{aligned} \tag{3.120}$$

Introducing the new variables

$$\psi_x = \epsilon_1 - w_{o,x}, \quad \psi_y = \epsilon_2 - w_{o,y}, \quad \phi_z = \epsilon_3/2, \tag{3.121}$$

the kinematic relations used by Hanna and Leissa (1994) are recovered:

$$u = z\psi_x - \frac{4z^3}{3h^2}(\psi_x + w_{o,x}) - \frac{z^3}{3}\phi_{z,x}, \tag{3.122a}$$

$$v = z\psi_y - \frac{4z^3}{3h^2}(\psi_y + w_{o,y}) - \frac{z^3}{3}\phi_{z,y}, \tag{3.122b}$$

$$w = w_o + z^2\phi_z. \tag{3.122c}$$

If the transverse normal strain is neglected, the kinematic relations used in the higher-order plate theory (HOPT) of Reddy (1984) are recovered:

$$u = u_o + z\left[\Psi_x - \frac{4}{3}\left(\frac{z}{h}\right)^2\left(\Psi_x + \frac{\partial w}{\partial x}\right)\right], \tag{3.123a}$$

$$v = v_o + z\left[\Psi_y - \frac{4}{3}\left(\frac{z}{h}\right)^2\left(\Psi_y + \frac{\partial w}{\partial y}\right)\right], \tag{3.123b}$$

$$w = w_o. \tag{3.123c}$$

The strain-displacement relations for the HOPT are

$$\epsilon_{xx} = \frac{\partial u_o}{\partial x} + z\left[\frac{\partial \Psi_x}{\partial x} - \frac{4}{3}\left(\frac{z}{h}\right)^2\left(\frac{\partial \Psi_x}{\partial x} + \frac{\partial^2 w}{\partial x^2}\right)\right] \tag{3.124}$$

$$\epsilon_{yy} = \frac{\partial v_o}{\partial y} + z\left[\frac{\partial \Psi_y}{\partial y} - \frac{4}{3}\left(\frac{z}{h}\right)^2\left(\frac{\partial \Psi_y}{\partial y} + \frac{\partial^2 w}{\partial y^2}\right)\right] \tag{3.125}$$

$$\epsilon_{zz} = 0 \tag{3.126}$$

$$\epsilon_{yz} = \frac{\partial w_o}{\partial y} + \Psi_y - 4\frac{z^2}{h^2}\left(\Psi_y + \frac{\partial w}{\partial y}\right) \tag{3.127}$$

$$\epsilon_{xz} = \frac{\partial w_o}{\partial x} + \Psi_x - 4\frac{z^2}{h^2}\left(\Psi_x + \frac{\partial w}{\partial x}\right) \tag{3.128}$$

$$\epsilon_{xy} = \frac{\partial u_o}{\partial y} + \frac{\partial v_o}{\partial x} + z\left[\frac{\partial \Psi_x}{\partial y} + \frac{\partial \Psi_y}{\partial x} - \frac{4}{3}\left(\frac{z}{h}\right)^2\left(\frac{\partial \Psi_x}{\partial y} + \frac{\partial \Psi_y}{\partial x} + 2\frac{\partial^2 w}{\partial x \partial y}\right)\right]. \tag{3.129}$$

These can be rewritten as

$$\epsilon_{xx} = \epsilon_{xx}^o + zK_x + z^3 K_x', \quad \epsilon_{yy} = \epsilon_{yy}^o + zK_y + z^3 K_y', \tag{3.130a,b}$$

$$\epsilon_{yz} = \epsilon_{yz}^o + z^2 K_{yz}', \quad \epsilon_{xz} = \epsilon_{xz}^o + z^2 K_{xz}', \tag{3.130c,d}$$

$$\epsilon_{xy} = \epsilon_{xy}^o + zK_{xy} + z^3 K_{xy}', \tag{3.130e}$$

where

$$K_x' = -\frac{4}{3h^2}\left(\frac{\partial \Psi_x}{\partial x} + \frac{\partial^2 w_o}{\partial x^2}\right), \quad K_y' = -\frac{4}{3h^2}\left(\frac{\partial \Psi_y}{\partial y} + \frac{\partial^2 w_o}{\partial y^2}\right), \tag{3.131a,b}$$

$$K_{yz}' = -\frac{4}{h^2}\left(\Psi_y + \frac{\partial w_o}{\partial y}\right), \quad K_{xz}' = -\frac{4}{h^2}\left(\Psi_x + \frac{\partial w_o}{\partial x}\right), \tag{3.131c,d}$$

$$K_{yz}' = -\frac{4}{3h^2}\left(\frac{\partial \Psi_x}{\partial y} + \frac{\partial \Psi_y}{\partial x} + 2\frac{\partial^2 w_o}{\partial x^2}\right). \tag{3.131e}$$

The stress resultants N_i, M_i, and R_i are defined by

$$(N_x, M_x, P_x) = \int_{-\frac{h}{2}}^{\frac{h}{2}} \sigma_{xx}(1, z, z^3)dz \tag{3.132a}$$

$$(N_y, M_y, P_y) = \int_{-\frac{h}{2}}^{\frac{h}{2}} \sigma_{yy}(1, z, z^3)dz \tag{3.132b}$$

$$(N_{xy}, M_{xy}, P_{xy}) = \int_{-\frac{h}{2}}^{\frac{h}{2}} \sigma_{xy}(1, z, z^3)dz \tag{3.132c}$$

$$(Q_2, R_2) = \int_{-\frac{h}{2}}^{\frac{h}{2}} \sigma_4(1, z^2)dz \tag{3.132d}$$

$$(Q_1, R_1) = \int_{-\frac{h}{2}}^{\frac{h}{2}} \sigma_5(1, z^2)dz. \tag{3.132e}$$

The inertias $I_1 - I_7$ are defined by

$$I_i = \int_{-\frac{h}{2}}^{\frac{h}{2}} \rho z^{i-1} dz, \quad i = 1, 7. \tag{3.133}$$

The equations of motion are

$$\frac{\partial N_x}{\partial x} + \frac{\partial N_y}{\partial y} = I_1 \ddot{u} + \bar{I}_2 \ddot{\Psi}_x - \frac{4}{3h^2} I_4 \frac{\partial^2 \ddot{w}}{\partial x}, \tag{3.134a}$$

$$\frac{\partial N_{xy}}{\partial x} + \frac{\partial N_y}{\partial y} = I_1 \ddot{v} + \bar{I}_2 \ddot{\Psi}_y - \frac{4}{3h^2} I_4 \frac{\partial^2 w}{\partial y^2}, \tag{3.134b}$$

$$\frac{\partial Q_x}{\partial x} + \frac{\partial Q_y}{\partial y} + q - \frac{4}{h^2}\left(\frac{\partial R_1}{\partial x} + \frac{\partial R_2}{\partial y}\right) + \frac{4}{3h^2}\left(\frac{\partial^2 P_x}{\partial x^2} + 2\frac{\partial^2 P_{xy}}{\partial y^2} + \frac{\partial^2 P_y}{\partial y^2}\right)$$
$$= I_1 \ddot{w} - \left(\frac{4}{3h^2}\right)^2 I_7 \left(\frac{\partial^2 \ddot{w}}{\partial x^2} + \frac{\partial^2 \ddot{w}}{\partial y^2}\right) + \frac{4}{3h^2} I_4 \left(\frac{\partial \ddot{u}}{\partial x} + \frac{\partial \ddot{v}}{\partial y}\right)$$
$$+ \frac{4}{3h^2} \bar{I}_5 \left(\frac{\partial \ddot{\Psi}_x}{\partial x} + \frac{\partial \ddot{\Psi}_y}{\partial y}\right), \tag{3.134c}$$

$$\frac{\partial M_x}{\partial x} + \frac{\partial M_{xy}}{\partial y} - Q_x + \frac{4}{h^2} R_x - \frac{4}{3h^2}\left(\frac{\partial P_x}{\partial x} + \frac{\partial P_{xy}}{\partial y}\right)$$
$$= \bar{I}_2 \ddot{u} + \bar{I}_3 \ddot{\Psi}_x - \frac{4}{3h^2} \bar{I}_5 \frac{\partial \ddot{w}}{\partial x}, \tag{3.134d}$$

$$\frac{\partial M_{xy}}{\partial x} + \frac{\partial M_y}{\partial y} - Q_y + \frac{4}{h^2} R_y - \frac{4}{3h^2}\left(\frac{\partial P_{xy}}{\partial x} + \frac{\partial P_y}{\partial y}\right)$$
$$= \bar{I}_2 \ddot{v} + \bar{I}_3 \ddot{\Psi}_y - \frac{4}{3h^2} \bar{I}_5 \frac{\partial \ddot{w}}{\partial y}, \tag{3.134e}$$

$$\bar{I}_2 = I_2 - \frac{4}{3h^2} I_4, \quad \bar{I}_5 = I_5 - \frac{4}{3h^2} I_7, \quad \bar{I}_3 = I_3 - \frac{8}{3h^2} I_5 + \frac{16}{9h^4} I_7. \tag{3.135}$$

The constitutive equations are

$$\{N\} = [A]\{\epsilon^o\}, \tag{3.136}$$

$$\begin{Bmatrix} \{M\} \\ \{P\} \end{Bmatrix} = \begin{bmatrix} [D] & [F] \\ [F]^{\mathrm{T}} & [H] \end{bmatrix} \begin{Bmatrix} \{K\} \\ \{K'_y\} \end{Bmatrix}, \tag{3.137}$$

$$\begin{Bmatrix} Q_y \\ Q_x \\ R_y \\ R_x \end{Bmatrix} = \begin{bmatrix} A_{44} & 0 & D_{44} & 0 \\ 0 & A_{55} & 0 & D_{55} \\ D_{44} & 0 & F_{44} & 0 \\ 0 & D_{55} & 0 & F_{55} \end{bmatrix} \begin{Bmatrix} \epsilon^o_{yz} \\ \epsilon^o_{xz} \\ K'_{yz} \\ K'_{xz} \end{Bmatrix}. \tag{3.138}$$

The complexity of this higher-order theory is obvious from these equations, but improvements in the results are relatively small; therefore, these theories are not used extensively.

3.3.4 Static Deflections of Plates

In order to evaluate the effect of several assumptions used in the development of the equations of motion, consider the static deflections of a simply supported, orthotropic, rectangular plate according to both the classical plate theory and the first-order shear deformation theory. This example is selected because it admits an exact solution called the Navier solution. Because no approximate solution procedure is involved, the two theories can be compared directly. Exact solutions are also available for rectangular plates with simple supports on two opposite sides and with any support conditions along the other two sides; these are called Levy-type solutions (Vinson and Sierakowski 1987).

The boundary conditions for a simply supported rectangular plate are written as

$$w(0, y) = w(a, y) = w(x, 0) = w(x, b) = 0 \tag{3.139}$$

$$M_x(0, y) = M_x(a, y) = M_y(x, 0) = M_y(x, b) = 0. \tag{3.140}$$

With the classical plate theory, we expand the transverse displacement and the transverse load into double Fourier series as

$$w(x, y) = \sum_{m,n}^{\infty} W_{mn} \sin \frac{m\pi x}{a} \sin \frac{n\pi y}{b} \tag{3.141}$$

$$q = \sum_{m,n} q_{mn} \sin \frac{m\pi x}{a} \sin \frac{n\pi y}{b}. \tag{3.142}$$

Substituting into (3.117), (3.139), and (3.140) shows that both the equations of equilibrium and the boundary conditions are satisfied when the bending-twisting coupling coefficients are zero ($D_{16} = D_{26} = 0$). Substituting (3.141) and (3.142) into (3.117), multiplying through by $\sin \frac{m'\pi x}{a} \sin \frac{n'\pi y}{b}$, integrating over Ω, and using the result

$$\int_0^a \sin \frac{m\pi x}{a} \sin \frac{m'\pi x}{a} dx = \begin{cases} \frac{a}{2} & \text{when } m = m' \\ 0 & \text{when } m \neq m' \end{cases} \tag{3.143}$$

yields the following expression for the coefficients in the double series expansion for the transverse displacements:

$$W_{mn}\frac{\pi^4}{a^4}\left[D_{11}m^4 + 2(D_{12} + 2D_{66})m^2n^2r^2 + D_{22}n^4r^4\right] = q_{mn} \tag{3.144}$$

where $r = a/b$ is the aspect ratio of the plate.

Using (3.141) and (3.144), a solution is found as long as q is expressed in a double series as in (3.142). Now, examine how this is done for three special cases: a concentrated force, a patch load, and a uniformly distributed load. When the external load is a concentrated force P applied at x_o and y_o,

$$q = P\delta(x - x_o)\,\delta(y - y_o) \tag{3.145}$$

where δ is the Dirac delta function. Multiplying (3.145) by $\sin(m\pi\frac{x}{a})\sin(n\pi\frac{y}{b})$ and integrating over Ω gives

$$q_{mn} = \frac{4P}{ab}\sin\left(m\pi\frac{x_o}{a}\right)\sin\left(n\pi\frac{y_o}{b}\right). \tag{3.146}$$

When P is applied in the center ($x_o/a = y_o/b = 1/2$), (3.146) indicates that $q_{mn} = 0$ when either m or n is even. Then, from (3.144), $W_{mn} = 0$, and therefore only odd values of m and n contribute to the plate deflection.

For a uniform pressure q distributed over a patch centered at (x_o, y_o),

$$q_{mn} = \frac{4q}{ab}\int_{x_o-\alpha}^{x_o+\alpha}\sin\frac{m\pi x}{a}dx \cdot \int_{y_o-\beta}^{y_o+\beta}\sin\frac{n\pi y}{b}dy, \tag{3.147}$$

$$q_{mn} = \frac{16q}{mn\pi^2}\sin\frac{m\pi x_o}{a}\sin\frac{n\pi y_o}{b}\sin\frac{m\pi\alpha}{a}\sin\frac{n\pi\beta}{b}. \tag{3.148}$$

When α/a and β/b are both very small (3.148) reduces to (3.146), since the total load applied on the patch is $4\alpha\beta q$.

To better simulate the normal pressures found during Hertzian contact, a cosine load is sometimes used. The pressure over a rectangular patch of size $2\alpha \times 2\beta$ centered at (x_o, y_o) is distributed according to

$$q(x, y) = q_o\cos\left(\frac{\pi}{2\alpha}(x - x_o)\right)\cos\left(\frac{\pi}{2\beta}(y - y_o)\right). \tag{3.149}$$

Then

$$q_{mn} = \frac{64P\sin\frac{m\pi x_o}{a}\sin\frac{n\pi y_o}{b}\cos\frac{m\pi\alpha}{a}\cos\frac{n\pi\beta}{b}}{ab\alpha^2\beta^2\left(\frac{m}{a} - \frac{1}{2\alpha}\right)\left(\frac{n}{b} - \frac{1}{2\beta}\right)\left(\frac{m}{a} + \frac{1}{2\alpha}\right)\left(\frac{n}{b} - \frac{1}{2\beta}\right)}$$

$$\text{when } \frac{m}{a} \neq \frac{1}{2\alpha},\quad \frac{n}{b} \neq \frac{1}{2\beta}, \text{ and}$$

$$\text{when } q_{mn} = 0 \text{ when either } \frac{m}{a} = \frac{1}{2\alpha} \quad \text{or} \quad \frac{n}{b} = \frac{1}{2\beta}. \tag{3.150}$$

When a uniformly distributed pressure is applied over the entire surface of the plate,

$$q_{mn} = \frac{4q_o}{ab} \int_\Omega \sin\frac{m\pi x}{a} \sin\frac{n\pi y}{b} d\Omega \tag{3.151}$$

$$q_{mn} = \frac{16q_o}{mn\pi^2} \quad \text{for } m, n = 1, 3, 5 \ldots. \tag{3.152}$$

$q_{mn} = 0$ when either m or n is even.

Using the FSDT, the solution is taken in the form

$$w_o = \sum_{m,n=1}^{\infty} W_{mn} \sin\left(m\pi\frac{x}{a}\right) \sin\left(n\pi\frac{y}{b}\right) \tag{3.153a}$$

$$\psi_x = \sum_{m,n=1}^{\infty} X_{mn} \cos\left(m\pi\frac{x}{a}\right) \sin\left(n\pi\frac{y}{b}\right) \tag{3.153b}$$

$$\psi_y = \sum_{m,n=1}^{\infty} Y_{mn} \sin\left(m\pi\frac{x}{a}\right) \cos\left(n\pi\frac{y}{b}\right). \tag{3.153c}$$

Substituting into the equations of motion (3.110)–(3.112) gives three coupled algebraic equations for each combination of m and n:

$$\begin{bmatrix} K_{11} & K_{12} & K_{13} \\ K_{12} & K_{22} & K_{23} \\ K_{13} & K_{23} & K_{33} \end{bmatrix} \begin{Bmatrix} W_{mn} \\ X_{mn} \\ Y_{mn} \end{Bmatrix} = \begin{Bmatrix} q_{mn} \\ 0 \\ 0 \end{Bmatrix} \tag{3.154}$$

where

$$K_{11} = A_{55}\left(\frac{m\pi}{a}\right)^2 + A_{44}\left(\frac{n\pi}{b}\right)^2, \quad K_{12} = A_{55}\frac{m\pi}{a},$$

$$K_{13} = A_{44}\frac{n\pi}{b}, \quad K_{22} = D_{11}\left(\frac{m\pi}{a}\right)^2 + D_{66}\left(\frac{n\pi}{b}\right)^2 + A_{55}, \tag{3.155}$$

$$K_{23} = D_{12}\frac{mn\pi^2}{ab} + D_{66}\frac{mn\pi^2}{ab}, \quad K_{33} = D_{66}\left(\frac{m\pi}{a}\right)^2 + D_{22}\left(\frac{n\pi}{b}\right)^2 + A_{44}.$$

For each (m, n) combination, the system of three equations (3.154) with three unknowns is solved for the coefficients W_{mn}, X_{mn}, and Y_{mn}. The solution to the problem is obtained by adding the contribution of all these terms in (3.153).

Example 3.4: Square SSSS Composite Plate Subjected to Concentrated Force in the Center To assess the influence of shear deformation on the deflections of composite plates, consider a 10-ply [0°,90°,0°,90°,0°]s graphite-epoxy plate with material properties

$$\begin{aligned} E_1 &= 120 \text{ GPa}, \quad E_2 = 7.9 \text{ GPa}, \quad \nu_{12} = 0.30, \\ G_{12} &= G_{23} = G_{13} = 5.5 \text{ GPa}, \quad \rho = 1580 \text{ kg/m}^3 \end{aligned} \tag{3.156}$$

Table 3.2. *Central deflection (10^{-3} mm) for simply supported composite plate subjected to a unit force applied in the center*

Plate size (mm)	Number of terms	CPT	FSDT
200	199 × 199	1.1766	7.2541
100	99 × 99	1.7941	1.8622
	199 × 199	1.7942	1.8712
50	99 × 99	.44852	.51627
	199 × 199	.44854	.52524
	499 × 499	.44854	.53707
25	99 × 99	.11213	.17950
	199 × 199		.18845
	499 × 499		.20028
	999 × 999		.20923

Table 3.3. *Central deflection (10^{-3} mm) for simply supported composite plate subjected to a unit patch load*

Plate size (mm)	Number of terms	CPT	FSDT
200	99 × 99	7.1643	7.2190
	199 × 199	7.1643	7.2187
100	99 × 99	1.7834	1.8286
	199 × 199	1.7834	
50	99 × 99	.43950	.47534
25	99 × 99	.10481	.13131
	199 × 199	.10481	.13131

and a laminate thickness of 2.69 mm. The plate is square with simple supports along all four edges and is first subjected to a unit force applied in the center. The central deflections determined using the CPT and the SDPT for different size plates are given in Table 3.2. As expected, as the ratio a/h becomes small, the effect of shear deformation becomes larger. It must be noted that with the SDPT, the double series solution converges very slowly.

Example 3.5: Square SSSS Plate Under Patch Loading Next, the unit load is assumed to be uniformly distributed over a square patch of size $2h \times 2h$ centered in the center of the plate. Table 3.3 shows that, while the effects of shear deformation are still present, the convergence problem has been eliminated because the singularity that exists with a concentrated force has been eliminated. For impact problems, the contact force is distributed over a small area which can be included in the model. To estimate the size of impact-induced damage in laminated composites, accurate solutions near the impact point are needed; accounting for the actual pressure distribution under the impactor, plate theories can be used.

3.3.5 Harmonic Wave Propagation

The inplane motion is governed by (3.105) and (3.106), and for a symmetric laminate, the constitutive equations are

$$\begin{Bmatrix} N_x \\ N_y \\ N_{xy} \end{Bmatrix} = \begin{bmatrix} A_{11} & A_{12} & A_{16} \\ A_{12} & A_{22} & A_{26} \\ A_{16} & A_{26} & A_{66} \end{bmatrix} \begin{Bmatrix} u_{,x} \\ v_{,y} \\ u_{,y} + v_{,x} \end{Bmatrix} \tag{3.157}$$

and $I_2 = 0$. For symmetric and balanced laminates, $A_{16} = A_{26} = 0$, and the equations of motion can be written as

$$\begin{aligned} A_{11} u_{,xx} + A_{66} u_{,yy} + (A_{12} + A_{66}) v_{,xy} &= \rho \ddot{u}, \\ A_{22} v_{,yy} + A_{66} v_{,xx} + (A_{12} + A_{66}) u_{,xy} &= \rho \ddot{v}. \end{aligned} \tag{3.158}$$

Consider the propagation of harmonic waves of the form

$$u = U e^{ik\{\eta - ct\}}, \qquad v = V e^{ik\{\eta - ct\}} \tag{3.159}$$

where $\eta = x \cos\phi + y \sin\phi$, k is the wavenumber and c is the phase velocity. Substituting (3.159) into (3.158) gives

$$\begin{bmatrix} M_{11} & M_{12} \\ M_{12} & M_{22} \end{bmatrix} \begin{Bmatrix} U \\ V \end{Bmatrix} = 0 \tag{3.160}$$

$$\begin{aligned} M_{11} &= \left(A_{11} \cos^2\phi + A_{66} \sin^2\phi - \rho c^2\right) k^2 \\ M_{12} &= (A_{12} + A_{66}) k^2 \sin\phi \cos\phi \\ M_{22} &= \left(A_{22} \sin^2\phi + A_{66} \cos^2\phi - \rho c^2\right) k^2. \end{aligned} \tag{3.161}$$

The phase velocity of the two waves propagating in a direction is determined from setting the determinant of the matrix M to zero. Equations (3.160) and (3.161) indicate that c is independent of the wavenumber k. An extensional wave and a shear wave can propagate in the 0° direction with phase velocities

$$c^2 = A_{11}/\rho \quad \text{and} \quad c^2 = A_{66}/\rho \tag{3.162}$$

respectively. Similarly, the extensional and shear waves propagating in the 90° direction have phase velocities

$$c^2 = A_{22}/\rho, \quad c^2 = A_{66}/\rho. \tag{3.163}$$

When $A_{11} = A_{22}$, a pure extensional wave ($U = V$) and a purely shear wave ($U = -V$) will propagate in the 45° with phase velocities

$$c^2 = (A_{11} + A_{12} + 2A_{66})/(2\rho), \quad c^2 = (A_{11} - A_{12})/(2\rho) \tag{3.164}$$

respectively.

Example 3.6: Inplane Wave Propagation in Graphite-Epoxy Laminates For a graphite-epoxy unidirectional laminate, with the following properties

$$Q_{11} = 157 \text{ GPa}, \quad Q_{22} = 10 \text{ GPa}, \quad Q_{12} = 3 \text{ GPa},$$

$$Q_{66} = 7.5 \text{ GPa}, \quad \rho = 1530 \text{ kg/m}^3 \tag{3.165}$$

(Veidt and Sachse 1994), the phase velocity of the two inplane waves vary with the direction of propagation as shown in Fig. 3.8 for a 0° unidirectional

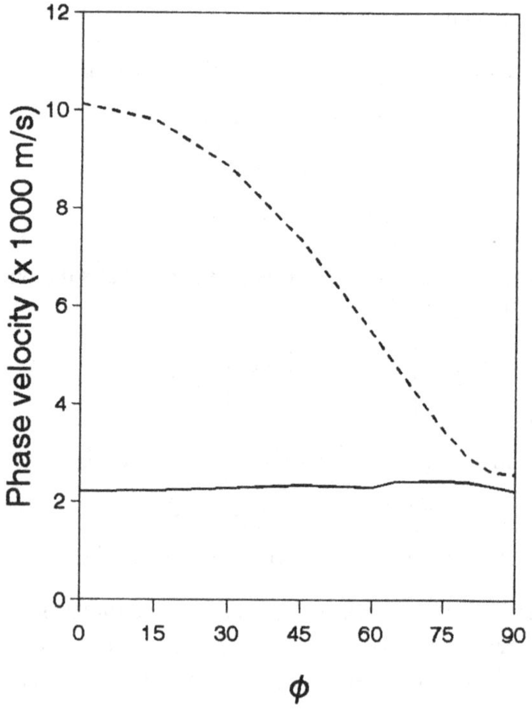

Figure 3.8. Dispersion curves for inplane wave propagation in unidirectional laminate (0° direction).

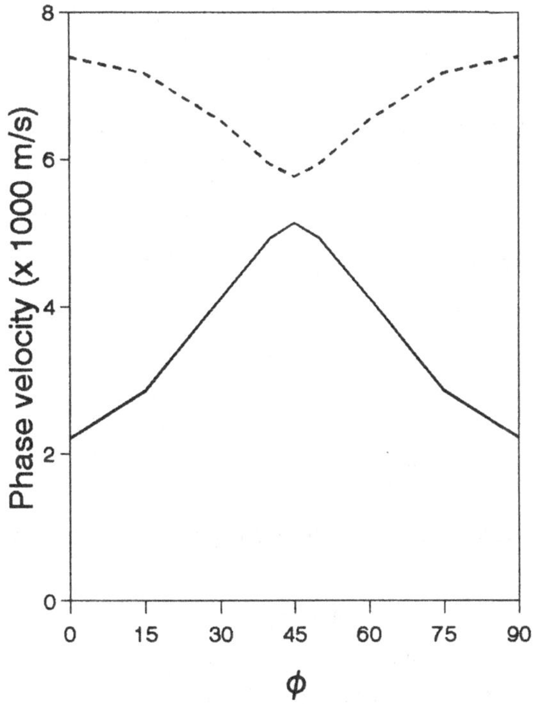

Figure 3.9. Dispersion curves for inplane wave propagation in $[0,90]_S$ laminate.

laminate. Similar results are obtained for a $[0,90]_S$ laminate (Fig. 3.9), and in this case $A_1 = A_2$ so that a pure extensional wave and pure inplane shear wave propagate in the 45° direction.

Now, consider the propagation of transverse waves in composite plates. For straight-crested waves propagating in a direction defined by a vector $\bar{n}$ oriented at an angle ϕ from the x-axis, the displacements are expressed as

$$w = A\mathrm{e}^{ik(\bar{n}\cdot\bar{r}-ct)} \tag{3.166}$$

where $\bar{r} = x\bar{i} + y\bar{j}$, $\bar{n} = \cos\phi\bar{i} + \sin\phi\bar{j}$. Using the CPT, substitution into the equation of motion (3.117) gives the dispersion relation

$$\begin{aligned} c^2 = \big[&D_{11}\cos^4\phi + 2(D_{12} + 2D_{66})\sin^2\phi\cos^2\phi + D_{22}\sin^4\phi \\ &+ 4D_{16}\sin\phi\cos^3\phi + 4D_{26}\sin^3\phi\cos\phi\big]k^2/I_1 \end{aligned} \tag{3.167}$$

which indicates that the phase velocity c varies with the direction of propagation. For isotropic plates ($D_{11} = D_{22} = D_{12} + 2D_{66} = D$, $I_1 = \rho h$), the phase velocity is the same in all directions and can be calculated from

$$c^2 = \frac{D}{\rho h} k^2. \tag{3.168}$$

Therefore, with the CPT, short waves propagate at very high velocities. Including rotatory inertia, the dispersion relation becomes

$$\begin{aligned} c^2 = \big[&D_{11} \cos^4 \phi + 2(D_{12} + 2D_{66}) \sin^2 \phi \cos^2 \phi + D_{22} \sin^4 \phi \\ &+ 4D_{16} \sin \phi \cos^3 \phi + 4D_{26} \sin^3 \phi \cos \phi\big] k^2 / \left(I_1 + I_3 k^2\right). \end{aligned} \tag{3.169}$$

As the wavenumber k becomes large, the phase velocity tends to the limit

$$\begin{aligned} c = \big[\{&D_{11} \cos^4 \phi + 2(D_{12} + 2D_{66}) \sin^2 \phi \cos^2 \phi + D_{22} \cos^4 \phi \\ &+ 4D_{16} \sin \phi \cos^3 \phi + 4D_{26} \sin^3 \phi \cos \phi\} / I_3\big]^{\frac{1}{2}}, \end{aligned} \tag{3.170}$$

and therefore the phase velocity of bending waves is bounded for short wavelength.

For wave propagation in shear deformable plates, consider harmonic waves in the form

$$w = W e^{ik(\eta - ct)}, \quad \psi_x = X e^{ik(\eta - ct)}, \quad \psi_y = Y e^{ik(\eta - ct)} \tag{3.171}$$

where W, X, and Y are amplitudes, k is the wavenumber, c is the phase velocity, and $\eta = \bar{r} \cdot \bar{n}$, where $\bar{r} = x\bar{i} + y\bar{j}$ is the position vector and $\bar{n} = \cos \phi \bar{i} + \sin \phi \bar{j}$ is the unit vector in the direction of propagation. Substituting (3.171) into (3.110)–(3.112) gives the complex eigenvalue problem

$$[a_{ij}] \begin{Bmatrix} W \\ \psi_x \\ \psi_y \end{Bmatrix} = 0. \tag{3.172}$$

The coefficients of the $[a_{ij}]$ matrix are defined as

$$\begin{aligned} a_{11} &= -k^2\left(A_{44} \sin^2 \phi + A_{55} \cos^2 \phi + 2A_{45} \sin \phi \cos \phi\right) + I_1 k^2 c^2, \\ a_{12} &= a_{21} = ik(A_{55} \cos \phi + A_{45} \sin \phi), \\ a_{13} &= a_{31} = ik(A_{44} \sin \phi + A_{45} \cos \phi), \\ a_{22} &= D_{11} k^2 \cos^2 \phi + 2D_{16} k^2 \sin \phi \cos \phi + D_{66} k^2 \sin^2 \phi \\ &\quad + A_{55} - I_3 k^2 c^2, \\ a_{23} &= a_{32} = (D_{12} + D_{66}) k^2 \sin \phi \cos \phi + D_{16} k^2 \cos^2 \phi \\ &\quad + D_{26} k^2 \sin^2 \phi + A_{45}, \\ a_{33} &= D_{22} k^2 \sin^2 \phi + D_{66} k^2 \cos^2 \phi + 2D_{26} k^2 \sin \phi \cos \phi \\ &\quad + A_{44} - I_3 k^2 c^2, \end{aligned} \tag{3.173}$$

where i denotes the complex number $i^2 = -1$. For a nontrivial solution, the determinant of the a_{ij} matrix must be equal to zero. For each value of the wavenumber k, there are three wave velocities or equivalently three values of the frequency $\omega = ck$.

For very long waves ($k \to 0$) the phase velocity of one of the waves tends to zero, while for the other two waves the phase velocity becomes infinite. The cutoff frequencies for those two waves are solutions of

$$\omega^4 I_3^2 - \omega^2 I_3(A_{44} + A_{55}) + A_{44}A_{55} - A_{45}^2 = 0. \tag{3.174}$$

It is interesting to note that, very often, G_{23} and G_{13} are taken to be equal to G_{12} for lack of better information. When $G_{13} = G_{23}$, then $A_{44} = A_{55}$ and $A_{45} = 0$, and from (3.174), the two cutoff frequencies are the same and are given by $\omega^2 = A_{44}/I_3$.

As the wavenumber increases, the phase velocity approaches a constant value. When $\phi = 0°$, the phase velocities for the three waves tend to

$$c^2 = A_{55}/I_1 \tag{3.175}$$

and to the solutions of the equation

$$c^4 I_3^2 - c^2 I_3(D_{11} + D_{66}) + D_{11}D_{66} - D_{16}^2 = 0. \tag{3.176}$$

When $\phi = 90°$, the phase velocities for the three waves tend to

$$c^2 = A_{44}/I_1 \tag{3.177}$$

and to the solutions of the equation

$$c^4 I_3^2 - c^2 I_3(D_{22} + D_{66}) + D_{22}D_{66} - D_{26}^2 = 0. \tag{3.178}$$

Equations (3.175)–(3.178) will be used later to analyze wave propagation in plates subjected to impact.

Example 3.7: Wave Propagation in Graphite-Epoxy Plates As an example, consider a 10-ply $[0°,90°,0°,90°,0°]_S$ graphite-epoxy plate with material properties given by (3.156) with a laminate thickness of 2.69 mm. Figure 3.10 shows the dispersion relations determined using the FSDT. Results are given for the first two modes for waves propagating in the 0° and 90° directions. This figure also shows results for the first mode propagation in the 45° direction. Phase velocities for the third mode are much higher than those for the first two modes, so this third mode will not be displayed here. As the wavenumber increases, the phase velocity for the first mode reaches an asymptotic value given by (3.175). Because the shear moduli are all the same in this case this asymptotic value is independent of the direction of propagation and

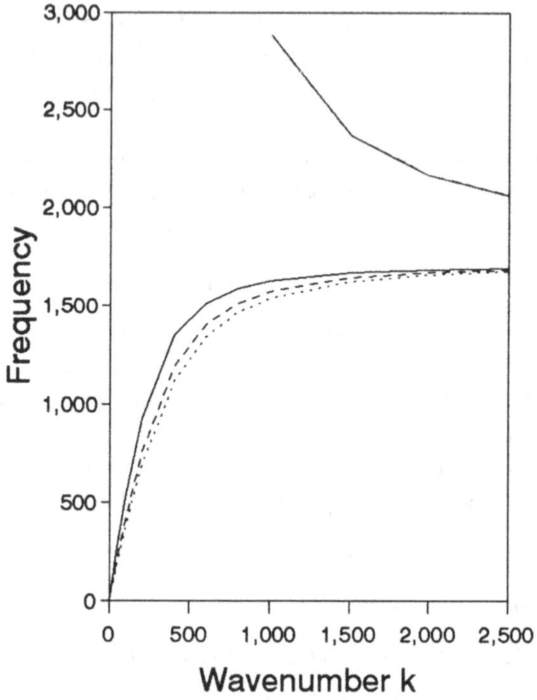

Figure 3.10. Wave propagation in 10-ply symmetric, specially orthotropic laminate.

is equal to $c = (G_{13}/I_1)^{\frac{1}{2}} = 1703$ m/s. For the second mode, since in this case $D_{16} = 0$, the phase velocity decreases to the asymptotic value given by $c = (D_{66}/I_3)^{\frac{1}{2}} = 1865$ m/s when $\phi = 0°$ and $90°$. For the third mode, the phase velocity decreases to an asymptotic value of $c = (D_{11}/I_3)^{\frac{1}{2}} = 1761$ m/s when $\phi = 0$, and $c = (D_{22}/I_3)^{\frac{1}{2}} = 5491$ m/s when $\phi = 90$. The effect of rotary inertia is small. For wave propagation in the 0° direction, neglecting rotary inertia ($I_3 = 0$) results in a maximum change in phase velocity of 0.6% as the wavenumber varies from 0 to 2500.

Figure 3.11 presents a comparison of the values of dispersion relations predicted by the classical plate theory and the first-order shear deformation theory for waves propagating in the 0 and 45° directions. Since the CPT account for only one mode, only the first mode predicted by the FSDT is plotted. The classical plate theory is seen to be adequate for long wavelengths. In the 0° direction, there is a 5.5% difference between the two curves when $k = 100$, which corresponds to a wavelength equal to approximately 23 times the thickness of the plate. This confirms the conclusions drawn earlier from the static analysis: the effect of shear deformation can be significant, and the classical

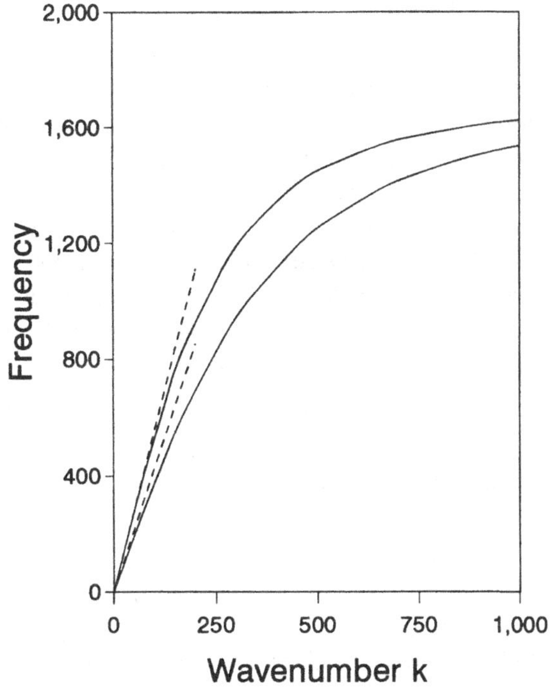

Figure 3.11. Wave propagation in 0° and 45° directions according to the CPT and the FSDT.

plate theory is valid only for thin plates. As noted earlier, including the effect of rotary inertia in the classical plate theory leads to bounded phase velocities for short waves, but that limit happens to coincide with that for the third mode obtained from the FSDT. In Fig. 3.11, dispersion curves for the CPT with and without rotary inertia are not distinguishable. Thus, including the effect of rotary inertia does not extend the range of applicability of the CPT.

Next, we determine the dispersion relations for a $[45,-45]_S$ and a $[45,-45]_{2S}$ graphite-epoxy laminate with material properties given by (3.156) and a laminate thickness of 2.69 mm. These two layups will have the same bending rigidities except for the D_{16} and D_{26} terms. Therefore, we will be able to study the effect of these bending-twisting coefficients on the dynamic behavior of the plate. Figure 3.11 shows that for the four-ply laminate, the effect of these coupling terms is significant, but also that as the number of plies is increased to eight, the effect of the D_{16} and D_{26} terms can be neglected. This result has been established in many studies and will be verified in the study of the free vibrations of composite plates.

3.3.6 Free Vibrations of Plates

In this section, we consider the free vibration of simply supported rectangular composite plates with symmetric layups and no bending-twisting coupling coefficients. In this case exact solutions are obtained using the classical plate theory and the first-order shear deformation theory. Therefore, the two theories can be compared directly without the use of variational approximation methods or finite element discretization. Levy-type solutions are available for rectangular plates with two opposite edges that are simply supported.

According to the classical plate theory, for a symmetrically laminated plate without bending-twisting coupling ($D_{16} = D_{26} = 0$), free vibrations are governed by the equation

$$D_{11}\frac{\partial^4 w}{\partial x^4} + 2(D_{12} + 2D_{66})\frac{\partial^4 w}{\partial x^2 \partial y^2} + D_{22}\frac{\partial^4 w}{\partial y^4} + I_1 \ddot{w} = 0. \tag{3.179}$$

For a rectangular plate with simple supports along the four edges, the double Fourier series

$$w(x, y, t) = \sum_{m,n=1}^{\infty} W_{mn} \sin\frac{m\pi x}{a} \sin\frac{n\pi y}{b} e^{i\omega_{mn} t} \tag{3.180}$$

satisfies the equation of motion (3.179) and the boundary conditions

$$\begin{aligned} w(0, y, t) = w(a, y, t) = w(x, 0, t) = w(x, b, t) = 0, \\ M_x(0, y, t) = M_x(a, y, t) = M_y(x, 0, t) = M_y(x, b, t) = 0. \end{aligned} \tag{3.181}$$

Substituting (3.180) into (3.179) multiplying by $\sin(\frac{m\pi x}{a})\sin(\frac{n\pi y}{b})$, and integrating over the domain occupied by the plate, the natural frequencies are obtained as

$$\omega_{mn}^2 = \frac{\pi^4}{a^4 I_1}\left[D_{11}m^4 + 2(D_{12} + 2D_{66})m^2n^2r^2 + D_{22}n^4r^4\right]. \tag{3.182}$$

For simply supported shear deformable plates with symmetric layups and no bending-twisting coupling, the displacement functions

$$\begin{aligned} w &= \sum_{m,n=1}^{\infty} W_{mn} \sin\frac{m\pi x}{a} \sin\frac{n\pi y}{b} e^{i\omega t} \\ \Psi_x &= \sum_{m,n=1}^{\infty} X_{mn} \cos\frac{m\pi x}{a} \sin\frac{n\pi y}{b} e^{i\omega t} \\ \Psi_y &= \sum_{m,n=1}^{\infty} Y_{mn} \sin\frac{m\pi x}{a} \cos\frac{n\pi y}{b} e^{i\omega t} \end{aligned} \tag{3.183}$$

satisfy the boundary conditions (3.139). Substituting (3.183) into the equations of motion (3.110)–(3.112) gives the eigenvalue problem

$$\left\langle [K] - \omega_{mn}^2 [M] \right\rangle \begin{Bmatrix} W_{mn} \\ X_{mn} \\ Y_{mn} \end{Bmatrix} = 0 \tag{3.184}$$

where the stiffness matrix $[K]$ is defined by (3.155) and the mass matrix $[M]$ is a diagonal matrix with

$$m_{11} = I_1, \quad m_{22} = m_{33} = I_3.$$

For nontrivial solutions, the determinant of the matrix $([k] - \omega_{mn}^2[m])$ must vanish.

Example 3.8: Free Vibrations of Laminated Plates In order to assess the effect of shear deformation and rotary inertia on the dynamics of composite plates, consider the example first studied by Chen and Sun (1985a,b) and Sun and Chen (1985) of a 10-ply $[0°,90°,0°,90°,0°]_S$ with material properties given by (3.156) and a laminate thickness of 2.69 mm. For square plates with $a = 25$, 50, 100, and 200 mm, the natural frequencies for modes (1,1), (1,3), (3,1), and (3,3) are determined using the classical plate theory and the first-order shear deformation theory. These three modes were selected because, as we shall see, for plates impacted in the center, modes with even values of m or n do not participate in the response of the plate. The ratio $\omega_{\mathrm{FSDT}}/\omega_{\mathrm{CPT}}$, determined from (3.182) and (3.184), is plotted versus the ratio a/h in Fig. 3.13. As expected from the previous study of beams, the effect of shear deformation and rotary inertia becomes significant for thick plates and becomes more pronounced as m and n increase. However, for the present case the CPT gives very accurate results for the three modes considered when $a = 200$ mm or $a/h = 74.34$, which is the size of the plate studied by Chen and Sun. It was verified that the effect of rotary inertia on the natural frequencies is small, and if the effect were neglected, the resulting curves would be indistinguishable from those in Fig. 3.12.

This example shows that the effect of rotatory inertia is negligible and that shear deformation is also negligible for thin plates.

3.3.7 Transient Response of a Plate

The exact solution for the transient response of an infinite plate was given by Medick (1961). For an isotropic plate of arbitrarily large radius subjected to a force $P \cdot f(t)$ in the center, the displacements are given as

$$w(r,t) = \frac{P}{4\pi(\rho h D)^{\frac{1}{2}}} \int_0^t \int_0^\lambda f(\zeta) \sin\left(\frac{ar^2}{t-\lambda}\right) \frac{d\lambda}{(t-\lambda)} d\zeta \tag{3.185}$$

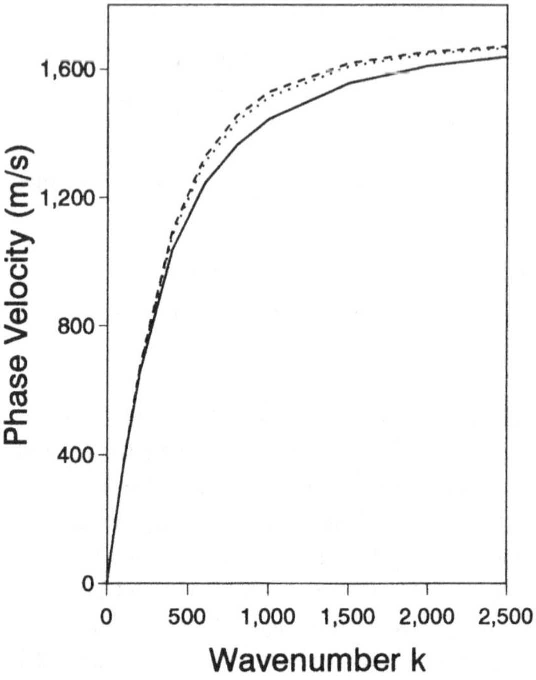

Figure 3.12. Effect of bending-twisting coupling terms on wave propagation. (solid line $[45,-45]_S$; dotted line: $[45,-45]_{2S}$; dashed line: $[45,-45]_{2S}$ with $D_{16} = D_{26} = 0$).

where $a = \frac{1}{4}(\frac{\rho h}{D})^{1/2}$. When $f(t)$ is a step function in time,

$$f(t) = \begin{cases} 1 & t > 0 \\ 0 & t > 0 \end{cases} \tag{3.186}$$

the displacement is given by

$$w_1(r,t) = \frac{P}{4\pi(\rho h D)^{\frac{1}{2}}} t H(ar^2/t) \tag{3.187}$$

where

$$H(x) = \frac{\pi}{2} - S_i(x) + xC_i(x) - \sin(x) \tag{3.188}$$

with $S_i(x) = \int_0^x \frac{\sin t}{t} dt$ and $C_i(x) = \int_0^x \frac{\cos t}{t} dt$.

When the loading is defined by

$$f(t) = \begin{Bmatrix} 1 & 0 < t \leq \beta \\ 0 & t > \beta \end{Bmatrix} \tag{3.189}$$

the response is given by

$$w_2(r,t) = w_1(r,t) - \begin{cases} 0 & t < \beta \\ w_1(r, t-\beta) & t > \beta \end{cases}. \tag{3.190}$$

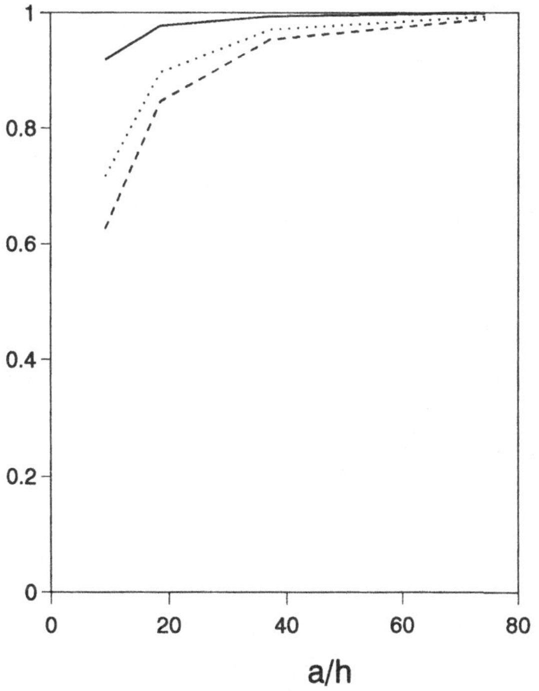

Figure 3.13. Effect of shear deformation and rotatory inertia on the natural frequencies of simply supported plates. Solid line: mode 1; dotted line: mode 2; dashed line: mode 3.

Example 3.9: Step Load on an Infinite Isotropic Plate Consider a steel plate, 5 mm thick with $E = 210$ GPa, $\nu = 0.3$, and $\rho = 7960$ kg/m^3. When subjected to a unit force ($P = 1$), the displacements given by (3.187) are shown in Fig. 3.14. This solution shows that the deformation is confined to a small region near the point of application of the force, and, because the plate is infinite, there are no waves reflected back from the boundary. With a finite-size plate, waves will eventually reach the boundaries and be reflected back, and then the incident and reflected waves will combine to produce the total plate deflections.

Since the classical plate theory is used to obtain this solution, high frequency waves can propagate at very high speeds. In fact, the theory of elasticity and the more advanced plate theories show that for short wavelengths, the phase velocity cannot exceed an asymptotic value given by (3.175). For the present case this value is $c = (G/\rho)^{1/2} = 3185$ m/s, which means that after 10 μs, the wave should have reached a radius equal to 6.37 times the plate thickness. Figure 3.14 shows that small oscillations are present beyond that radius.

In practice, plates have finite dimensions, and waves that initially propagate away from the impact point will eventually be reflected back from the

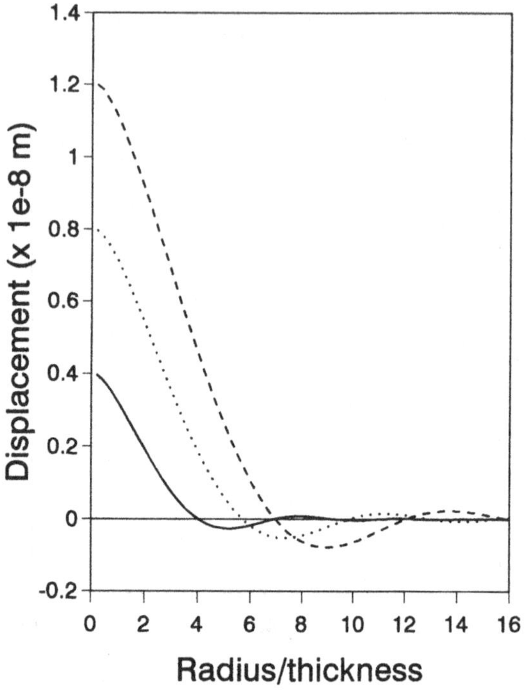

Figure 3.14. Displacements of infinite plate subjected to a step load.

boundaries. After that time, the response will be affected by the boundary conditions. A solution for finite-sized composite plates is presented now to investigate the effect of finite size and anisotropy. The distributed loading and the displacement functions for a shear deformable plate (FSDT) can be expanded into the double Fourier series

$$
\begin{aligned}
q &= \sum_{m,n=1}^{\infty} q_{mn}(t) \sin \frac{m\pi x}{a} \sin \frac{n\pi y}{b} \\
w &= \sum_{m,n=1}^{\infty} W_{mn}(t) \sin \frac{m\pi x}{a} \sin \frac{n\pi y}{b} \\
\Psi_x &= \sum_{m,n=1}^{\infty} X_{mn}(t) \cos \frac{m\pi x}{a} \sin \frac{n\pi y}{b} \\
\Psi_y &= \sum_{m,n=1}^{\infty} Y_{mn}(t) \sin \frac{m\pi x}{a} \cos \frac{n\pi y}{b}.
\end{aligned}
\tag{3.191}
$$

In this way, for a simply supported plate with symmetric layup and no bending-twisting coupling, the boundary conditions (3.139)–(3.140) are satisfied.

Substituting (3.191) into the equations of motion (3.110)–(3.112) yields for each mode the set of three ordinary differential equations

$$[M]\begin{Bmatrix} \ddot{W}_{mn} \\ \ddot{X}_{mn} \\ \ddot{Y}_{mn} \end{Bmatrix} + [K]\begin{Bmatrix} W_{mn} \\ X_{mn} \\ Y_{mn} \end{Bmatrix} = \begin{Bmatrix} q_{mn} \\ 0 \\ 0 \end{Bmatrix} \tag{3.192}$$

with the generalized mass and stiffness matrices given by (3.184).

3.3.8 Variational Models

In general, exact solutions for the dynamic response of laminated plates are not available and variational approximation methods can be used. According to the first-order shear deformation plate (FSDT) theory, the strain energy in a laminated plate is given by (3.90). The constitutive equations (3.99) and (3.103) can be written in matrix form as

$$\begin{Bmatrix} N \\ M \end{Bmatrix} = \begin{bmatrix} A & B \\ B & D \end{bmatrix}\begin{Bmatrix} \epsilon_o \\ \kappa \end{Bmatrix}$$

$$Q = \bar{A}\gamma. \tag{3.193}$$

With this notation, the strain energy in the plate can be written as

$$U = \frac{1}{2}\int_\Omega \left[\epsilon_o^{\mathrm{T}} A \epsilon_o + \kappa^{\mathrm{T}} D \kappa + 2\epsilon_o^{\mathrm{T}} B \kappa\right] d\Omega + \frac{1}{2}\int_\Omega \gamma^{\mathrm{T}} \bar{A} \gamma \, d\Omega. \tag{3.194}$$

For a symmetric laminate, the B matrix is zero and the inplane and transverse motions are uncoupled. In addition, if the effects of shear deformation are neglected (classical plate theory), the last term in (3.194) can be neglected and the strain energy becomes

$$U = \frac{1}{2}\int_\Omega \kappa^{\mathrm{T}} D \kappa \, d\Omega \tag{3.195}$$

when only out-of-plane deformations are considered. Using (3.94), the kinetic energy for a symmetric plate subjected to transverse deformations can be written as

$$T = \frac{1}{2}\int_\Omega \left\{ I_1 \dot{w}_o^2 + I_3\left(\dot{\psi}_x^2 + \dot{\psi}_y^2\right) \right\} d\Omega \tag{3.196}$$

when the effect of rotatory inertia are neglected, $I_3 = 0$, and only the first term is left in the integrand of (3.196). For plates subjected to a distributed pressure q in the transverse direction only, the potential energy of the external forces can be written as

$$V = -\int_\Omega q w_o \, d\Omega. \tag{3.197}$$

With the Rayleigh-Ritz variational approximation method, the transverse displacements are taken as

$$w = \sum_{i=1}^{N} c_i \phi_i(x, y) \tag{3.198}$$

where the approximation functions ϕ_i must satisfy the essential boundary conditions and where the c_i are constants to be determined. Substituting (3.198) into (3.195)–(3.197) yields

$$U = \frac{1}{2} c^{\mathrm{T}} \int_\Omega \Phi''^{\mathrm{T}} D \Phi'' d\Omega c \tag{3.199}$$

$$T = \frac{1}{2} \int_\Omega \rho \dot{c}^{\mathrm{T}} \Phi^{\mathrm{T}} \Phi \dot{c} d\Omega \tag{3.200}$$

$$V = - \int_\Omega c^{\mathrm{T}} \Phi^{\mathrm{T}} q d\Omega. \tag{3.201}$$

Using Hamilton's principle, the equations of motion are obtained as

$$[M]\{\ddot{c}\} + [K]\{c\} = \{F\} \tag{3.202}$$

where

$$[M] = \int_\Omega \rho \Phi^{\mathrm{T}} \Phi d\Omega \tag{3.203a}$$

$$[K] = \int_\Omega \Phi''^{\mathrm{T}} D \Phi'' d\Omega \tag{3.203b}$$

$$\{F\} = \int_\Omega^{\mathrm{T}} q \Phi^{\mathrm{T}} d\Omega. \tag{3.203c}$$

$[M]$, $[K]$, and F are the generalized mass matrix, stiffness matrix, and force vector respectively.

The approximation functions in (3.198) must satisfy the essential boundary conditions along the edges of the plate; if those functions form a complete set, convergence to the exact solution is guaranteed. Along a free edge, no essential condition needs to be satisfied. Along a simply supported edge $w = 0$ and along a clamped edge, both the transverse displacement and the slope in a direction normal to the edge ($\frac{\partial w}{\partial n} = 0$) must vanish. Polynomial functions can be used very effectively because it is easy to select a set of functions that is complete and satisfies the boundary conditions. Using polynomial functions facilitates the evaluation of the integrals in (3.203).

For example, for a rectangular plate, a typical approximation function will be of the form

$$\phi_i(x, y) = x^{m+\alpha} y^{n+\beta} (x - a)^\gamma (y - b)^\delta \tag{3.204}$$

where m and n can vary from 1 to p and 1 to q, respectively ($p \cdot q = N$). The coefficients $\alpha, \beta, \gamma, \delta$ account for the support conditions along the edges $x = 0, y = 0, x = a$, and $y = b$, respectively. α, β, γ, and δ are taken to be –1, 0, 1 for free edges, simply supported edges, and clamped edges respectively. Support conditions for a rectangular plate are designated by four letters for the $x = 0$, $y = 0$, $x = a$, and $y = b$ edges. F denotes a free edge, S a simply supported edge, and C a clamped edge. With this notation, the approximation functions take the form

$$\phi_i(x, y) = x^{m-1}y^{n-1} \tag{3.205a}$$

$$\phi_i(x, y) = x^m y^n (x - a)(y - b) \tag{3.205b}$$

$$\phi_i(x, y) = x^{m+1} y^n (y - b) \tag{3.205c}$$

for FFFF, SSSS, and CSFS plates respectively. Abrate (1995e) study the free vibrations of rectangular plate with intermediate line supports. With r supports located along the lines $x = a_1, a_2, \ldots, a_r$ and s supports along the lines $y = b_1$, $b_2, \ldots, b_s$ the approximation functions were taken as

$$\phi_i(x, y) = x^{m+\alpha} y^{n+\beta} (x - a)^\gamma (y - b)^\delta \prod_{j=1}^{r} (x - a_j) \prod_{k=1}^{s} (y - b_k). \tag{3.206}$$

Following the same approach, polynomial approximation can be developed for other geometries. For example, for triangular plates with one edge along the x-axis, the approximation functions can be taken as (Abrate 1996a)

$$\phi_i(x, y) = x^{m-1} y^{n+\alpha} \langle cy - bx \rangle^\beta \, [bx + (a - c)y - ab]^\gamma \tag{3.207}$$

where α, β, and γ account for the support conditions along the sides $y = 0$, $cy - bx = 0$, and $bx + (a - c)y - a = 0$ and are taken to be –1, 0, 1 for free, simply supported, and clamped edges, respectively.

Example 3.10: Free Vibration of Plate with Intermediate Line Supports Consider a graphite-epoxy square plate with simple supports along the four edges and line supports along the lines $x = a/4$ and $y = a/2$, where a is the length of the side. The elastic properties of the material are $E_1 = 181.0$ GPa, $E_2 = 10.30$ GPa, $\nu_{12} = 0.28$, $G_{12} = 7.17$ GPa. The natural frequencies are nondimensionalized as follows: $\Omega^2 = \omega^2 \rho h a^4/(D_{11}D_{22})^{1/2}$. The first six natural frequencies for angle-ply laminates are given in Table 3.4 as a function of the fiber orientation. The approximation functions are taken as

$$\phi_i(x, y) = x^\alpha y^\beta (x - a)(y - b)\left(x - \frac{a}{4}\right)\left(y - \frac{a}{2}\right).$$

Table 3.4. *Natural frequencies of graphite-epoxy angle-ply, two-way-two-span square plates*

Fiber orientation (degrees)	Natural frequencies (rad/s)					
	Ω_1	Ω_2	Ω_3	Ω_4	Ω_5	Ω_6
0	57.52	63.00	103.3	124.2	168.6	171.4
15	63.74	70.53	123.3	145.7	170.6	175.4
30	70.80	81.41	158.3	161.4	170.5	188.7
45	75.09	92.65	147.0	162.0	198.8	237.5
60	81.64	110.4	135.3	159.4	202.4	224.4
75	86.16	116.5	127.1	152.3	160.8	186.8
90	85.17	101.5	130.5	132.1	142.9	154.4

Table 3.5. *Natural frequencies of clamped, elliptical, isotropic plate*

1	9 × 9	10.2158	21.2604	21.2604	34.8772	34.8770	39.771
	Lam*	10.2160	21.2601	21.2601	34.8786	34.8786	39.7712
2	9 × 9	27.3774	39.4974	55.9758	69.8580	77.0367	88.0479
	Lam	27.4773	39.4976	55.9773	69.8557	77.0443	88.0472
3	9 × 9	56.8004	71.5912	90.2352	113.2753	140.7441	150.0907
	Lam	56.8995	71.5902	90.2380	113.2661	140.7461	150.0889
4	9 × 9	97.5989	115.6089	137.2739	164.3236	195.3101	255.1315
	Lam	97.5984	115.6084	137.2686	164.3248	195.3398	255.0946
5	9 × 9	149.6378	171.0975	196.2046	229.8105	266.2457	350.4704
	Singh**	149.66	171.10	198.55	229.81		

*Lam et al. (1992).
**Singh and Chakraverty (1994).

The first six natural frequencies have converged for the first four significant figures when an 8 × 8 approximation is used. That is, α and β vary from 1 to 8, and $N = 64$ in the displacement approximation (3.198).

Example 3.11: Free Vibration of Elliptical Plate For clamped, isotropic, elliptical plates, the approximation functions are taken as

$$\phi_j(x, y) = x^m y^n \left[\left(\frac{x}{a} \right)^2 + \left(\frac{y}{b} \right)^2 - 1 \right]^2$$

where a and b are the lengths of the semi-axes in the x- and y-directions respectively. The natural frequencies obtained using the present approach are in excellent agreement with those presented recently by Lam et al. (1992) and Singh and Chakraverty (1994) as shown in Table 3.5.

3.3.9 Effect of Boundary Conditions

To accurately predict the contact force history and the transient response of the structure to an impact, very sophisticated models of the structure can be used. However, in most analyses the edges of the beam or the plate are assumed to have classical boundary conditions: free, simply supported, or clamped. In practice, it is difficult to achieve such ideal conditions, and some flexibility remains in the supports. One study (Palazotto et al. 1991) showed that, while shear deformations and geometric nonlinearities are present, the fact that the edges were neither perfectly simply supported nor perfectly clamped produced a much larger effect. Experimental values of the contact force history fell between those predicted assuming simply supported boundary conditions and those obtained assuming that the panel was perfectly clamped.

A simple approach was developed (Abrate et al. 1996) for the identification of the boundary conditions of beams. The transverse vibrations of Bernoulli-Euler beams are governed by the equation of motion (3.20), and for free vibration, following Section 3.2.7, the displacements are taken in the form given by (3.42). The mode shapes are given by (3.45). At each end, the actual support conditions are modeled by a translational and a rotational spring, and the boundary conditions can be written as

$$\begin{aligned}
V(0,t) &= -EI\frac{\partial^3}{\partial x^3}w(0,t) = k_1^t w(0,t)\\
V(L,t) &= -EI\frac{\partial^3}{\partial x^3}w(L,t) = -k_2^t w(L,t)\\
M(0,t) &= EI\frac{\partial^2}{\partial x^2}w(0,t) = k_1^r\frac{\partial}{\partial x}w(0,t)\\
M(L,t) &= EI\frac{\partial^2}{\partial x^2}w(L,t) = -k_2^r\frac{\partial}{\partial x}w(L,t)
\end{aligned} \tag{3.208}$$

where V is the shear force, M is the bending moment, and L is the length of the beam. Superscripts t and r refer to translational and rotational spring constants, respectively. For each natural frequency ω_i, the boundary conditions (3.208) yield four equations that must be satisfied simultaneously:

$$\begin{bmatrix}
1 & -\bar{k}_1^t & -1 & -\bar{k}_1^t\\
\left(-\bar{k}_2^t\cdot \mathrm{SS}-\mathrm{CC}\right) & \left(-\bar{k}_2^t\cdot \mathrm{CC}-\mathrm{SS}\right) & \left(-\bar{k}_2^t\cdot \mathrm{SH}+\mathrm{CH}\right) & \left(-\bar{k}_2^t\cdot \mathrm{CH}+\mathrm{SH}\right)\\
\bar{k}_1^r & 1 & \bar{k}_1^r & -1\\
\left(\bar{k}_2^r\cdot \mathrm{CC}-\mathrm{SS}\right) & \left(-\bar{k}_2^r\cdot \mathrm{SS}-\mathrm{CC}\right) & \left(\bar{k}_2^r\cdot \mathrm{CH}+\mathrm{SH}\right) & \left(\bar{k}_2^r\cdot \mathrm{SH}+\mathrm{CH}\right)
\end{bmatrix}
\begin{Bmatrix} A_i\\ B_i\\ C_i\\ D_i\end{Bmatrix} = 0 \tag{3.209}$$

Table 3.6. *Natural frequencies (rad/s) of aluminum beam*

Mode (1)	Experimental (2)	Cantilevered beam (3)	H&J* (4)	Elastically supported (5)
1	122.7	148.912	131.5	122.683
2	766.86	933.203	790.1	797.393
3	2131.6	2612.93	2148	2240.33
4	4157.8	5120.11	4176	4294.40
5	6818.64	8463.48	6868	6818.64
6	10019.1	12642.2	10065	9890.68

*Hassiotis and Jeong (1995).

where

$$\bar{k}_1^t = \frac{k_1^t}{EI\beta_i^3}, \quad \bar{k}_1^r = \frac{k_1^r}{EI\beta_i}, \quad \bar{k}_2^t = \frac{k_2^t}{EI\beta_i^3}, \quad \bar{k}_2^r = \frac{k_2^r}{EI\beta_i}, \tag{3.210}$$

and

$$\mathrm{SS} = \sin(\beta_i L), \quad \mathrm{CC} = \cos(\beta_i L), \quad \mathrm{SH} = \sinh(\beta_i L), \quad \mathrm{CH} = \cosh(\beta_i L). \tag{3.211}$$

For nontrivial solutions, the determinant of the matrix of the coefficients must vanish. For a given set of spring constants, the natural frequencies and mode shapes can be determined by finding the zeroes of a 4×4 determinant. The support stiffnesses are selected to minimize some objective measure of the difference between predicted and measured values of a set of natural frequencies:

$$\Lambda = \sum_{i=1}^{N} \left| \frac{\omega_i}{\Omega_i} - 1 \right| \tag{3.212}$$

where Ω_i is the measured natural frequency, ω_i is the predicted value, and N is the number of modes considered.

Example 3.12: Identification of Boundary Conditions for an Aluminum Beam
In this example, a 25.4-mm-wide, 6.35-mm-thick, 495.3-mm-long aluminum beam of uniform cross section is assumed to be clamped at one end and free at the other. The modulus of elasticity and the density of the material are $E = 71$ MPa and $\rho = 2210$ kg/m^3, respectively. The natural frequencies predicted using classical cantilever boundary conditions (Table 3.6, column 3) are significantly higher that the experimental values (column 2). At $x = 0$, the support provides

Table 3.7. *Natural frequencies (rad/s) of composite plate*

Mode (1)	Experimental (2)	Finite element (3)	Cantilever beam (4)	Elastically supported (5)
1	399.3	453.6	453.254	399.300
2	1113	—	—	—
3	2507.3	2826.4	2840.50	2527.58
4	3992.1	4193.2	—	—
5	7012.0	7859.6	7953.47	7011.96

some flexibility, and the best values for the rotational and translational spring constants are found to be $k_1^t = 732{,}560$ N/m, $k_1^r = 668$ N · m/rad.

Example 3.13: Identification of Boundary Conditions for a Composite Plate Proper modeling of the support conditions for composite plates subjected to impact loading has long been recognized as a problem needing attention. Experiments were conducted on 3-layer graphite-epoxy specimens with a ply thickness of 0.015 in. and a total thickness of 0.045 in. The width was 2 in., the length 6 in., and the measured material properties are $E_1 = 19.832$ Msi, $E_2 = 1.603$ Msi, $\nu_{12} = 0.314$, $G_{12} = 0.926$ Msi, $\rho = 0.058$ lb/in.^{3}. Specimens were clamped at one end, and the natural frequencies and mode shapes were determined experimentally. Free vibrations were studied using a finite element model with an 8×17 mesh of 4 noded isoparametric plate elements, assuming the specimen to be clamped along the edge $x = 0$. The predicted natural frequencies (Table 3.7, column 3) are significantly higher than those measured (column 2). Since the material properties are known accurately, the discrepancy is expected to come from imperfect boundary conditions. Experiments and finite element results indicate that, for this example, modes 1, 3, and 5 are pure bending modes and modes 2 and 4 involve torsional motion. Results obtained using the Bernoulli-Euler beam theory with cantilever boundary conditions (column 4) show that the bending modes are predicted accurately and that shear deformation effects are small because the specimens are thin. The difference between the finite element results (column 3) and the cantilever beam results (column 4) is due to the effect of shear deformation. Using the present approach, the support at the root is modeled by two linear springs whose stiffnesses were determined by minimizing the error measure E_B for the first three bending modes. Using $k_1^t = 3595$ lb/in. and $k_1^r = 684.0$ lb · in./rad, excellent agreement is obtained between experimental results in column 2 and the predicted frequencies in column 5.

3.3.10 Summary

In this section, the derivation of three plate theories were reviewed: the classical plate theory, the first-order shear deformation theory, and a higher-order plate theory. One static example illustrated the fact that as the ratio between the length of the side and the thickness of the plate becomes small, shear deformation effects become significant. Loading a plate with a concentrated force creates a singularity at the point of application of that force, which results in slow convergence in the solution procedure. In practice, loads are introduced through a small patch, and under such a loading, convergence problems are avoided and realistic distributions of transverse shear forces are obtained. Harmonic wave propagation analysis confirms the effect of shear deformation and indicates that the effects of rotary inertia are small in the range in which the plate theory is applicable. The bending-twisting coupling terms have a significant effect with four-ply laminates but are negligible when there are eight plies or more.

Since there are only a few cases of plates for which closed-form solutions can be found, a variational approximation method capable of handling many complex cases was presented. Actual boundary conditions are often different than the ideal conditions and will significantly affect the dynamics of the structure. A procedure to determine the actual boundary conditions has been presented.

3.3.11 Exercise Problems

3.6 Use a one-term polynomial approximation in the Rayleigh-Ritz method to determine the fundamental natural frequency of a rectangular, simply supported plate. Compare the results with the exact solution given by (3.182).

3.7 Using the classical plate theory, derive an exact solution for the deflection of a simply supported plate subjected to uniformly distributed static load.

3.8 Using the first-order shear deformable plate theory, derive an exact solution for the deflection of a simply supported rectangular plate subjected to uniformly distributed static load. Assume that the laminate layup is specially orthotropic. Investigate the effect of shear deformation as a function of the ratio between the length of the side of the plate and the thickness.

3.9 Derive a one-term Rayleigh-Ritz approximation for the fundamental natural frequency of an orthotropic rectangular plates with all four edges fully clamped. Use a polynomial approximation function.

3.10 Develop an exact solution for the deflection of a rectangular, simply supported plate subjected to a force applied in its center. Assume that force is applied as (1) a concentrated force and (2) a uniform pressure applied on a small square patch centered at the center of the plate. Plot the displacements along one line of symmetry of the plate, and investigate the effect of patch size and the number of terms retained in the solution.

3.11 Develop a solution for the dynamic response of a rectangular, simply supported plate subjected to step load applied in its center. Compare this solution with the solution given by Medick (Section 3.3.7) for the transient response of an infinite plate to the same load. Show that the solution for an infinite plate is valid for very short times. That is when the disturbance has not yet reached the boundary and reflected back.

3.4 Impact Models

Several types of mathematical models are used to study the impact of a structure by a foreign object. In certain cases, the contact force history can be accurately determined by modeling the structure by an equivalent spring-mass system. When the structure behaves quasi-statically, it is also possible to calculate the maximum contact force by assuming that when the contact force reaches its maximum, the sum of the strain energy in the structure and the energy required for indentation is equal to the initial kinetic energy of the projectile. This energy-balance approach is useful when one is interested not in the complete contact force history but rather only in maximum contact force. When the dynamics of the structure must be more accurately accounted for, more sophisticated models are needed.

In this section, the various models used for impact analysis are discussed, and examples are presented to illustrate various features of the impact event.

3.4.1 Spring-Mass Models

Spring-mass models are simple and provide accurate solutions for some types of impacts often encountered during tests on small-size specimens. The most complete model (Fig. 3.15) consists of one spring representing the linear stiffness of the structure (K_{bs}), another spring for the nonlinear membrane stiffness, a mass M_2 representing the effective mass of the structure, the nonlinear contact stiffness, and M_1, the mass of the projectile. From the free body diagrams of the two masses M_1 and M_2, we obtain the equations of motion:

$$\begin{aligned} M\ddot{x}_1 + P &= 0, \\ m\ddot{x}_2 + K_{bs}x_2 + K_m x_2^3 - P &= 0. \end{aligned} \tag{3.213}$$

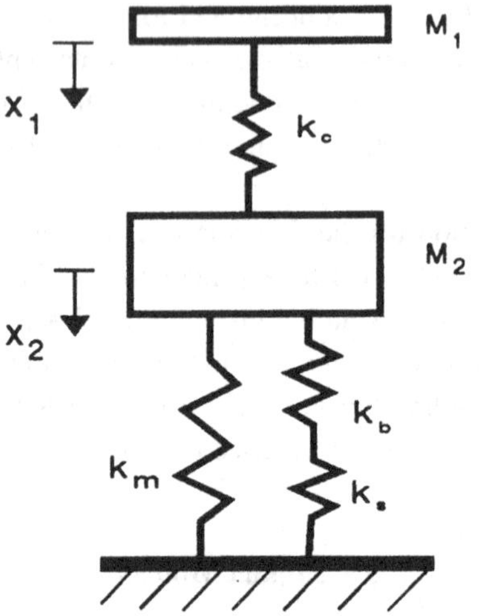

Figure 3.15. Spring-mass model.

P is the contact force which, as discussed in Chapter 2, is a highly nonlinear function of the indentation $x_1 - x_2$, which can be expressed as

$$P = P(x_1 - x_2) \tag{3.214}$$

when $x_1 > x_2$. When $x_1 < x_2$, contact ceases, $P = 0$, and (3.208) govern the free vibration of the model. The dynamics of the system described by (3.208) and (3.209) and the initial conditions

$$\ddot{x}_1(0) = V, \quad x_1(0) = x_2(0) = 0 \tag{3.215}$$

can be studied numerically.

When geometrical nonlinearities and the indentation are negligible, the model can be significantly simplified to a single degree of freedom system with the equation of motion

$$M\ddot{x} + K_{bs}x = 0. \tag{3.216}$$

In that model, the effective mass of the structure is neglected and the structure and the projectile move together as soon as contact is made ($x_1 = x_2 = x$). The general solution to (3.216) is

$$x = A \sin \omega t + B \cos \omega t \tag{3.217}$$

where $\omega^2 = K_{bs}/M_1$. The constants A and B are determined from the initial conditions (3.215), which give the result

$$x = \frac{V}{\omega} \sin \omega t. \tag{3.218}$$

Since the contact force P is equal to the force in the linear spring K_{bs}, the contact force history is given by

$$F = K_{bs}x = V(K_{bs}M_1)^{\frac{1}{2}} \sin \omega t \tag{3.219}$$

for $\omega t < \pi$. When $\omega t > \pi$ (3.214) predicts a negative contact force, which is impossible because contact between the projectile and the structure is unilateral. Therefore, separation occurs for $t = T_c = \pi/\omega$. The contact duration T_c increases with the mass of the projectile and decreases as the stiffness of the structure increases:

$$T_c = \pi(M_1/K_{bs})^{\frac{1}{2}}. \tag{3.220}$$

Equation (3.219) shows that the maximum contact force is directly proportional to the impact velocity and also increases with both the square root of the stiffness of the structure and the square root of the mass of the projectile. The maximum contact force is often plotted versus the initial kinetic energy of the projectile. For a given projectile (3.219) predicts that the maximum contact force increases with the square root of the kinetic energy.

Example 3.14: Impact on a Half-Space Consider the impact on a half-space made out of steel by a 10-mm-radius steel ball with an initial velocity of 10 mm/s. The mass of the impactor is 32.24 g, Young's modulus is taken as 215.8 GPa, the density as 7966 kg/m^3, and the contact stiffness as 1.538×10^{10} N/m$^{3/2}$. In this case, there is no overall deflection of the target, so the motion of the projectile is governed by (3.213), where x_1 represents both the displacement of the projectile and the indentation of the half-space and $P = k_c x_1^{3/2}$. This nonlinear equation of motion was solved using Newmark's step-by-step integration method with a time step of 2 μs. The contact force history is given by the solid line in Fig. 3.16. The contact duration is 170 μs, and the maximum contact force is 6.96 N.

Some investigators linearize the contact law in order to avoid numerically solving a nonlinear differential equation. In this case, taking $P = K\alpha$, the equation of motion of the projectile is given by (3.216), the contact force is given by (3.219), and the contact duration is given by (3.220), where K_{bs} is replaced by K. If K is determined from (3.219) so that the maximum contact force matches that determined numerically ($K = 1.503 \times 10^7$ N/m), the contact duration is significantly underestimated (Fig. 3.16). If K is determined from (3.220) so that the contact duration matches the numerically determined contact duration ($K = 1.101 \times 10^7$ N/m), the maximum contact force is severely underestimated (Fig. 3.16). Therefore, for impacts dominated by local indentation, the use of linearized contact laws leads to significant errors.

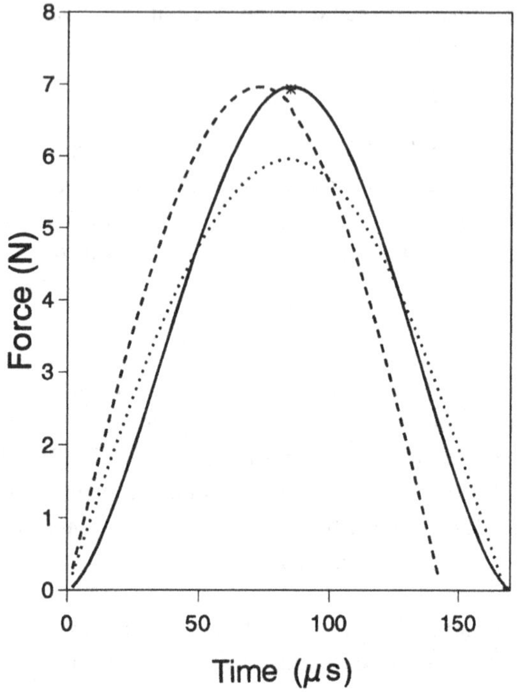

Figure 3.16. Impact of a sphere on a half-space.

3.4.2 Energy-Balance Models

Another approach for analyzing the impact dynamics is to consider the balance of energy in the system. The initial kinetic energy of the projectile is used to deform the structure during impact. Assuming that the structure behaves quasi-statically, when the structure reaches its maximum deflection, the velocity of the projectile becomes zero and all the initial kinetic energy has been used to deform the structure. The overall deformation of the structure usually involves bending, shear deformation, and for large deformations, membrane stiffening effects. Local deformations in the contact zone also are to be considered. For impacts that induce only small amounts of damage, the energy needed to create damage can be neglected. Therefore, the energy-balance equation can be written as

$$\frac{1}{2}MV^2 = E_b + E_s + E_m + E_c \tag{3.221}$$

where the subscripts b, s, and m refer to the bending, shear, and membrane components of the overall structural deformation, and E_c is the energy stored in the contact region during indentation. It is always possible to express the force-deflection relation in the form

$$P = K_{bs} W + K_m W^3 \tag{3.222}$$

where K_{bs} is the linear stiffness including bending and transverse shear deformation effects, K_m is the membrane stiffness, and W is the deflection at the impact point. Then

$$E_b + E_s + E_m = \frac{1}{2} K_{bs} W_{\max}^2 + \frac{1}{4} K_m W_{\max}^4. \tag{3.223}$$

Both experimental and analytical studies of contact between smooth indentors and laminated composites indicate that during the loading phase, the contact law can be written as

$$P = n\alpha^{\frac{3}{2}} \tag{3.224}$$

where α represents the relative motion or indentation of the structure by the projectile:

$$E_c = \int_0^{\alpha_{\max}} P \, d\alpha = \frac{2}{5} n \alpha_{\max}^{\frac{5}{2}} \tag{3.225}$$

Using (3.222), (3.224), and (3.225), the maximum indentation can be expressed in terms of the maximum displacement of the structure at the impact location:

$$\alpha_{\max} = \left(\frac{P}{n}\right)^{\frac{2}{3}} = n^{-\frac{2}{3}} \left\langle K_{bs} W_{\max} + K_m W_{\max}^3 \right\rangle^{\frac{2}{3}} \tag{3.226}$$

After substitution into (3.225), the contact energy becomes

$$E_c = \frac{2}{5} n^{-\frac{2}{3}} \left(K_{bs} W_{\max} + K_m W_{\max}^3\right)^{\frac{5}{3}}. \tag{3.227}$$

Using (3.223) and (3.227), the energy-balance equation becomes

$$\frac{1}{2} M V^2 = \frac{1}{2} K_{bs} W_{\max}^2 + \frac{1}{4} K_m W_{\max}^4 + \frac{2}{5} n^{-\frac{2}{3}} \left(K_{bs} W_{\max} + K_m W_{\max}^3\right)^{\frac{5}{3}}. \tag{3.228}$$

This equation can be solved numerically for $W_{\max}$, and the maximum contact force can then be obtained after substitution into (3.222). In order to examine the relative effects of the bending, shear, membrane, and indentation components of the deformation, we rewrite (3.228) as

$$\frac{1}{2} M V^2 = \frac{1}{2} K_{bs} W_{\max}^2 \left[1 + \frac{1}{2} \frac{K_m}{K_{bs}} W_{\max}^2 + \frac{4}{5} \left(\frac{K_{bs} W_{\max}}{n}\right)^{\frac{2}{3}} \cdot \left(1 + \frac{K_m}{K_{bs}} W_{\max}^2\right)^{\frac{5}{3}}\right]. \tag{3.229}$$

Neglecting the membrane effects ($K_m = 0$), this expression simplifies to

$$\frac{1}{2}MV^2 = \frac{1}{2}K_{bs}W_{\max}^2\left[1 + \frac{4}{5}\left(\frac{K_{bs}W_{\max}}{n}\right)^{\frac{2}{3}}\right] \tag{3.230}$$

or, in terms of the maximum contact force,

$$\frac{1}{2}MV^2 = \frac{1}{2}\frac{P_{\max}^2}{K_{bs}} + \frac{4}{5}\frac{P_{\max}^{\frac{5}{3}}}{n^{\frac{2}{3}}}. \tag{3.231}$$

If we further neglect the effect of local deformation in the contact zone, represented by the second term on right-hand side,

$$P_{\max} = V(K_{bs}M)^{\frac{1}{2}}. \tag{3.232}$$

In this case, the maximum contact force increases linearly with the initial velocity of the projectile.

For the special case of impact on a thick specimen, if the deformation of the structure is negligible, the maximum indentation is reached when the initial kinetic energy of the projectile has been used entirely to indent the structure. That is,

$$\frac{1}{2}MV^2 = \frac{2}{5}n\alpha^{\frac{5}{2}} = \frac{2}{5}\frac{F^{\frac{5}{3}}}{n^{\frac{2}{3}}} \tag{3.233}$$

from which we obtain the maximum contact force

$$F = \left(\frac{5}{4}\right)^{\frac{3}{5}}(M^3V^6n^2)^{\frac{1}{5}}. \tag{3.234}$$

An approximate value of the contact duration can be calculated, replacing the nonlinear contact stiffness by a linear spring with stiffness

$$k = \frac{F_{\max}}{\alpha_{\max}} = \left(\frac{5}{4}\right)^{\frac{1}{5}}M^{\frac{1}{5}}V^{\frac{2}{5}}n^{\frac{4}{5}}. \tag{3.235}$$

The contact duration is then half the period for free vibrations of a single degree of freedom system with stiffness k and mass M.

$$\begin{aligned} T_c &= \frac{T}{2} = \frac{\pi}{\omega} = \pi\left(\frac{M}{k}\right)^{\frac{1}{2}} = \pi\left(\frac{4}{5}\right)^{\frac{1}{10}}(M^2V^{-1}n^{-2})^{\frac{1}{5}} \\ &= 3.0723(M^2V^{-1}n^{-2})^{\frac{1}{5}}. \end{aligned} \tag{3.236}$$

The actual contact duration is determined by considering the energy-balance equation for an arbitrary time t during impact

$$\frac{1}{2}MV^2 = \frac{1}{2}M\dot{\alpha}^2 + \frac{2}{5}n\alpha^{\frac{5}{2}}. \tag{3.237}$$

Separating the variables t and α, we obtain

$$dt = \frac{d\alpha}{\left(V^2 - \frac{4}{5}M^{-1}n\alpha^{\frac{5}{2}}\right)^{\frac{1}{2}}}. \tag{3.238}$$

The contact duration is double the time needed to reach the maximum indentation:

$$T_c = 2\int_0^{\alpha_{\max}} \frac{d\alpha}{\left(V^2 - \frac{4}{5}M^{-1}n\alpha^{\frac{5}{2}}\right)^{\frac{1}{2}}}. \tag{3.239}$$

With the change of variables $x = \alpha/\alpha_{\max} = \alpha/(\frac{5}{4}\frac{MV^2}{n})^{2/5}$, we obtain

$$\begin{aligned} T_c &= \frac{2\alpha_{\max}}{V}\int_0^1 \frac{dx}{\langle 1 - x^{\frac{5}{2}}\rangle^{\frac{1}{2}}} = 2.94\,\frac{\alpha_{\max}}{V} \\ &= 3.2145\langle M^2V^{-1}n^{-2}\rangle^{\frac{1}{5}}. \end{aligned} \tag{3.240}$$

The approximate value of the contact duration (3.236) is close to the exact result (3.240).

Example 3.15: Impact on a Half-Space by the Energy-Balance Method For the case considered in Example 3.12, the maximum contact force can be calculated by (3.234), and the maximum contact force can be calculated using (3.240). Excellent agreement is obtained between the numerical solution from Example 3.14 and the prediction of the energy-balance model given by (3.234), (3.240) as shown on Fig. 3.16.

3.4.3 Response of a Bernoulli-Euler Beam to the Impact by a Mass

In this section we present a concrete example of complete model for predicting the impact dynamics. In this context, a complete model is one in which the dynamics behavior of the structure is described accurately. This means that all the vibration modes that participate in the response have to be predicted accurately and must be retained in the model. For a simply supported beam, an analytical solution can be found for the natural frequencies and mode shapes. The transient response is then expressed in terms of these mode shapes, and all participating modes can be included. This simple example first treated by Timoshenko (1913) will be used to show how to develop a complete model in which the dynamics of the structure, the dynamics of the projectile, and the contact behavior are all modeled accurately. This approach can then be generalized to more complex situations.

For a beam governed by (3.20), the mode shapes are given by (3.45) with the parameters β and the coefficients A–D being selected so that the boundary conditions at $x = 0, L$ are satisfied. The transverse displacement of the beam is expanded in the series

$$w(x,t) = \sum_{j=1}^{N} \alpha_j(t)\phi_j(x). \tag{3.241}$$

N is the number of modes retained in the dynamic model of the beam, determined by trial and error. Substituting (3.241) into the equation of motion, multiplying by $\phi_i(x)$, and integrating over the length of the beam gives the modal equations

$$m_i\ddot{\alpha}_i + k_i\alpha_i = f_i(t) \tag{3.242}$$

where the modal stiffness, modal mass, and modal force are defined as

$$k_i = \int_0^L \phi_{i,xx}^2\,dx, \quad m_i = \int_0^L \phi_i^2 dx, \quad f_i(t) = \int_0^L p(x,t)\phi_i(x)\,dx.$$

When a concentrated force F is applied at $x = a$,

$$p(x,t) = P(t)\delta(x-a) \tag{3.243}$$

where δ is the Dirac delta function. Then,

$$f_i(t) = P\phi_i(a). \tag{3.244}$$

The motion of the projectile is governed by

$$M\ddot{x} + P = 0. \tag{3.245}$$

The magnitude of the contact force is determined using Hertz contact law:

$$P = k_c\langle x(t) - w(a,t)\rangle^{\frac{3}{2}}. \tag{3.246}$$

The $N + 1$ equations of motion (3.242) and (3.246) can be written in matrix form as

$$[M]\{\ddot{X}\} + [K]\{X\} = \{F\} \tag{3.247}$$

where in this case the mass matrix $[M]$ is a diagonal matrix with the first N terms being the m_i defined in (3.242) and the last diagonal term being the mass of the projectile. The stiffness matrix is also a diagonal matrix here. Its first N diagonal terms are the modal stiffness parameters k_i defined in (3.242) and the last diagonal term is zero. The pseudo-force vector $\{F\}$ consists of the N terms f_i defined in (3.244) and the last term is $-P$ obtained from (3.246). The initial conditions are that for $t = 0$, the beam is at rest and the displacement of the projectile is also zero, so $\{X(0)\} = 0$. The velocity of every point along the beam is zero initially, and the initial velocity of the projectile is V. Therefore, the only nonzero term in the initial velocity vector $\{\dot{X}(0)\}$ is the last term and it is equal to V.

The equations of motion (3.247) with the initial conditions just described are solved using Newmark's time integration technique. Assuming that the solution is known for step n (3.247) are solved to determine displacements, velocities, and accelerations at the end of step $n + 1$. With this method, the solution at the $(n + 1)$th time step is calculated by solving the system of linear algebraic equations

$$[\tilde{A}]\{X\}_{n+1} = \{\tilde{F}\} \tag{3.248}$$

where

$$\begin{aligned}
&[\tilde{A}] = [K] + a_o[M] \\
&\{\tilde{F}\} = \{F\}_{n+1} + [M]\langle a_o\{X\}_n + a_1\{\dot{X}\}_n + a_2\{\ddot{X}\}_n\rangle \\
&a_o = \frac{1}{\beta \Delta t^2}, \quad a_1 = a_o \Delta t, \quad a_2 = \frac{1}{2\beta} - 1, \\
&a_3 = (1-\alpha)\Delta t, \quad a_4 = \alpha \Delta t.
\end{aligned} \tag{3.249}$$

Once (3.248) is solved and $\{X\}$ is known at time $(n+1)\Delta t$, the first and second derivatives are computed from

$$\begin{aligned}
\{\ddot{X}\} &= a_o\langle\{X\}_{n+1} - \{X\}_n\rangle - a_1\{\dot{X}\}_n - a_2\{\ddot{X}\}_n \\
\{\dot{X}\}_{n+1} &= \{\dot{X}\}_n + a_3\{\ddot{X}\}_n + a_4\{\ddot{X}\}_{n+1}
\end{aligned} \tag{3.250}$$

The difficulty is that $\{F\}$ is a nonlinear function of the displacements at the end of step $n + 1$. $\{F\}$ is first estimated from the displacements at step n, (3.247) are solved for the displacements at the end of step $n + 1$, and a new estimate of $\{F\}$ is obtained. After several iterations, this procedure converges to the solution at the end time step $n + 1$.

Example 3.16: Impact of a Steel Sphere on a Simply Supported Steel Beam
This example first proposed by Timoshenko (1913) is considered a benchmark problem and has been studied by many authors (e.g., Goldsmith 1960, Lee 1940, Sun and Huang 1975). A 10 × 10 × 153.5 mm simply supported steel beam is impacted by a 10-mm-radius steel ball with an initial velocity of 10 mm/s. The mass of the impactor is 32.24 g, Young's modulus is taken as 215.8 GPa, the density as 7966 kg/m^3, and the contact stiffness as 1.538×10^{10} N/m$^{3/2}$. The results in Fig. 3.17 are in excellent agreement with those presented by previous investigators. The maximum contact force is 5.0 N and the contact duration of 158 μs. During contact, the kinetic energy of the projectile is reduced while the kinetic energy and the strain energy stored into the target increase (Fig. 3.18). After contact ceases, the kinetic energy of the projectile

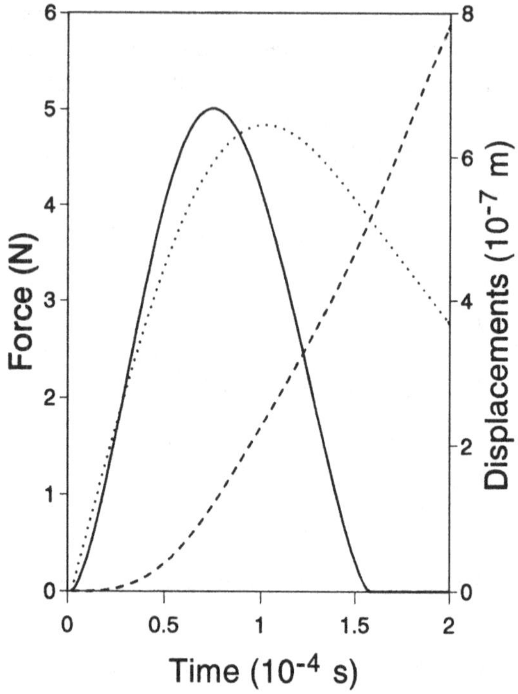

Figure 3.17. Impact on a simply supported beam. Solid line: contact force; dotted line: displacement of projectile; dashed line: central beam displacement.

remains constant and the sum of the kinetic and strain energies of the target also remains constant since both bodies are free of external loads. After contact ends, the target undergoes free vibrations and there is a continuous exchange between the kinetic energy and the strain energy, the sum being constant as shown in Fig. 3.18. From Figs. (3.17) and (3.18), we see that the velocity of the projectile is not zero when the contact force reaches its maximum and at that time the kinetic energy stored into the beam is larger than the strain energy and represents a significant portion of the total energy in the system. This indicates that the energy-balance approach would not be successful for this problem. Figure 3.19 shows the displacements of the beam for $t = 76\ \mu s$, that is, when the contact force is near its maximum; Figure 3.20 shows the velocity distribution along the beam at the same instant. Under the same force, the static deflection would be 2.1529×10^{-6} m, while for the impact case at $t = 76\ \mu s$, the central deflection is 1.24×10^{-7} m. Therefore, it is not appropriate to assume a quasi-static behavior for the beam, as is done in the energy-balance approach.

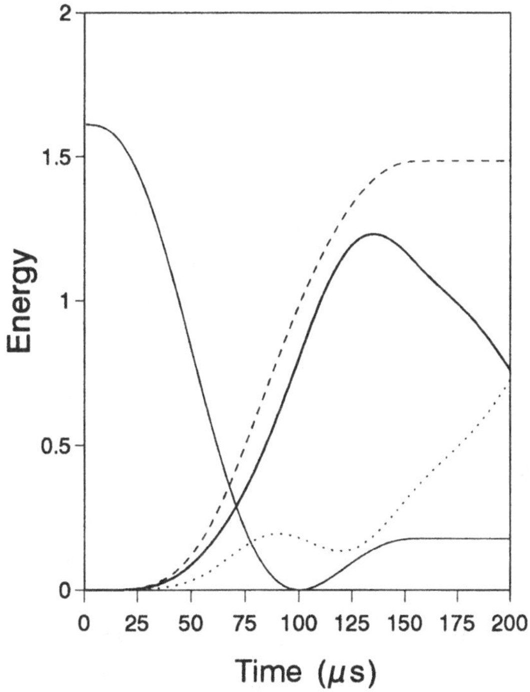

Figure 3.18. Kinetic and strain energies in simply supported beam during impact. Thin line: kinetic energy of impactor; thick line: kinetic energy of the beam; dotted line: strain energy in the beam; dashed line: total energy stored in beam.

Example 3.17: Impact of a Steel Sphere on a Steel Beam Another example first treated by Goldsmith (1960) and then by Sun and Huang (1975) consists of a 12.7 × 12.7 × 645.2 mm simply supported steel beam impacted in the center by a 12.7-mm-diameter steel sphere. Both the beam and the sphere have the following material properties: $E = 206.8$ GPa, $\nu = 0.3$, $\rho = 7837$ kg/m^3. When the contact is assumed to follow Hertz's law ($k = 1.208 \times 10^{10}$ N/m$^{3/2}$), the contact force and the displacements of the projectile and the center of the beam are as shown in Fig. 3.21. The effects of permanent indentation are included by taking the contact law to be

$$F = 3.6149 \times 10^8 \alpha^{1.128} \tag{3.251}$$

during the loading phase and

$$F = F_{\max}\left(\frac{\alpha - \alpha_p}{\alpha_{\max} - \alpha_p}\right)^{\frac{3}{2}} \tag{3.252}$$

for the unloading phase, where $F_{\max}$ and $\alpha_{\max}$ are the maximum force and

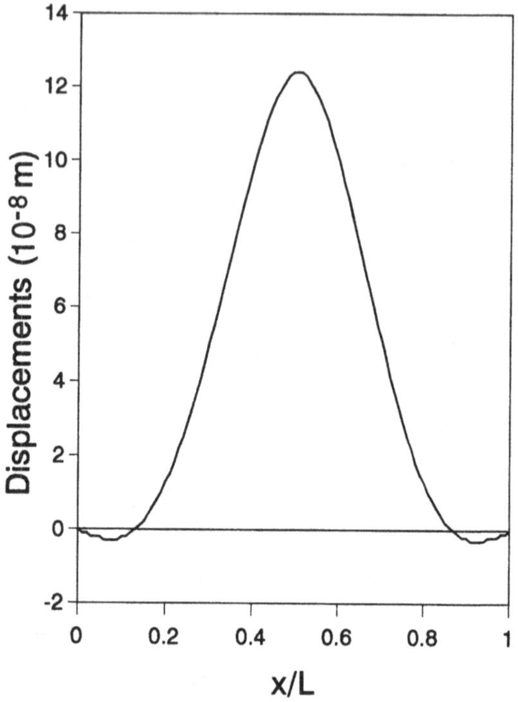

Figure 3.19. Transverse beam displacements when the contact force reaches its maximum ($t = 76\ \mu$s).

maximum indentation reached during the loading phase and α_p is the permanent indentation, taken to be 0.0003124 m. First consider the case when the contact law is given by (3.247) for both the loading and unloading phase. Figure 3.22 shows that the magnitude of the contact force is reduced compared to the case with elastic contact (Fig. 3.21). When the contact law is given by (3.251) for the loading phase and (3.252) for the unloading phase, the results in Fig. 3.23 are the same as in Fig. 3.22 until the contact force reaches its maximum. After that, the contact force drops quickly. Considering the evolution of the kinetic and strain energies of the beam and the kinetic energy of the projectile (Fig. 3.24), we see that after impact the kinetic energy of the projectile is very small and that about 71% of the initial kinetic energy has been lost. Figure 3.25 shows that with this type of contact behavior, most of the energy stored in the local deformation of the beam is not recovered upon unloading.

This approach can be generalized to cases where a simple closed-form solution for transient response of the target is not available. In most cases, an

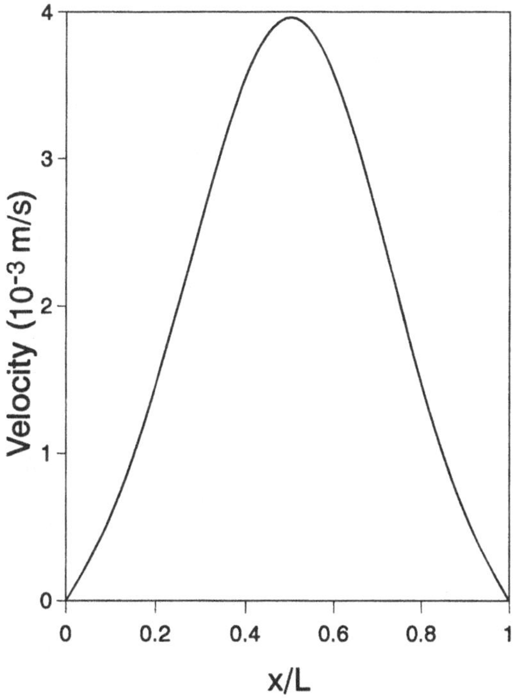

Figure 3.20. Velocities of the beam when the contact force reaches its maximum (t = 76 μs).

approximate solution must be used. If the geometry remains simple, a variational approximation method such as the Rayleigh-Ritz method (Section 3.3.8) can be used very efficiently. The finite element method can be used especially for cases with complex geometry. In either case, the target will be modeled by a set of N coupled ordinary differential equations of motion, and the motion of the projectile is still governed by (3.240). Those $N + 1$ equations can be written as (3.247) again, but now the stiffness and mass matrices are not diagonal matrices anymore. However, the general procedure for formulating the problem and solving the equations of motion remains the same.

The number of modes to be retained in the model for representing the dynamics of the structure is usually determined by trial and error. For the simply supported beams in examples 3.16 and 3.17 the natural frequencies and mode shapes are known exactly and all that is needed is to increase N, the number of modes in the model. For more complex situations in which approximate solutions are used, the mode shapes are known only approximately, and the number of degrees of freedom in the finite element model or the number of

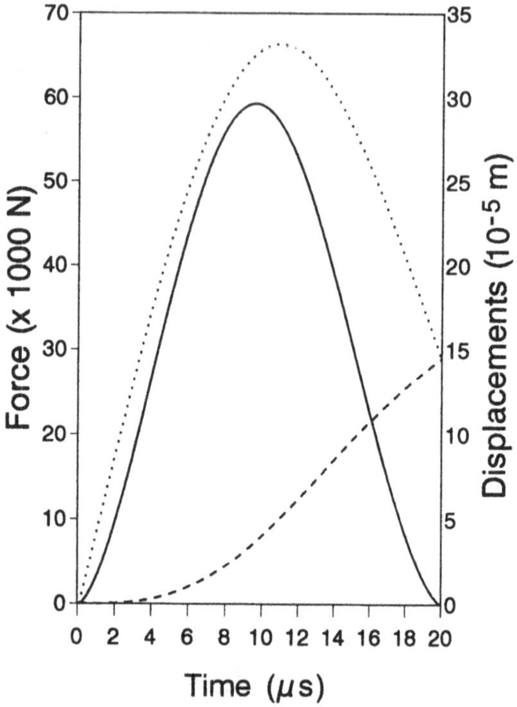

Figure 3.21. Projectile and target displacements for impact of simply supported beam by a steel sphere. Hertzian contact (Goldsmith's example). Solid line: contact force; dotted line: displacement of projectile; dashed line: central beam displacement.

displacement approximation functions in the variational model has to be increased significantly in order to accurately predict the required modes. A rule of thumb is that in order to accurately predict N modes, $2N$ degrees of freedom are needed in the model. Another problem is that as more modes participate in the response, the accuracy is limited by the validity of the structural theory used. For example, as shown in Section 3.2.8, the lower natural frequencies and mode shapes of slender beams are predicted accurately with the Bernoulli-Euler beam theory, but for higher modes, this theory will be in error and a more refined theory such as the Timoshenko beam theory must be used.

3.4.4 Impact on a Simply Supported Plate (Classical Plate Theory)

For rectangular plates with simple supports along the edges, the Navier solution can be used to obtain a closed-form solution for the transient response to a given loading. For this simple case, a simple complete model can be developed for

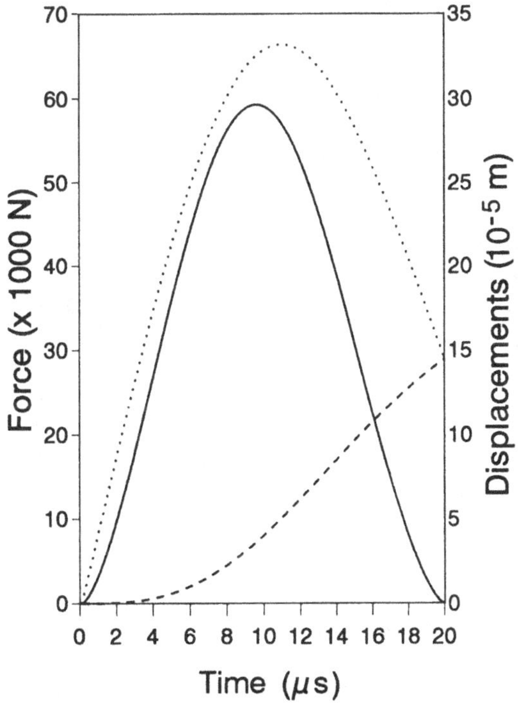

Figure 3.22. Impact of simply supported beam by a steel sphere. Inelastic contact (Eq. 3.243). Solid line: contact force; dotted line: displacement of projectile; dashed line: central beam displacement.

the impact problem. For other geometries or boundary conditions, variational models or finite elements must be used, but nevertheless this model is useful in the analysis of many test situations; it will be shown that in many practical cases the contact duration is so short that the plate experiences only local deformation around the point of impact. In this case the deformation is not affected by the presence of the boundaries, and so the present model is applicable.

The motion of an orthotropic plate is governed by (3.179), according to the classical plate theory. The boundary conditions for a rectangular SSSS plate (3.181) and the equation of motion plate are satisfied when the displacements are expanded into the double series

$$w(x, y, t) = \sum_{m,n=1}^{\infty} \alpha_{mn} \sin \frac{m\pi x}{a} \sin \frac{n\pi y}{b}. \tag{3.253}$$

For a concentrated force applied at (x_o, y_o),

$$q = F\delta(x - x_o)\delta(y - y_o). \tag{3.254}$$

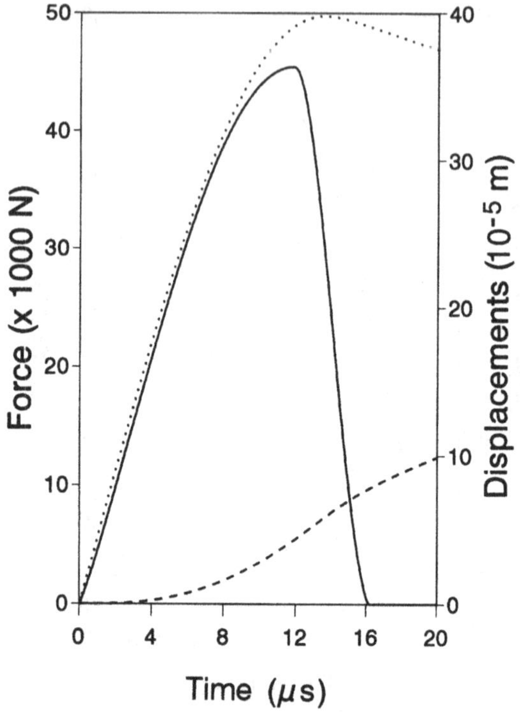

Figure 3.23. Impact of simply supported beam by a steel sphere. Inelastic contact with different loading and unloading curves (Eqs. 3.243, 3.244). Solid line: contact force; dotted line: displacement of projectile; dashed line: central beam displacement.

After substitution into the equation of motion, we find that for each mode,

$$\ddot{\alpha}_{mn} + \omega_{mn}^2 \alpha_{mn} = \frac{4F}{abI_1} \sin \frac{m\pi x_o}{a} \sin \frac{n\pi y_o}{b} \tag{3.255}$$

where ω_{mn} is the natural frequency given by (3.182). If m and n vary respectively from 1 to p and 1 to q, the motion of the plate is then described by $N = p \cdot q$ equations like (3.255).

The motion of the impactor is governed by (3.245) with the contact force given by

$$F = k_c \alpha^{\frac{3}{2}} = k_c \langle x - w(x_o, y_o) \rangle^{\frac{3}{2}} \tag{3.256}$$

and the initial conditions

$$\begin{aligned} \alpha_{mn}(0) &= \dot{\alpha}_{mn}(0) = 0 \\ x(0) = 0, &\quad \dot{x}(0) = V. \end{aligned} \tag{3.257}$$

The $N + 1$ differential equations can be written in matrix form as in (3.247) and integrated using Newmark's step-by-step time integration method.

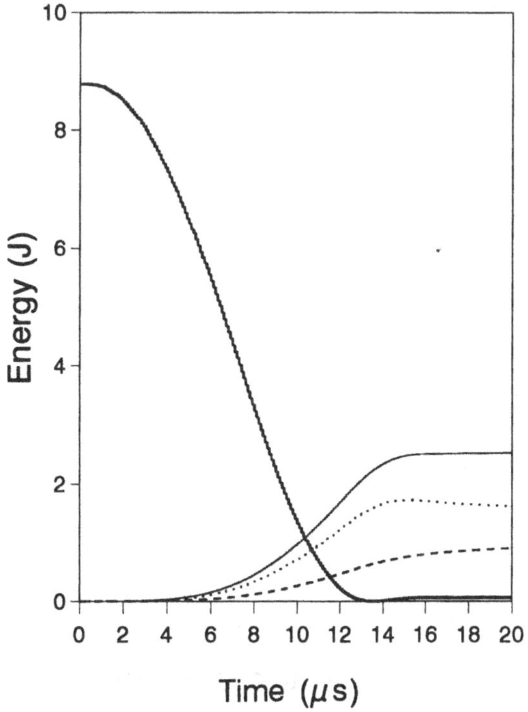

Figure 3.24. Kinetic energy of the projectile during impact of a simply supported beam by a steel sphere (Goldsmith's example). Thin line: kinetic energy of impactor; thick line: kinetic energy of the beam; dotted line: strain energy in the beam; dashed line: total energy stored in beam.

3.4.5 Impact on a Simply Supported Plate (SDPT)

The problem of impact on a simply supported plate governed by the first-order shear deformation plate theory can be formulated in a similar fashion. The displacement can be taken in the form

$$w = \sum_{m,n=1}^{p,q} W_{mn} \sin \frac{m\pi x}{a} \sin \frac{n\pi y}{b}$$

$$\psi_x = \sum_{m,n=1}^{p,q} X_{mn} \cos \frac{m\pi x}{a} \sin \frac{n\pi y}{b} \tag{3.258}$$

$$\psi_y = \sum_{m,n=1}^{p,q} Y_{mn} \sin \frac{m\pi x}{a} \cos \frac{n\pi y}{b}.$$

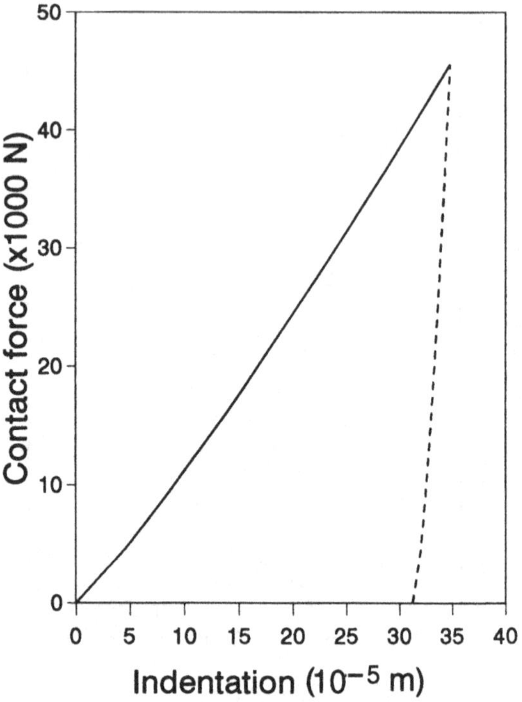

Figure 3.25. Indentation during impact of a simply supported beam by a steel sphere.

Substitution into the equations of motion (3.110)–(3.112) yields the set of three coupled ordinary differential equations

$$[m]\begin{Bmatrix} \ddot{W}_{mn} \\ \ddot{X}_{mn} \\ \ddot{Y}_{mn} \end{Bmatrix} + [k]\begin{Bmatrix} W_{mn} \\ X_{mn} \\ Y_{mn} \end{Bmatrix} = \begin{Bmatrix} f_{mn} \\ 0 \\ 0 \end{Bmatrix} \tag{3.259}$$

where

$$f_{mn} = \frac{4F}{abI_1}\sin\frac{m\pi x_o}{a}\sin\frac{n\pi y_o}{b}$$

and the matrices K and M are the same as in (3.184). Therefore, there will be three equations for each mode considered in the response of the plate. The motion for the projectile is again governed by (3.245). The initial conditions are defined by

$$\begin{gathered} x(0) = 0, \quad \dot{x}(0) = V, \quad w(x, y, 0) = 0, \quad \dot{w}(x, y, 0) = 0, \\ \psi_x(x, y, 0) = \psi_y(x, y, 0) = \dot{\psi}_x(x, y, 0) = \dot{\psi}_y(x, y, 0) = 0. \end{gathered} \tag{3.260}$$

As with simply supported plates governed by the classical plate theory, the solution can be obtained by step-by-step integration using Newmark's method.

Example 3.18: Impact of a Sphere on an Isotropic Square Plate To assess the accuracy of various models for impact dynamics analysis, we first consider the case of a steel square plate clamped along all four edges and impacted by a steel ball. The dimensions of the plate are 7.874 in. $\times$ 7.874 in. $\times$ 0.315 in. (200 mm $\times$ 200 mm $\times$ 8 mm). Young's modulus is 30×10^6 psi, Poisson's ratio is 0.30, and the specific weight is 0.282 lb/in.^{3}. The sphere has a radius of 0.3937 in. (10 mm) and an initial velocity of 39.37 in./s (1 m/s). This example was first analyzed by Karas (1939) and subsequently by many other investigators, such as Bachrach (1988), Wu and Chang (1989), Bachrach and Hansen (1989), and Choi and Chang (1992).

This example is first analyzed as a simply supported plate using the classical plate theory. The contact force history (Fig. 3.26), with a maximum force of 311.4 lb occurring after 33 μs and a contact duration of approximately 73 μs, is in excellent agreement with the results given by previous investigators. Figure 3.27 shows that, while initially small, plate deflections are significant. Figure 3.28 indicates that the velocity of the center of the plate reaches a maximum of

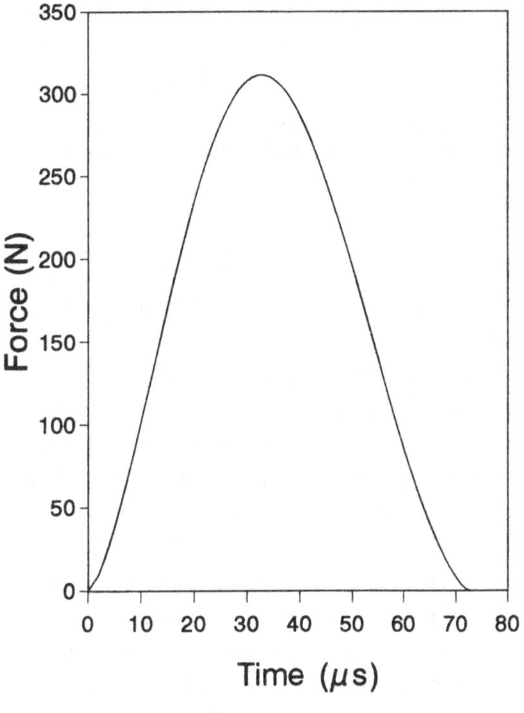

Figure 3.26. Impact of a simply supported plate by a steel sphere (Karas' example). Contact force history.

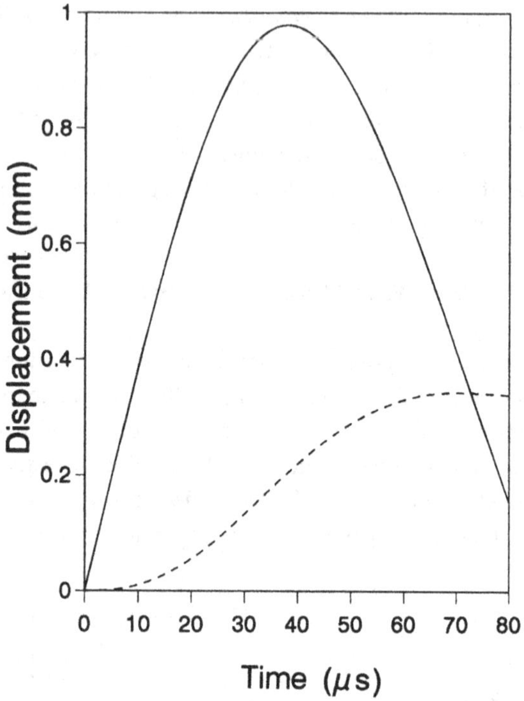

Figure 3.27. Impact of a simply supported plate by a steel sphere (Karas' example). Solid line: displacement of projectile; dashed line: plate displacement at the impact point.

8.774 in./s after 33 μs, precisely when the contact force reaches its maximum. After impact, the velocity of the projectile is –26.128 in./s, which implies that the final kinetic energy of the projectile is only 44% of its kinetic energy before impact. Conversely, 66% of the kinetic energy of the projectile has been transferred to the plate. These results are in excellent agreement with those presented by other investigators who considered the plate to be clamped along all four edges. Figure 3.29 shows the deflections of the plate along the line $y/b = 0.5$ for $t = 33$ μs, that is, when the contact force reaches its maximum. For such times deformation near the edges of the plate is very small, and it can be said that this is a case of wave propagation–controlled impact and that the deformation wave has not yet sensed the presence of the boundary. This explains why accurate results can be obtained while modeling the plate as being simply supported instead of clamped. Figure 3.29 shows that, at least in the first half of the contact duration, the plate does not deform in a quasi-static manner or according to the first free vibration mode. The dynamics of the

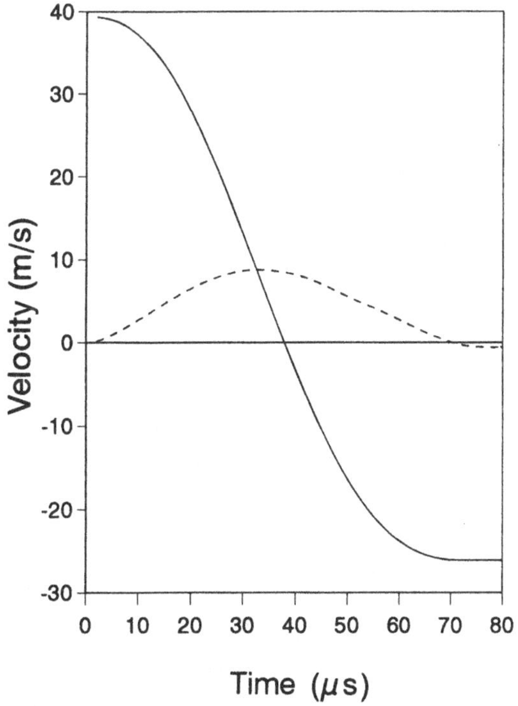

Figure 3.28. Impact of a simply supported plate by a steel sphere (Karas' example). Solid line: velocity of projectile; dashed line: plate velocity at the impact point.

target must be modeled accurately since a large fraction of the impact energy is absorbed by the target and since its response involves the participation of several modes. Therefore, the energy-balance or spring mass models developed under the assumption of a quasi-static or single mode approximation for the behavior of the structure are expected to be inaccurate for this example.

Now, let us use the energy-balance approach to find the maximum contact force and the contact duration by first neglecting the overall deflection of the plate (3.234). As expected, this approach overestimates the maximum contact force and underestimates the contact duration. The second approach is to assume that the structure behaves quasi-statically and that it can be modeled by a spring with stiffness

$$k_2 = \frac{Eh^3}{0.0611a^2}. \tag{3.261}$$

The maximum contact force predicted using (3.231) is slightly underestimated, which can be explained by the fact that because the mass of the plate is large

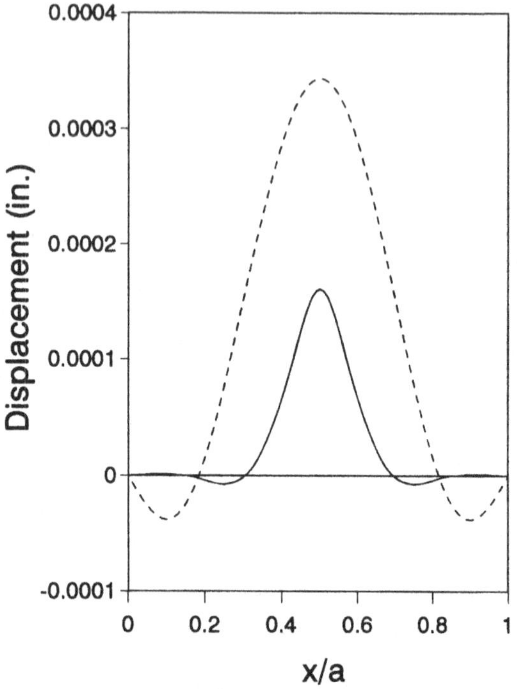

Figure 3.29. Impact of a simply supported plate by a steel sphere (Karas' example). Displacement along $y/b = 0.5$. Solid line: $t = 33$ μs; dashed line: $t = 73$ μs.

compared to the mass of the projectile, the inertia of the plate will make it appear stiffer and therefore increase the contact force. A plot of the contact force history computed using the two-degree-of-freedom model described in Section 3.4.1 shows that the solution obtained by such a simple model compares well with that obtained with the more complex finite element model used by Bachrach (1988).

Example 3.19: Impact on a Circular Aluminum Plate This example was studied by Shivakumar et al. (1985a) and Wang and Yew (1990). A clamped circular aluminum plate with a 38 mm radius is impacted by a 19-mm-radius steel impactor with an initial velocity of 2.54 m/s. We seek to determine how the contact duration and the maximum contact force vary as the thickness of the plate increases. The energy-balance method is used, and the effects of plate deflection and geometrical nonlinearities are examined. Young's modulus for aluminum is taken as 68.95 GPa. For a clamped circular plate with immovable edges, the bending and membrane stiffness are given by (Shivakumar et al. 1985)

$$K_b = \frac{4\pi E_r h^3}{3(1-\nu^2)a^2}, \tag{3.262a}$$

$$K_m = \frac{(353-191\nu)\pi E h}{648(1-\nu)a^2}. \tag{3.262b}$$

Using the energy-balance model (3.228), the maximum contact force is shown to increase as plate thickness increases and to reach a limit when the deflections of the plate becomes negligible compared to the indentation by the projectile (Fig. 3.30a). The contact force history and the history of projectile and central plate deflections are determined using the two-degree-of-freedom model described in Section 3.4.1. If inertia effects in the plate are neglected ($m_2 = 0$), a smooth contact force history is obtained and the maximum value matches that obtained using the energy-balance model, which also neglects the

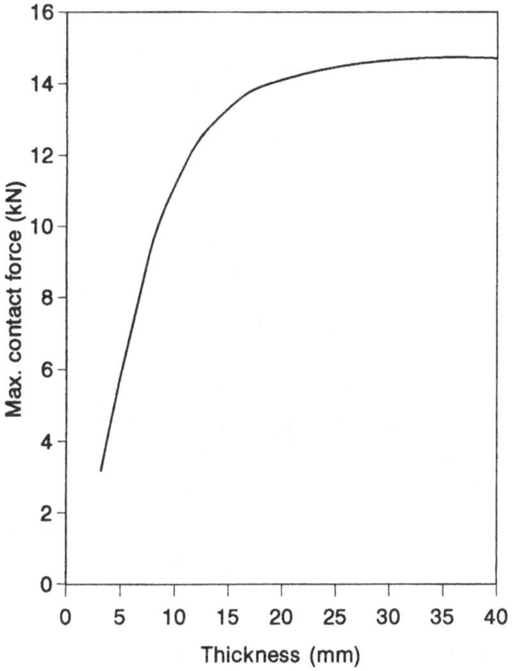

(a)

Figure 3.30. Impact on aluminum circular plate. a) Evolution of maximum contact force with laminate thickness. b) Contact force history (solid line: $m_2 = M_p/4$; dashed line: $m_2 = 0$). c) Solid line: projectile displacement; dashed line: central displacement of plate ($m_2 = M_p/4$).

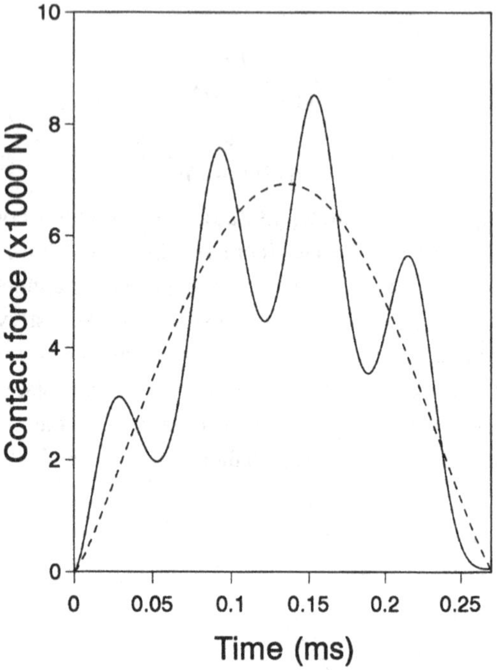

(b)

Figure 3.30. *(cont.)*

kinetic energy of stored in the target. However, if the effective mass of the plate is taken as $\frac{1}{4}$ of the total mass of the plate, as suggested by Shivakumar et al. (1985), strong oscillations in the contact force history are observed (Fig. 3.30b). The contact duration is not affected. Additional insight is gained by looking at the displacement histories (Fig. 3.30c). The projectile displacement remains smooth, whereas the central displacement of the plate is seen to oscillate. The displacements of the plate are larger than the indentation as shown in Fig. 3.30c and therefore cannot be neglected. Results in Fig. 3.30b,c are in excellent agreement with those reported by Wang and Yew (1990), who performed a detailed finite analysis.

Example 3.20: Impact on Square Composite Plate The following example has been studied by many investigators, including Chen and Sun (1985), Bachrach (1988), Cairns and Lagace (1989). The 200 mm by 200 mm graphite epoxy plate with a $[90,0,90,0,90]_S$ layup is impacted in the center by a 12.7-mm-diameter steel sphere with an initial velocity of 3 m/s. The properties of the

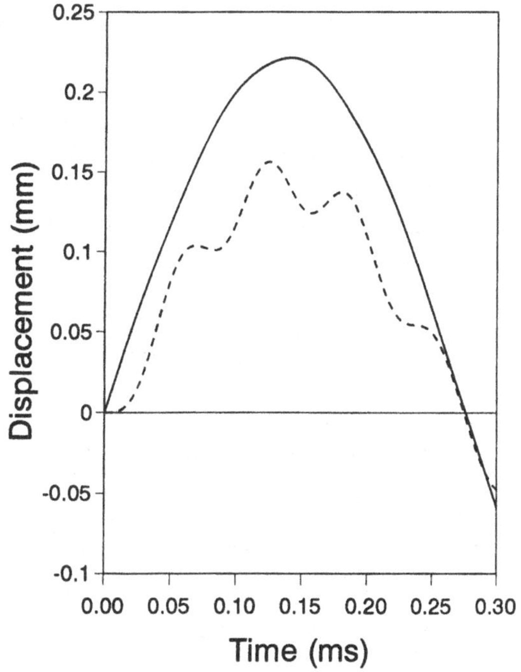

(c)

Figure 3.30. *(cont.)*

T300/934 graphite epoxy are

$$E_1 = 141.2 \text{ GPa}, \quad E_2 = 9.72 \text{ GPa}, \quad \nu_{12} = 0.30, \quad \nu_{23} = 0.30,$$
$$G_{12} = 5.53 \text{ GPa}, \quad G_{23} = 3.74 \text{ GPa}, \quad \rho = 1536 \text{ kg/m}^3. \tag{3.263}$$

The ply thickness is 0.269 mm, and the plate is assumed to be simply supported along all four edges. The contact stiffness is taken as 9.8488×10^8, and the mass of the impactor is 8.537 g. The contact force history found using the classical plate theory (Fig. 3.31a) is in good agreement with the results of previous investigators. Two impacts occur during the first 400 μs. The first one ends when the displacement of the target exceeds that of the projectile (Fig. 3.31b). Roughly 100 μs later, a second impact occurs.

Example 3.21: Impact on Square Composite Plate A slightly different example was studied in detail by Qian and Swanson (1990). A 200 mm by 200 mm graphite epoxy plate with a $[0,90,0,90,0]_S$ layup is impacted in the center by a 12.7-mm-diameter steel sphere with an initial velocity of 3 m/s. The properties of the T300/934 graphite epoxy are given by (3.156). The ply thickness is 0.269 mm, and the plate is assumed to be simply supported along all four edges. The contact stiffness is taken as 8.394×10^8 N/m$^{3/2}$, and the mass of the impactor is 8.537 g.

The natural frequencies for the first five modes that will participate to the response of the plate were determined according to the classical plate theory (3.203) and the first-order shear deformation theory (3.205), assuming simply supported boundary conditions. The results in Table 3.8 indicate that for this example the effect of shear deformation and rotary inertia are small.

Table 3.8. *First five natural frequencies (rad/s) for the 10-ply example by Qian and Swanson (1990)*

m	n	CPT	FSDT	% difference
1	1	1900.6	1897.7	0.15
1	3	9856.5	9781.2	0.76
3	1	12617	12462	1.23
3	3	17105	16884	1.29
1	5	26633	26084	2.06

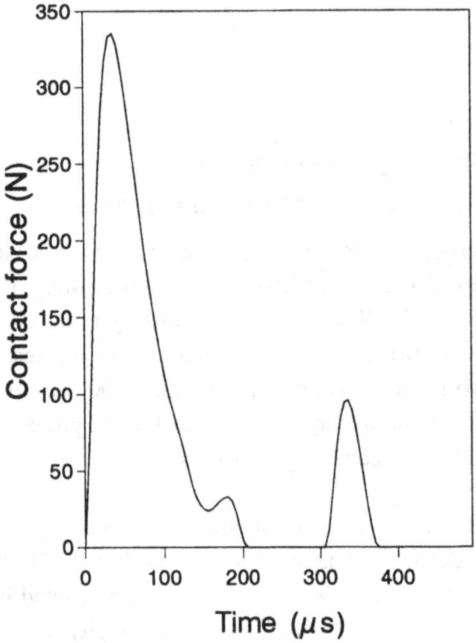

(a)

Figure 3.31. Impact on 10-plied graphite-epoxy laminated plate, $V = 3$ m/s. a) Contact force history. b) Solid line: projectile displacement; dashed line: plate deflection.

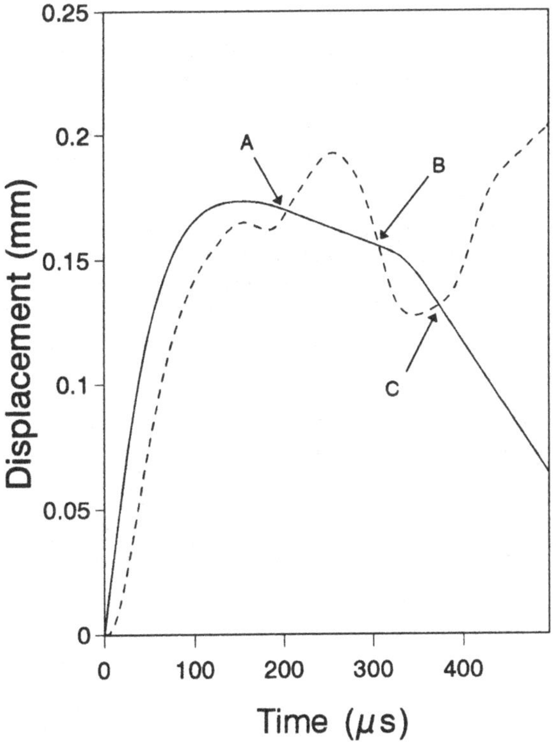

(b)

Figure 3.31. *(cont.)*

The contact force history found using the classical plate theory (Fig. 3.32a) is in good agreement with the results of previous investigators. Multiple impacts occur in this example, too. Contact ends at point A at a time $t_A = 225\ \mu$s, and a second impact starts at point B at time $t_B = 311\ \mu$s. Between points A and B, the central displacement of the plate becomes larger than the displacement of the projectile (Fig. 3.32b), while both the plate and the projectile are still moving in the positive direction. As the center of the plate starts moving backwards, contact is made with the impactor at point B, and a second impact occurs. The second impact ends at point C at time t_C. Right after point C, the projectile moves in the negative direction, and the center of the plate moves in the positive direction. Since no external force is applied to the projectile between A and B, Newton's law implies that the velocity of the projectile is constant during that time. Therefore, the displacement versus time curve (Fig. 3.32b) is a straight line between A and B, and the same observation can be made for the portion of the curve after t_C.

Qian and Swanson (1990) also looked at the effect of laminate thickness by considering two additional cases in which all parameters remained the same except that the thickness of each ply was multiplied by 2 and 4, respectively. Figures 3.33a,b indicate that in those two cases, a single impact occurs. A comparison between the results obtained using the classical plate theory and those

Table 3.9. *Maximum contact force predicted using the classical plate theory and the shear deformation plate theory for* $[0,90,0,90,0]_s$ *graphite-epoxy laminates with two different thicknesses*

	Contact force (N)	
Total thickness (mm)	Classical plate theory	Shear deformation plate theory
2.69	320.2	286.8
5.38	627.6	562.8

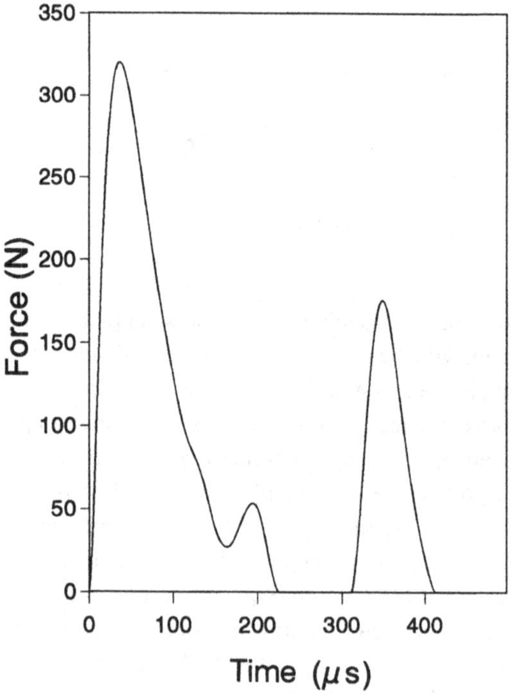

(a)

Figure 3.32. Impact on 10-plied graphite-epoxy laminated plate, $V = 3$ m/s. a) Contact force history. b) Solid line: projectile displacement; dashed line: plate deflection.

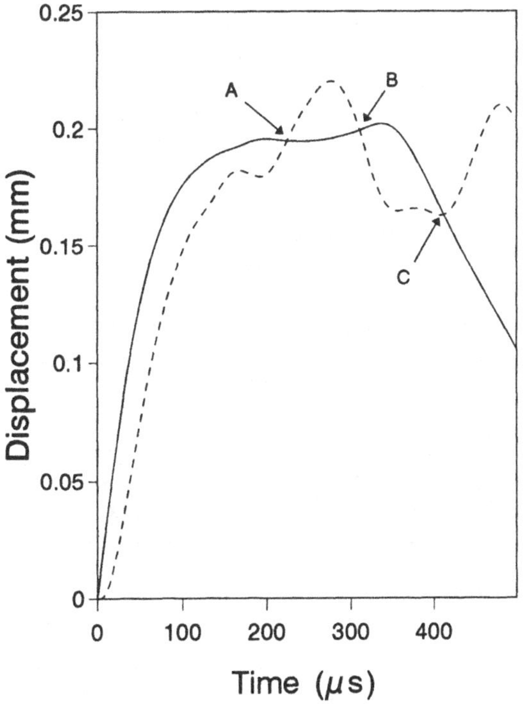

(b)

Figure 3.32. *(cont.)*

obtained by Qian and Swanson using the first-order shear deformation theory indicate that the maximum contact force is overestimated with the classical plate theory (Table 3.9). This is confirmed by a comparison between the results obtained from the two plate theories presented by Byun and Kapania (1992).

Chen and Sun (1985b) studied the same example of a $200 \times 200 \times 2.69$ mm graphite-epoxy plate with material properties given by (3.156) and by simply increasing the initial velocity of the 12.7-mm-diameter steel impactor to 10 m/s. Results (Fig. 3.34) are similar to those for a 3 m/s impact (Fig. 3.32) except, of course, for the magnitude of the results. Two impacts occur in the first 500 μs.

Example 3.22: Impact on Rectangular Composite Plates Chen and Sun (1985) also considered a 20-plied $[0,45,0,-45,0]_{2S}$ laminate with dimensions $152.4 \times 101.6 \times 2.69$ mm^3 impacted by a 12.7-mm-diameter steel ball with an initial velocity of 30 m/s. The material properties of the graphite-epoxy system used are again given by (3.156). The natural frequencies of the first five modes that will participate to the response of the plate were again determined according to the classical plate theory (3.203) and the first-order shear deformation theory (3.205), assuming simply supported boundary conditions. The results in Table 3.10 indicate that for this example, the effect of shear deformation and rotary inertia are still small. Therefore, the classical plate theory is expected to be adequate for this example.

Table 3.10. *First five natural frequencies (rad/s) for the 20-ply example by Chen and Sun (1985b)*

m	n	CPT	FSDT	% difference
1	1	5141.7	5120.2	0.42
3	1	24973	24370	2.41
1	3	26265	25843	1.61
3	3	46275	44631	3.55
5	1	64594	60681	6.06

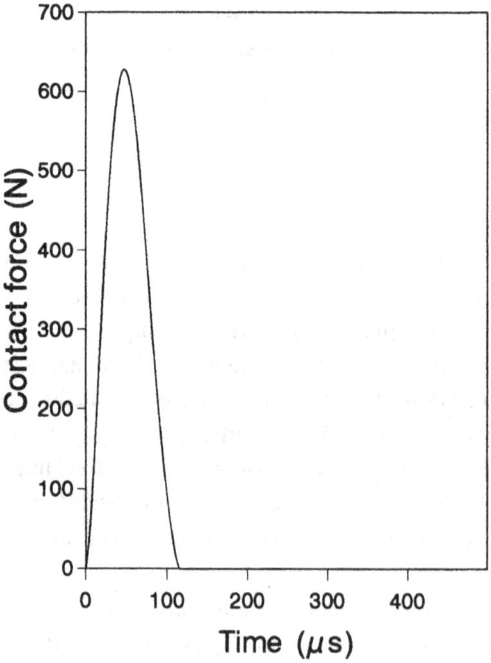

(a)

Figure 3.33. Impact on 10-plied graphite-epoxy laminated plate with double thickness, $V = 3$ m/s. a) Contact force history. b) Solid line: projectile displacement; dashed line: plate deflection.

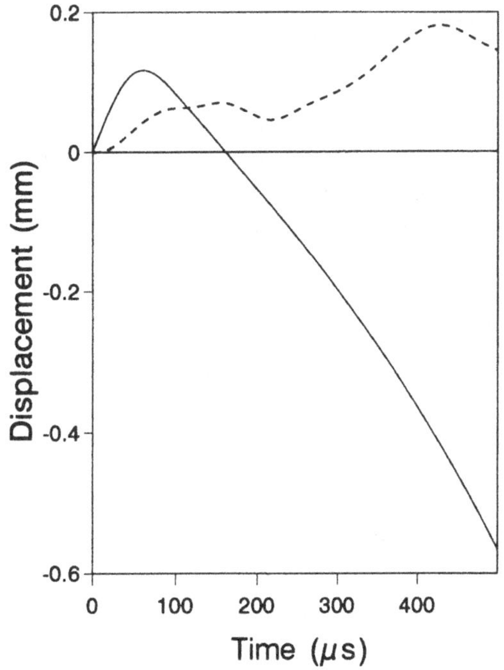

(b)

Figure 3.33. *(cont.)*

Three impacts occur during the first 800 μs (Fig. 3.35). As contact is lost between points A and B, and the central deflection of the plate becomes slightly larger that the thickness of the plate. This suggests that the geometrical nonlinearities should be considered. Chen and Sun (1985b), Kant and Mallikarjuna (1991), and Byun and Kapania (1992) showed that the response predicted by the linear model are very accurate for the first phase of the impact (up to point A in (Fig. 3.35). Discrepancies between the results between the linear and nonlinear approaches are seen in the time of arrival of the second and third impacts. Figure 3.36 shows that higher modes of vibration participate in the velocity response of the plate.

3.4.6 Approximate Solution for Wave-Controlled Impacts

An approximate solution was presented by Olsson (1992) for wave-controlled impacts. For this class of problems, the contact duration is short enough so that the deformation of the plate is confined to a small region near the impactor. Waves emanating from the impact point do not have time to reach the boundaries of the plate and reflect back. This situation is similar to an impact on an infinite plate.

In orthotropic plates, waves propagate at different speeds in different directions, and the wavefront will have an almost elliptical shape centered at the impact point. The analysis assumes that the area affected by impact can be approximated by a rectangular simply supported plate with side lengths a and b. When the origin of the coordinate system is at the center of the plate, the eigenfunctions can be written as

$$W_{jk} = \cos\frac{j\pi x}{a}\cos\frac{k\pi y}{b}. \tag{3.264}$$

The response to a force $F(t)$ applied at the center of the plate is of the form

$$w(x, y) = \sum_{j,k=1}^{\infty} W_{jk}\alpha_{jk} \tag{3.265}$$

where the αs are the modal participation factors. Substituting (3.265) and $q = F(t)\delta(x)\delta(y)$ into the equation of motion for a specially orthotropic classical

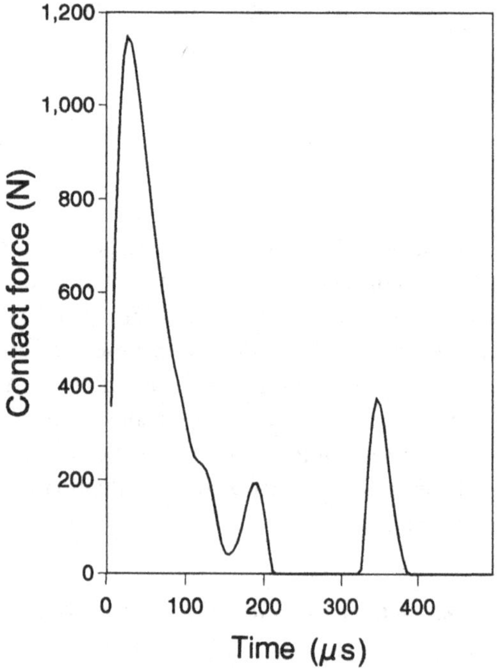

(a)

Figure 3.34. Impact on 10-plied graphite-epoxy laminated plate, $V = 10$ m/s. a) Contact force history. b) Solid line: projectile displacement; dashed line: plate deflection.

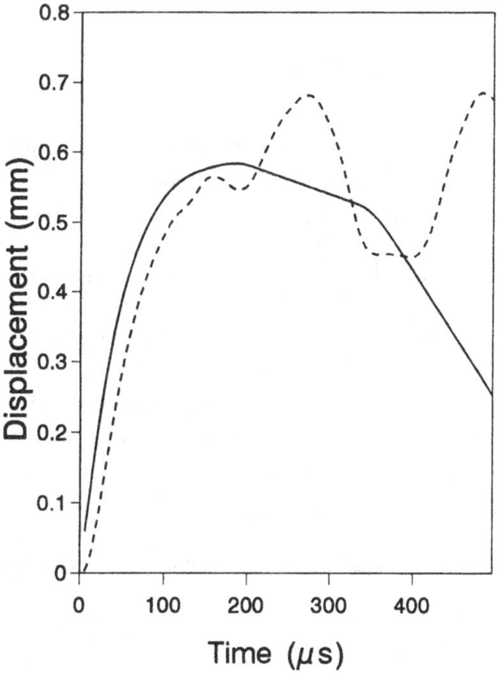

(b)

Figure 3.34. *(cont.)*

plate (3.117), multiplying by W_{jk}, and integrating over the area of the plate give the modal equations

$$\ddot{\alpha}_{jk} + \omega_{jk}^2 \alpha_{jk} = \frac{4F}{I_1 ab} \tag{3.266}$$

where

$$\omega_{jk}^2 = \left[D_{11}\left(\frac{j\pi}{a}\right)^4 + 2(D_{12} + 2D_{66})\left(\frac{j\pi}{a}\right)^2\left(\frac{k\pi}{b}\right)^2 + D_{22}\left(\frac{k\pi}{b}\right)^4 \right] \Big/ I_1$$

when j and k are odd. When either j or k is even, the right-hand side of (3.266) is equal to zero, and $\alpha_{jk} = 0$. Considering (3.266) at time τ, multiplying it by $\sin[(\omega_{jk}(t-\tau))]$ and integrating from 0 to t, it can be shown that

$$\alpha_{jk} = \frac{4}{I_1 ab\omega_{jk}} \int_0^t F(\tau)\sin[\omega_{jk}(t-\tau)]\,d\tau. \tag{3.267}$$

After substitution into (3.265), the displacement of the plate at the point of impact is found to be

$$w_1 = \frac{4}{I_1 ab}\int_0^t F(\tau)\sum_{j=1}^{\infty}\sum_{k=1}^{\infty}\frac{1}{\omega_{jk}}\sin[\omega_{jk}(t-\tau)]\,d\tau. \tag{3.268}$$

The double sum in this equation can be approximated by a continuous integration over j and k. Since only odd values of j and k are used in the summations, only half of the continuous integrals should be retained. Then,

$$w_1 = \frac{1}{mab}\int_0^t F(\tau)\int_0^{\infty}\int_0^{\infty}\frac{1}{\omega_{jk}}\sin[\omega_{jk}(t-\tau)]\,dj\,dk\,d\tau. \tag{3.269}$$

From (3.168) we know that for waves propagating in the x-direction, $\omega^2 = D_{11}k^4/I_1$, and the group velocity in that direction is $c_{gx} = d\omega/dk = 2\omega^{1/2}(D_{11}/I_1)^{1/4}$. Similarly, the group velocity in the y-direction is given by $c_{gy} = d\omega/dk = 2\omega^{1/2}(D_{22}/I_1)^{1/4}$. The lengths of the sides of the deformed region are proportional to those velocities so that

$$a^2/b^2 = (D_{11}/D_{22})^{\frac{1}{2}}. \tag{3.270}$$

It is further assumed that, since the waves resemble closed ellipses, only terms

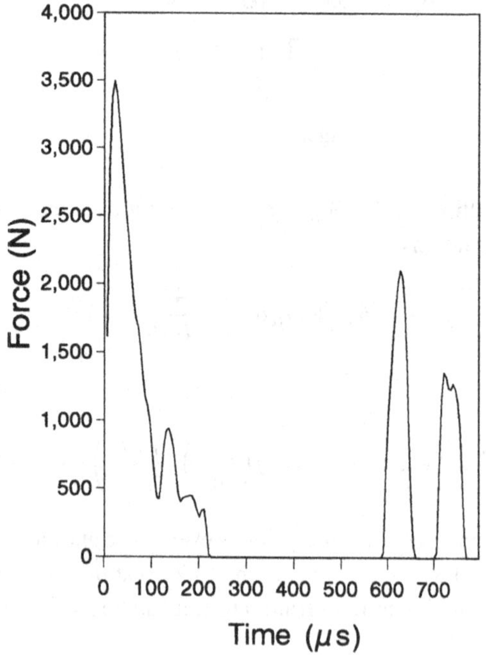

(a)

Figure 3.35. Impact on 20-plied graphite-epoxy laminated plate, $V = 30$ m/s. a) Contact force history. b) Solid line: projectile displacement; dashed line: plate deflection.

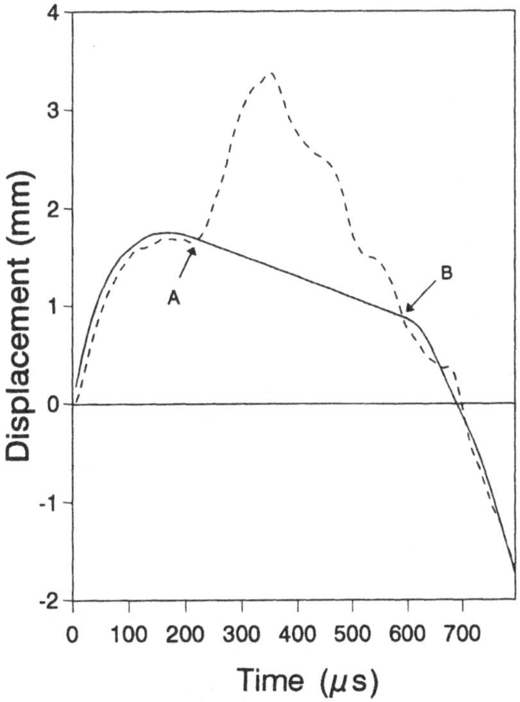

(b)

Figure 3.35. (*cont.*)

where $j = k$ should be retained in the modal expansion (3.265). Then the natural frequencies can be expressed as

$$ab\omega_{jk} = 2\pi^2(D^*/m)^{\frac{1}{2}}\, jk \tag{3.271}$$

where

$$D^* = \frac{A+1}{2}(D_{11}D_{22})^{\frac{1}{2}}, \quad A = (D_{12} + 2D_{66})/(D_{11}D_{22})^{\frac{1}{2}}.$$

The plate displacement at the impact point can be written as

$$w_1 = \frac{1}{8(I_1 D^*)^{\frac{1}{2}}} \int_0^t F(\tau)\, d\tau. \tag{3.272}$$

Example 3.23: Dynamic Response of an Infinite Plate Using Olsson's Approximation In order to determine the validity of Olsson's approach for calculating the transient response of the plate to force $F(t)$ using (3.272), we consider the case studied in Example 3.9, for which an exact solution was obtained. In this case, $I_1 = 39.8$ kg/m^3, $D = 2403.85$ Nm, and the applied force is a unit step

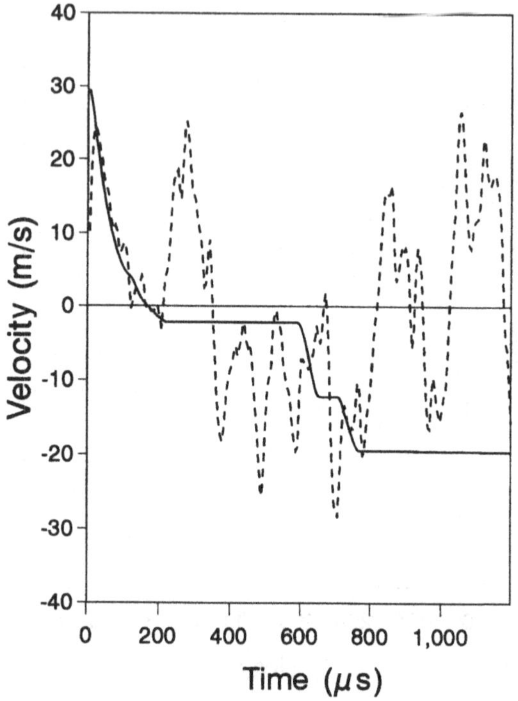

Figure 3.36. Impact on 20-plied graphite-epoxy laminated plate, $V = 30$ m/s (solid line: projectile velocity; dashed line: plate velocity).

function. The central displacements of the plate are 0.4041×10^{-8}, 0.8082×10^{-8}, 1.212×10^{-8} m after 10, 20, and 30 μs, respectively. Equation (3.272) indicates that under this type of loading, the central displacement of the plate increases linearly with time. This result is not obvious from the exact solution given by Medick (1961) (Section 3.3.7).

Applying Newton's law, the equation of motion of the projectile is found to be

$$M\ddot{w}_2 + F = 0 \tag{3.273}$$

where M is the mass of the projectile. After integrating twice, the displacement of the projectile is

$$w_2 = Vt - \frac{1}{M}\int_0^t \int_0^\tau F(\tau')\, d\tau'\, d\tau \tag{3.274}$$

where V is the initial velocity of the projectile. The indentation of the plate is defined by

$$\delta + w_1 - w_2 = 0. \tag{3.275}$$

Differentiating twice with respect to time, using (3.272) and (3.274) and the contact law $F = k_c\delta^{3/2}$, the indentation is found to be governed by the nonlinear differential equation

$$\frac{d^2\delta}{dt^2} + \frac{1}{8(I_1 D^*)^{\frac{1}{2}}} \cdot \frac{3}{2} k_c \delta^{\frac{1}{2}} \frac{d\delta}{dt} + \frac{k_c}{M}\delta^{\frac{3}{2}} = 0. \tag{3.276}$$

k_c is the contact stiffness, which can be estimated from the properties of the target and the indentor (see Chap. 2). Introducing the nondimensional variables

$$\bar{\delta} = \frac{\delta}{TV}, \quad \bar{t} = \frac{t}{T}, \tag{3.277}$$

the nondimensional indentation is governed by the nonlinear differential equation

$$\frac{d^2\bar{\delta}}{d\bar{t}^2} + \lambda\frac{3}{2}\bar{\delta}^{\frac{1}{2}}\frac{d\bar{\delta}}{d\bar{t}} + \bar{\delta}^{\frac{3}{2}} = 0 \tag{3.278}$$

with the initial conditions

$$\bar{\delta}(0) = 0, \quad \frac{d\bar{\delta}(0)}{d\bar{t}} = 1 \tag{3.279}$$

where

$$T = \left[M/\left(k_c V^{\frac{1}{2}}\right)\right]^{\frac{2}{5}}, \tag{3.280}$$

$$\lambda = k_c^{\frac{2}{5}} V^{\frac{1}{5}} M^{\frac{3}{5}} / \left[8(I_1 D^*)^{\frac{1}{2}}\right]. \tag{3.281}$$

The contact force is given by

$$F = \left[k_c^2 M^3 V^6\right]^{\frac{1}{5}} \bar{\delta}^{\frac{3}{2}} \tag{3.282}$$

and the displacement at the point of impact is given by

$$w(0,0,t) = \frac{MV}{8(I_1 D^*)^{\frac{1}{2}}} \int_0^{\bar{t}} \bar{\delta}^{\frac{3}{2}}\, d\tau. \tag{3.283}$$

Equation (3.278) was solved numerically for several values of the parameter λ, and Fig. 3.37 shows the different behaviors obtained. The highest contact force is obtained for $\lambda = 0$, in which case the plate is very rigid and the problem is that of an impact on a half-space. As the impact parameter increases, the contact force history becomes more and more asymmetrical; the contact duration increases as the deformation of the target becomes more significant. The maximum contact force and the evolutions of the time needed to reach the maximum contact force are plotted as a function of λ in Fig. 3.38. Results in this figure can be used to determine the maximum impact force without having to solve (3.278).

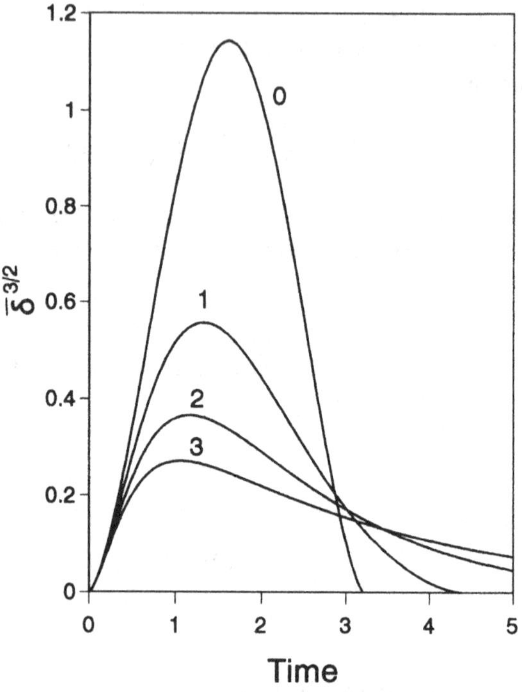

Figure 3.37. Nondimensional contact force as a function of the inelasticity parameter λ.

This approach brings out the fact that for an infinite plate, the impact is governed by a single parameter λ that combines the effect of the properties of the plate, those of the impactor, the contact stiffness, and the impact velocities. From the value of λ, one can anticipate what type of impact to expect and whether the indentation of the plate will be significant or not.

Example 3.24: Impact on Composite Plate Using Olsson's Approach The same problem treated in Example 3.20 is studied here using the approximate method presented above. In this case, $\lambda = 2.08$, the maximum nondimensional force $\bar{\delta} = 0.35505$ is reached when $\bar{t} = 1.15$. Knowing that for this problem $E_c = 9.72$ GPa, $R = 6.35$ mm, $M = 8.537$ g, and $V = 3$ m/s, the maximum force is predicted to be 307 N and to occur 33.7 μs after the beginning of the impact. The total impact duration is predicted to be 261.6 μs. A comparison of contact forces predicted by the complete model used in Example 3.20 and the approximate method proposed by Olsson indicates that the latter provides good predictions of the contact force history (Fig. 3.39). The maximum contact force is slightly underestimated, but the overall behavior is estimated well by the approximate method. Near the end of the first contact, the complete

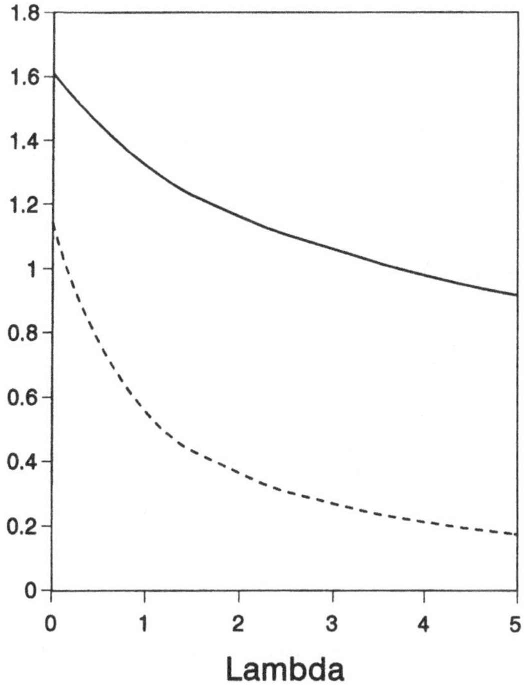

Figure 3.38. Peak contact force and time to reach the peak contact force as a function of the parameter λ (solid line: nondimensional time t; dashed line: nondimensional force $\bar{\delta}^{3/2}$).

model predicts a small rise in the contact force. This is attributed to waves reflected back from the boundaries that have reached the impact point. The approximate model does not account for these reflections and thus predicts a smooth decrease of the contact force. The approximate model does not predict the second impact, even though it is possible to study the dynamics of the projectile and the dynamics of the plate; if a second impact is to occur, the approximate model can be applied again.

Karas' example of impact of steel sphere on a steel plate was treated in Example 3.18 using a complete model. In that case, $\lambda = .2346$, and for such small values the contact force history is expected to be nearly symmetrical as shown in nondimensional form in Fig. 3.37. The maximum contact force occurs for $\bar{\delta}^{3/2} = 0.9186$ for $\bar{t} = 1.52$, which gives a maximum contact force of 301.6 lb at a time $t = 34.0\ \mu$s, which is to be compared with the maximum force of 311.4 lb at 33 μs obtained in Example 3.18. Therefore, this approximate method gives excellent results for this type of problem and is much more efficient.

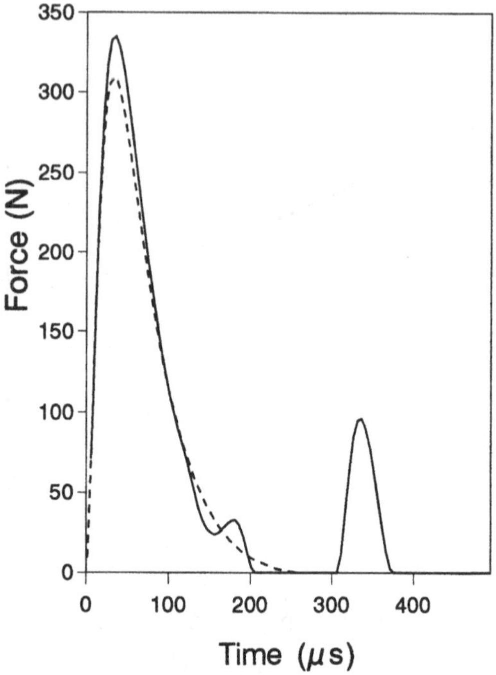

Figure 3.39. Comparison of contact force histories predicted by the approximate method and complete model.

The impulse transferred to the plate is $I = \int_0^t F(\tau)\,d\tau = MV \int_0^{\bar{t}} \bar{\delta}^{3/2}\,d\bar{\tau}$. When $\lambda = 0$, multiplying (3.278) by $d\bar{\delta}/d\bar{t}$ and integrating from 0 to the time when the indentation reaches its maximum, it can be shown that $\bar{\delta} = (5/4)^{2/5}$ at that time. Substituting this value of the nondimensional indentation into (3.282), the expression for the maximum contact force for an impact on a half-space (3.229), obtained using the energy-balance approach, is recovered.

When $\lambda > 0$, (3.278) can be integrated directly, and when the indentation reaches its maximum, $\bar{\delta}^{3/2} = \frac{1}{\lambda}[1 - \int_0^{\bar{t}} \bar{\delta}^{3/2}\,d\tau]$. The integral in that expression is a positive quantity between 0 and 1 that represents the ratio between the impulse applied to the plate up to that time and the initial momentum of the projectile. Therefore, $\bar{\delta}^{3/2} < 1/\lambda$, and the maximum contact force decreases as λ becomes larger.

Numerical integration of (3.278) shows that when $\lambda = 0$, $I = 2MV$ after the impact is completed. Since the total momentum is conserved, the momentum of the projectile after impact is $-MV$, and the rebound velocity of the projectile is $-V$. In this case, we have an elastic impact. When $\lambda > 1$, I tends to MV, which implies that the rebound velocity of the projectile is near zero and that the impact

Table 3.11. *Effect of plate thickness on contact force history*

h (mm)	λ	$\bar{\delta}_{max}^{3/2}$	P_{max} (N)	T_c (μs)
2.69	1.380	0.4647	363.7	171
5.38	0.4849	0.7569	592.4	114.5
10.76	0.1725	0.9693	758.6	105.6

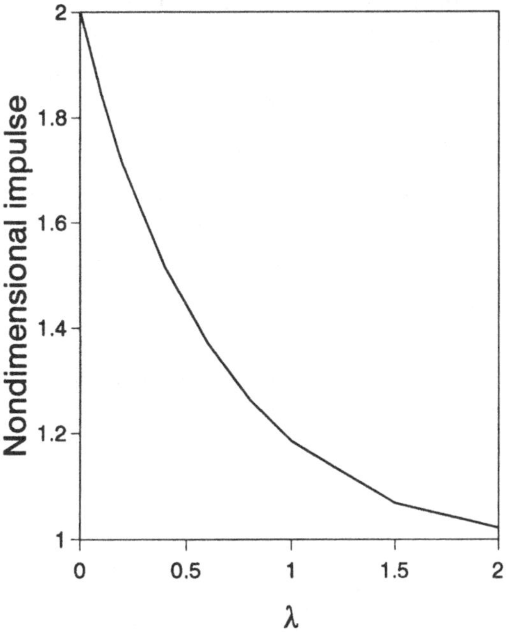

Figure 3.40. Momentum absorbed by plate as a function of λ.

appears to be inelastic. Most of the initial kinetic energy of the projectile is absorbed by the target. A plot of I/MV versus λ is shown in Fig. 3.40 to illustrate how the impulse varies with what is sometimes called the inelasticity parameter.

Example 3.25: Effect of Plate Thickness on Contact Force History For the $[0,90,0,90,0]_S$ plate studied in Example 3.30, with a total plate thickness of 2.69 mm, $\lambda = 1.38$. Increasing the total thickness to 5.38 and 10.76 mm by multiplying the thickness of each ply by 2 and 4 respectively gives $\lambda = 0.4879$ and 0.1725. Therefore, increasing the stiffness of the laminate reduces the value of the inelasticity parameter λ. The results of that analysis are summarized in Table 3.11.

As the plate thickness is increased, the overall deflections of the plate become negligible; for $\lambda = 0$, the maximum contact force calculated using (3.234) is $P_{max} = 894.8$ N, and the contact duration calculated using (3.240) is $T_c = 102.3\ \mu$s. This limit case gives the "hardest" impact, that is, the case when the contact force is maximum and the contact duration is the shortest.

This approach can be generalized to cases with different contact laws such that

$$F = k_c \delta^\alpha. \tag{3.284}$$

According to the Hertz theory of contact between two elastic solids, $\alpha = 3/2$. For laminated composite materials, $\alpha = 3/2$ also; for sandwich structures, $\alpha = 1$ before core damage is introduced. The nondimensional indentation is governed by the nonlinear differential equation

$$\frac{d^2\bar{\delta}}{d\bar{t}^2} + \lambda\alpha\bar{\delta}^{\alpha-1}\frac{d\bar{\delta}}{d\bar{t}} + \bar{\delta}^\alpha = 0 \tag{3.285}$$

where

$$T = \left[M/\left(k_c V^{\alpha-1}\right)\right]^{\frac{1}{\alpha+1}} \tag{3.286}$$

$$\lambda = \left[k_c V^{\alpha-1} M^\alpha\right]^{\frac{1}{\alpha+1}} / \left[8(mD^*)^{\frac{1}{2}}\right]. \tag{3.287}$$

For Hertzian contact, the results obtained by Olsson are recovered. With a linear contact law ($\alpha = 1$), (3.285) becomes

$$\frac{d^2\bar{\delta}}{d\bar{t}^2} + \lambda\frac{d\bar{\delta}}{d\bar{t}} + \bar{\delta} = 0 \tag{3.288}$$

where

$$\lambda = \frac{1}{8}\left(\frac{k_c M}{mD^*}\right)^{\frac{1}{2}}, \tag{3.289}$$

$$T = (M/k_c)^{\frac{1}{2}}. \tag{3.290}$$

Equation (3.288) is the equation of motion for single degree of freedom system with viscous damping. Introducing the variable $\eta = \lambda/2$ for convenience, we distinguish between the underdamped case for which $\eta < 1$ and the overdamped case with $\eta > 1$. The solution for the underdamped case is

$$\bar{\delta} = \frac{e^{-\eta\bar{t}}}{(1-\eta^2)^{\frac{1}{2}}}\sin\left[(1-\eta^2)^{\frac{1}{2}}\bar{t}\right]. \tag{3.291}$$

Figure 3.41 shows that as the damping increases, the maximum contact force decreases, the contact force becomes more asymmetrical, and the contact duration increases. This phenomenon is explained by the SDOF system analogy where, initially, when the velocity of the mass is high, the dashpots apply a restraining force proportional to the velocity. The system appears rigid, but as the velocity decreases, the restraining force becomes smaller, making the system appear more flexible. This explains the different shapes of the curves for the loading and unloading phases when $\eta > 0$. Stronger dashpots provide more damping, which explains why the indentation reaches its maximum earlier, its maximum value decreases, and λ becomes larger.

The solution for the overdamped case is

$$\bar{\delta} = \frac{e^{-\eta \bar{t}}}{2(\eta^2 - 1)^{\frac{1}{2}}} \left[e^{(\eta^2-1)^{\frac{1}{2}} \bar{t}} - e^{-(\eta^2-1)^{\frac{1}{2}} \bar{t}} \right]. \tag{3.292}$$

The response for $\eta = 1.01, 1.5, 2, 4, 6$ are shown in Fig. 3.41b. For very large values of the damping factor, the contact force appears to reach a maximum value and then remain constant. When $\eta \gg 1$, the response for the overdamped case can be approximated by

$$\bar{\delta} = \frac{1}{2\eta}[1 - e^{-2\eta t}] \tag{3.293}$$

which indicates that the indentation increases monotonically and reaches a maximum value of $1/(2\eta)$ as time increases. The maximum nondimensional contact force calculated from either (3.292) or (3.293) decreases rapidly as η increases. The approximate solution from (3.293) is seen to be accurate when $\eta > 3$.

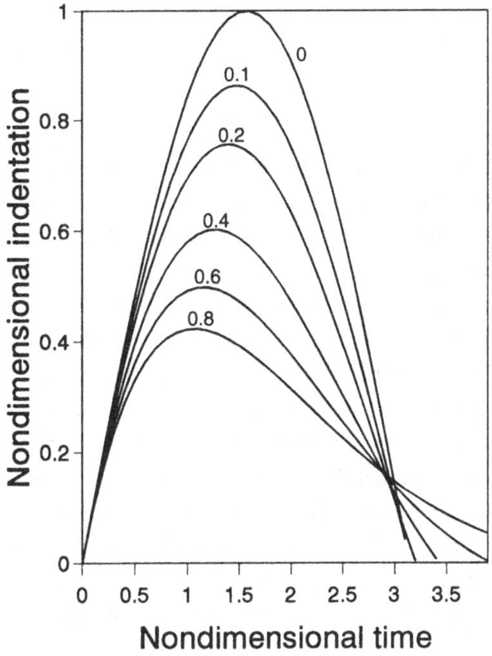

(a)

Figure 3.41. Nondimensional contact force history as a function of λ: a) $\eta < 1$, b) $\eta > 1$.

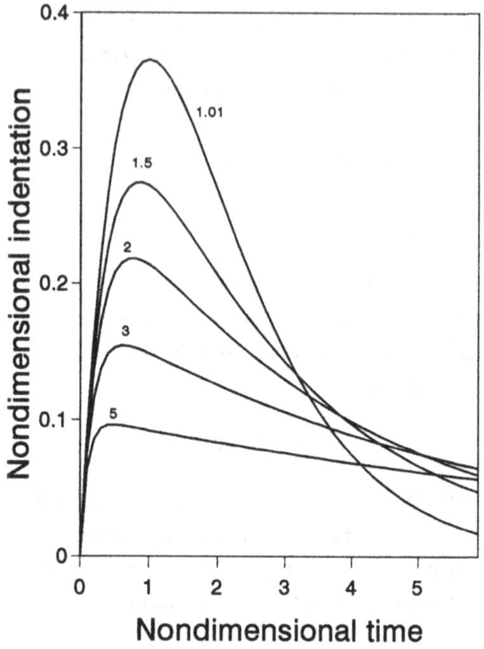

(b)

Figure 3.41. *(cont.)*

3.4.7 Summary

In this section four methods for studying impact dynamics have been presented: (1) spring-mass models, (2) energy-balance models, (3) complete models; and (4) approximate model for wave-controlled impacts. A complete model is one that fully accounts for the dynamics of the structure, the dynamics of the projectile, and the local indentation in the contact zone. Depending on the number of modes participating in the response, the geometry of the target, and the boundary conditions, a complete model may result in a large number of degrees of freedom and can be computationally expensive. In addition, with numerical solutions it is usually more difficult to get insight into the problem than with analytical solutions. The other three approaches were developed based on some simplifying assumptions. The spring-mass models are based on the assumption that the dynamics of the target can be represented by a single-degree-of-freedom system with equivalent mass and stiffness. With the energy-balance models, the target is assumed to deform in a quasi-static manner, the maximum contact force is assumed to occur when the deflection of the target is maximum, and that at that time the kinetic energy of the projectile and the target are zero. However, examples have shown that there are some cases for which the response of the structure is not quasi-static, the contact force reaches a maximum at a different time than at the deflection of the structure, and the velocity of the structure is not zero at that time. Therefore, the applicability of this method is limited.

The fourth approach for studying impact dynamics is based on a simplified analysis of solution for the impact on an infinite orthotropic plate. This solution is expected to be valid as long as the plate is large enough so that the deformation wave does not sense the presence of the boundary. In that case the problem is said to involve a wave-controlled impact. When the contact duration is large enough for the deformation wave to reach the boundary and be reflected back to the point of impact at least once, that approach cannot be used. In that case, we have a boundary-controlled impact. With the solution for wave-controlled impacts presented by Olsson, the problem is reduced to the solution of a single nondimensional nonlinear ordinary differential equation with a single nondimensional parameter λ. This parameter λ determines the shape of the contact force history. Several examples showed that this approach allows for a very efficient and accurate determination of the contact force history.

3.4.8 Exercise Problems

3.12 Use (3.272) to calculate the central deflection of an infinite plate at the end of an impulse of (a) sinusoidal, (b) rectangular, and (c) triangular shape. Show that the deflection will be the same provided that the impulse $I = \int_0^t F(\tau)\,d\tau$ is the same.

3.13 Write a computer program to solve (3.278) using Newmark's time integration method. Check results with Figure 3.37.

3.14 The Rayleigh-Ritz method can be used to develop simplified impact models. Use a one-term polynomial approximation to determine the effective stiffness and effective mass of an orthotropic rectangular plate fully clamped along all four edges.

3.15 Write a computer program to analyze the dynamic response of a simply supported, symmetrically laminated rectangular plate subjected to the impact of a mass using the approach given in Section 3.4.4.

3.16 Write a computer program to analyze the dynamic response of a fully clamped, symmetrically laminated rectangular plate subjected to the impact of a mass. The plate should be modeled using the Rayleigh-Ritz method with polynomial approximation functions as discussed in Section 3.3.8.

3.17 Write a computer program to analyze the dynamic response of infinite orthotropic plates subjected to the impact of a mass using the solution presented in Section 3.4.6.

3.18 For the case discussed in Example 3.35, use the computer programs developed in Exercises 3.15–3.17 to determine the effects of boundary conditions on the impact response and when the plate can be considered to be infinite. Vary the size of the plate to investigate these effects.

3.5 Theory of Shells

While the basic problems associated with impact on composite can be studied on flat specimens, in many applications curved panels are used and mathematical models for analyzing the effects of impacts on such structures must be analyzed. A brief review of shell theory is presented in this section, and then a solution for the dynamic response of a simply supported panel is developed. This solution can be used to develop a solution for the impact problem. For more complex geometries and boundary conditions, variational or finite element models can be used to model the structure. Relatively few studies have addressed the modeling of impact on composite shells.

3.5.1 Equations of Motion

The distance between points $(\xi_1, \xi_2, 0)$ and $(\xi_1 + d\xi_1, \xi_2 + d\xi_2, 0)$ on the midsurface is given by

$$(ds)^2 = d\bar{r} \cdot d\bar{r} = \alpha_1^2 (d\xi_1)^2 + \alpha_2 (d\xi_2)^2 \tag{3.294}$$

where $\bar{r}$ is the position vector for a point on the midsurface. The vectors $\bar{r}_1, \bar{r}_2$ are tangent to the coordinate lines, and the surface metrics are defined as

$$\alpha_1^2 = \bar{r}_1 \cdot \bar{r}_1, \quad \alpha_2^2 = \bar{r}_2 \cdot \bar{r}_2. \tag{3.295}$$

The elemental surface area is

$$dA = \alpha_1 \alpha_2 d\xi_1 d\xi_2 \tag{3.296}$$

and the unit vector normal to the midsurface is

$$\bar{n} = \frac{\bar{r}_1 x \bar{r}_2}{\alpha_1 \alpha_2}. \tag{3.297}$$

The position of an arbitrary point located at a distance z from the midsurface is defined by the vector

$$\bar{R} = \bar{r} + z\bar{n}. \tag{3.298}$$

The distance between two arbitrary points is found from

$$dS^2 = d\bar{R} \cdot d\bar{R} \tag{3.299}$$

with

$$d\bar{R} = \frac{\partial \bar{R}}{\partial \xi_1} d\xi_1 + \frac{\partial \bar{R}}{\partial \xi_2} d\xi_2 + \frac{\partial \bar{R}}{\partial z} dz. \tag{3.300}$$

After differentiation, (3.288) gives

$$\frac{\partial \bar{R}}{\partial \xi_1} = \frac{\partial \bar{r}}{\partial \xi_1} + z\frac{\partial \bar{n}}{\partial \xi_1}, \quad \frac{\partial \bar{R}}{\partial \xi_2} = \frac{\partial \bar{r}}{\partial \xi_2} + z\frac{\partial \bar{n}}{\partial \xi_2}. \tag{3.301}$$

Using Rodrigues' theorem,

$$\frac{\partial \bar{n}}{\partial \xi_1} = \frac{1}{R_1}\frac{\partial \bar{r}}{\partial \xi_1}, \quad \frac{\partial \bar{n}}{\partial \xi_2} = \frac{1}{R_2}\frac{\partial \bar{r}}{\partial \xi_2} \tag{3.302}$$

we find

$$\frac{\partial \bar{R}}{\partial \xi_1} = \left(1 + \frac{z}{R_1}\right)\frac{\partial \bar{r}}{\partial \xi_1}, \quad \frac{\partial \bar{R}}{\partial \xi_2} = \left(1 + \frac{z}{R_2}\right)\frac{\partial \bar{r}}{\partial \xi_2} \tag{3.303}$$

and

$$(dS)^2 = d\bar{R} \cdot d\bar{R} = L_1^2(d\xi_1)^2 + L_2^2(d\xi_2)^2 + L_3^2(dz)^2 \tag{3.304}$$

with

$$L_1 = \alpha_1\left(1 + \frac{z}{R_1}\right), \quad L_2 = \alpha_2\left(1 + \frac{z}{R_2}\right), \quad L_3 = 1. \tag{3.305}$$

For thin shells, the force and moment resultants are defined as

$$(N_i, M_i) = \int_{-\frac{h}{2}}^{\frac{h}{2}} \sigma_i(1, z)\, dz \tag{3.306}$$

$$(Q_1, Q_2) = K_i^2 \int_{-\frac{h}{2}}^{\frac{h}{2}} (\sigma_5, \sigma_4)\, dz. \tag{3.307}$$

The first-order shear deformation theory is obtained assuming that the displacement varies as

$$u = \frac{L_1}{\alpha_1}u_1 + z\phi_1, \quad v = \frac{L_2}{\alpha_2}u_2 + z\phi_2, \quad w = u_3 \tag{3.308}$$

where u, v, w are the displacements of an arbitrary point with coordinates (ξ_1, ξ_2, z), and where u_1, u_2, u_3 are the displacements of the point on the midsurface with coordinates $(\xi_1, \xi_2, 0)$. The strain-displacement relations are obtained as

$$\begin{Bmatrix} \varepsilon_1 \\ \varepsilon_2 \\ \varepsilon_6 \end{Bmatrix} = \begin{Bmatrix} \varepsilon_1^0 \\ \varepsilon_2^0 \\ \varepsilon_6^0 \end{Bmatrix} + z \begin{Bmatrix} \kappa_1 \\ \kappa_2 \\ \kappa_6 \end{Bmatrix} \tag{3.309}$$

$$\varepsilon_4 = \varepsilon_4^0, \quad \varepsilon_5 = \varepsilon_5^0 \tag{3.310}$$

where the midsurface strains are

$$\begin{aligned}
\varepsilon_1^0 &= \frac{1}{\alpha_1}\frac{\partial u_1}{\partial \xi_1} + \frac{u_3}{R_1} \\
\varepsilon_2^0 &= \frac{1}{\alpha_2}\frac{\partial u_2}{\partial \xi_2} + \frac{u_3}{R_2} \\
\varepsilon_6^0 &= \frac{1}{\alpha_1}\frac{\partial u_2}{\partial \xi_1} + \frac{1}{\alpha_2}\frac{\partial u_1}{\partial \xi_2} \\
\varepsilon_4^0 &= \frac{1}{\alpha_2}\frac{\partial u_3}{\partial \xi_2} + \phi_2 - \frac{u}{R_2} \\
\varepsilon_5^0 &= \frac{1}{\alpha_1}\frac{\partial u_3}{\partial \xi_1} + \phi_1 - \frac{u_1}{R_1}
\end{aligned} \tag{3.311}$$

and the curvatures are

$$\begin{aligned}
\kappa_1^0 &= \frac{1}{\alpha_1}\frac{\partial \phi_1}{\partial \xi_1} \\
\kappa_2^0 &= \frac{1}{\alpha_2}\frac{\partial \phi_2}{\partial \xi_2} \\
\kappa_6^0 &= \frac{1}{\alpha_1}\frac{\partial \phi_2}{\partial \xi_1} + \frac{1}{\alpha_2}\frac{\partial \phi_1}{\partial \xi_2} + \frac{1}{2}\left(\frac{1}{R_2} - \frac{1}{R_1}\right)\left(\frac{1}{\alpha_1}\frac{\partial u_2}{\partial \xi_1} - \frac{1}{\alpha_2}\frac{\partial u_1}{\partial \xi_2}\right).
\end{aligned} \tag{3.312}$$

The equations of motion are

$$\begin{aligned}
\frac{\partial N_1}{\partial x_1} + \frac{\partial}{\partial x_2}(N_6 + c_0 M_6) + \frac{Q_1}{R_1} &= \left(I_1 + \frac{2}{R_1}I_2\right)\frac{\partial^2 u_1}{\partial t^2} + \left(I_1 + \frac{I_3}{R_1}\right)\frac{\partial^2 \phi_1}{\partial t^2} \\
\frac{\partial N_2}{\partial x_2} + \frac{\partial}{\partial x_1}(N_6 - c_0 M_6) + \frac{Q_2}{R_2} &= \left(I_1 + \frac{2}{R_2}I_2\right)\frac{\partial^2 u_2}{\partial t^2} + \left(I_1 + \frac{I_3}{R_2}\right)\frac{\partial^2 \phi_2}{\partial t^2} \\
\frac{\partial Q_1}{\partial x_1} + \frac{\partial Q_2}{\partial x_2} - \left(\frac{N_1}{R_1} + \frac{N_2}{R_2} - q\right) &= I_1 \frac{\partial^2 u_3}{\partial t^2} \\
\frac{\partial M_1}{\partial x_1} + \frac{\partial M_6}{\partial x_2} - Q_1 &= I_3 \frac{\partial^2 \phi_1}{\partial t^2} + \left(I_1 + \frac{I_3}{R_1}\right)\frac{\partial^2 u_1}{\partial t^2} \\
\frac{\partial M_6}{\partial x_1} + \frac{\partial M_2}{\partial x_2} - Q_2 &= I_3 \frac{\partial^2 \phi_2}{\partial t^2} + \left(I_1 + \frac{I_3}{R_2}\right)\frac{\partial^2 u_2}{\partial t^2} \\
c_0 &= \frac{1}{2}\left(\frac{1}{R_1} - \frac{1}{R_2}\right), \quad dx_i = \alpha_i \, d\xi_i .
\end{aligned} \tag{3.313}$$

For the classical thin shell theory,

$$\phi_1 = -\frac{\partial u_3}{\partial x_1} + \frac{u_1}{R_1}, \quad \phi_2 = -\frac{\partial u_3}{\partial x_2} + \frac{u_2}{R_2} \tag{3.314}$$

and the equations above simplify to some extent.

3.5.2 Simply Supported Panels

For laminates with no extension-shear and bending-twisting coupling,

$$A_{16} = A_{26} = B_{16} = B_{26} = D_{16} = D_{26} = A_{45} = 0 \tag{3.315}$$

with the boundary conditions

$$\begin{aligned}
&N_1(0, x_2) = N_1(a, x_2) = M_1(0, x_2) = M_1(a, x_2) = 0 \\
&N_2(x_1, 0) = N_2(x_1, b) = M_2(x_1, 0) = M_2(x_1, b) = 0 \\
&u_3(0, x_2) = u_3(a, x_2) = u_2(0, x_2) = u_2(a, x_2) = 0 \\
&u_3(x_1, 0) = u_3(x_1, b) = u_1(x_1, 0) = u_1(x_1, b) = 0 \\
&\phi_2(x_1, 0) = \phi_2(x_1, b) = \phi_1(x_1, 0) = \phi_1(x_1, b) = 0,
\end{aligned} \tag{3.316}$$

and with the transverse load expressed as

$$q = \sum_{m,n=1}^{\infty} Q_{mn}(t) \sin\left(\frac{m\pi x_1}{a}\right) \sin\left(\frac{n\pi x_2}{b}\right), \tag{3.317}$$

a solution is sought in the form

$$\begin{aligned}
u_1(x_1, x_2, t) &= \sum_{m,n=1}^{\infty} U_{mn}(t) \cos\left(\frac{m\pi x_1}{a}\right) \sin\left(\frac{n\pi x_2}{b}\right) \\
u_2(x_1, x_2, t) &= \sum_{m,n=1}^{\infty} V_{mn}(t) \sin\left(\frac{m\pi x_1}{a}\right) \cos\left(\frac{n\pi x_2}{b}\right) \\
u_3(x_1, x_2, t) &= \sum_{m,n=1}^{\infty} W_{mn}(t) \sin\left(\frac{m\pi x_1}{a}\right) \sin\left(\frac{n\pi x_2}{b}\right) \\
\phi_1(x_1, x_2, t) &= \sum_{m,n=1}^{\infty} X_{mn}(t) \cos\left(\frac{m\pi x_1}{a}\right) \sin\left(\frac{n\pi x_2}{b}\right) \\
\phi_2(x_1, x_2, t) &= \sum_{m,n=1}^{\infty} Y_{mn}(t) \sin\left(\frac{m\pi x_1}{a}\right) \cos\left(\frac{n\pi x_2}{b}\right).
\end{aligned} \tag{3.318}$$

Substitution into the equations of motion yields a set of five differential equations of the form

$$[M]\{\ddot{\Delta}\} + [K]\{\Delta\} = \{F\} \tag{3.319}$$

where

$$\{\Delta\}^{\mathrm{T}} = \{U_{mn}, V_{mn}, W_{mn}, X_{mn}, Y_{mn}\}$$

for each m, n combination.

Once N, the number of modes to be retained, has been selected, N sets of equations of the form given by (3.319) are assembled along with the equation of motion of the projectile to form a $5N+1$ set of differential equations of the form given by (3.240). The contact appears on the right-hand side and is usually a nonlinear function of the displacements. The contact force is determined using an appropriate contact law. Then, the solution of the impact problem can be solved using the same approach used previously for beams and plates.

3.5.3 Exercise Problems

3.19 Write a computer program to determine the impact response of a simply supported panel using the solution presented in Section 3.5.2 for the dynamic response of the panel.

3.6 Scaling

It is often required to conduct impact experiments on a small-size model and extrapolate the results to a full size prototype. One might also like to compare analytical results obtained for different size specimen. Two basic approaches are used to develop scaling rules. The first approach is strictly a dimensional analysis method using Buckingham's Pi theorem, and the other starts with the equations of motion of the system in order to develop a set of scaling rules.

Morton (1988) developed scaling laws for homogeneous, isotropic beams subjected to transverse impact. A total of 13 parameters were identified: 1) geometrical parameters: l, h, b for beam length, height, and width, respectively; 2) the material properties of the beam: Young's modulus E and Poisson's ratio ν; 3) impactor characteristics ν_i, E_i, ρ_i, R_i, V_i; 4) the central deflection w, and time t. Applying Buckingham's Pi theorem, 10 nondimensional parameters were formed:

$$\begin{aligned} &\pi_1 = w/h, \quad \pi_2 = l/h, \quad \pi_3 = b/h, \quad \pi_4 = R_i/h, \\ &\pi_5 = E_i/E, \quad \pi_6 = \nu, \quad \pi_7 = \nu_i, \quad \pi_8 = \rho_i/\rho, \\ &\pi_9 = \rho_i V_i^2/E, \quad \pi_{10} = tV_i/h \end{aligned} \tag{3.320}$$

If these 10 π-terms form a complete set, then there exists a function f such that

$$f(\pi_1, \pi_2, \ldots, \pi_{10}) = 0. \tag{3.321}$$

If the linear dimensions of a model are scaled by a common factor s with respect to the prototype, the first four π-terms are the same as in the prototype. If the same materials are used, the next four π-terms also remain the same. For the π_9 term to remain scaled, since the materials used in the model and the prototype are already assumed to be the same, the impactor velocity V has to remain the same. Finally, the last parameter implies that

$$\left(\frac{tV_i^2}{h}\right)_m = \left(\frac{t'V_i^2}{sh}\right)_p. \tag{3.322}$$

Equation (3.264) indicates that time must be scaled in the model. If the deflection of the prototype beam reaches a maximum after time t, a one-tenth-scale model ($s = .1$) will reach its maximum deflection after $0.1t$. For such a beam it is possible to make an scale model and extrapolate the results to predict the impact response of the full-size prototype. However, if the material is strain rate sensitive,

$$\sigma = E(\epsilon + \tau\dot{\epsilon}) \tag{3.323}$$

where τ is a new parameter reflecting the strain rate sensitivity of the material, and a new nondimensional parameter is considered:

$$\pi_{11} = t/\tau. \tag{3.324}$$

If, as already assumed, the same materials are used for the model and the prototype, an additional condition for similarity is obtained:

$$t = t'. \tag{3.325}$$

This new condition is in conflict with (3.322); therefore, scale models cannot be developed when rate-sensitive materials are used. Fortunately, with the commonly used material systems, strain rate effects are negligible during impact. Morton (1988) showed how these scaling laws are applied to graphite-epoxy beams.

Swanson and coworkers (1993) took a different approach to develop use of scaling rules for plates and shells. The equations of motion for a symmetrically specially orthotropic laminated plate are

$$A_{55}\left(\frac{\partial \psi_x}{\partial x} + \frac{\partial^2 w}{\partial x^2}\right) + A_{44}\left(\frac{\partial \psi_y}{\partial y} + \frac{\partial^2 w}{\partial y^2}\right) + q = I_1 \ddot{w}_o \tag{3.326a}$$

$$D_{11}\frac{\partial^2 \psi_x}{\partial x^2} + (D_{12} + D_{66})\frac{\partial^2 \psi_y}{\partial y \partial x} + D_{66}\frac{\partial^2 \psi_x}{\partial y^2} - A_{55}\left(\psi_x + \frac{\partial w}{\partial x}\right) = I_3 \ddot{\psi}_x \tag{3.326b}$$

$$(D_{12} + D_{66})\frac{\partial^2 \psi_x}{\partial y\, \partial x} + D_{66}\frac{\partial^2 \psi_y}{\partial x^2} + D_{22}\frac{\partial^2 \psi_y}{\partial y^2} - A_{44}\left(\psi_y + \frac{\partial w}{\partial y}\right) = I_3 \ddot{\psi}_y. \tag{3.326c}$$

Equations (3.326) govern the behavior of both the model and a prototype. Assume that the layup can be scaled and that the variables are related by

$$T_p = \lambda T_m \tag{3.327}$$

where T_p is a typical variable for the prototype and T_m is the corresponding variable for the model. λ is the scale factor for that variable. Assuming that the same materials are used for both the model and the prototype, the equation of motion for the prototype becomes

$$c_1 D_{11}\frac{\partial^2\psi_x}{\partial x^2}+c_2(D_{12}+D_{66})\frac{\partial^2\psi_y}{\partial y\partial x}+c_3 D_{66}\frac{\partial^2\psi_x}{\partial y^2}$$
$$-c_4 A_{55}\left(\psi_x+\frac{\partial w}{\partial x}\right)=c_5 I_3\ddot{\psi}_x \tag{3.328}$$

where

$$c_1=\lambda_h^3\lambda_{\psi_x}/\lambda_x^2,\quad c_2=\lambda_h^3\lambda_{\psi_x}/\lambda_y^2,\quad c_3=\lambda_h^3\lambda_{\psi_y}/(\lambda_x\lambda_y),$$
$$c_4=\lambda_h\lambda_{\psi_x},\quad c_5=\lambda_h\lambda_w/\lambda_x,\quad c_6=\lambda_\rho\lambda_h^3\lambda_{\psi_x}/\lambda_t^2. \tag{3.329}$$

The necessary condition for similitude between the prototype and the model is that

$$c_1=c_2=c_3=c_4=c_5=c_6. \tag{3.330}$$

This gives the similitude rules

$$\lambda_x/\lambda_y=1,\quad \lambda_{\psi_x}/\lambda_{\psi_y}=1,\quad \lambda_x/\lambda_h=1,$$
$$\lambda_w/\left(\lambda_h\lambda_{\psi_x}\right)=1,\quad \lambda_\rho\lambda_h^2/\lambda_t^2=1. \tag{3.331}$$

The other two equations of motion (3.326b,c) give

$$\lambda_y/\lambda_y=1,\quad \lambda_w/\left(\lambda_h\lambda_{\psi_y}\right)=1,\quad \lambda_\rho\lambda_x/\lambda_w=1. \tag{3.332}$$

Requiring strain similitude, we also have

$$\lambda_{\psi_x}=1,\quad \lambda_{\psi_y}=1. \tag{3.333}$$

The impact force is determined from the contact pressure using

$$F(t)=\int_A q\,dx\,dy. \tag{3.334}$$

Since $\lambda_\rho=1$ from the previous results, if the contact area is scaled by λ^2, then the contact force F is scaled by λ^2. The impact energy is scaled by λ^3, and since the mass of the impactor is scaled also by λ^3, the impact velocities must be the same. Experiments conducted on laminated plates with scale factors of 1,2,3,4,5 showed that during impact, the strain histories at similar locations follow the same time history provided that time is scaled properly.

In summary, two approaches for developing scaling rules are available. The first method is based on dimensional analysis; the second makes use of the equations of motion of the system. In general, it is possible to scale the displacement, contact forces, and strains between the model and the prototype, but usually it is not possible to scale the damage size.

4
Low-Velocity Impact Damage

This chapter deals broadly will the subject of *impact resistance*, which is the study of damage induced by foreign object impact in a laminate and the factors affecting it. Methods for predicting impact damage are discussed in Chapter 5, and the study of damage tolerance – that is, the effect of impact damage on the stiffness, strength, fatigue life, and other properties of the laminate – will be presented in Chapter 6.

An understanding of impact damage development, the failure modes involved, and the various factors affecting damage size has been gained through extensive experimental studies. In this chapter, several of the most commonly used impact test procedures will be discussed. Experimental techniques for impact damage detection and detailed mapping of the damage zone after impact are reviewed. Of the many different techniques discussed in the literature, some are nondestructive, and others are destructive. Some techniques are used extensively, and others have seen only limited applications. A few experimental techniques have been developed to observe damage development during the impact event. While not attempting to give an exhaustive description of the techniques used, the objectives are to briefly describe each one and to give a general idea of what the most commonly used techniques are.

Understanding the process of impact damage initiation and growth and identifying the governing parameters are important for the development of mathematical models for damage prediction, for designing impact resistant structures, and for developing improved material systems. The basic morphology of impact damage, its development, and the parameters affecting its initiation, growth, and final size will be described. Qualitative models providing intuitive explanations for the observed damage patterns will be presented here, but the presentation of detailed mathematical models for damage prediction will be delayed until Chapter 5.

4.1 Impact Tests

To simulate actual impact by a foreign object, a number of test procedures have been suggested (Table 4.1). The initial kinetic energy of the projectile is an important parameter to be considered, but several other factors also affect the response of the structure. A large mass with low initial velocity may not cause the same amount of damage as a smaller mass with higher velocity, even if the kinetic energies are exactly the same. In one case, the impact might induce an overall response of the structure, while in the other the response might be localized in a small region surrounding the point of impact. Therefore, the selection of the appropriate test procedure must be made very carefully to ensure that test conditions are similar to the impact conditions to be experienced by the actual structure.

At the moment, two types of tests are used by most investigators, although many details of the actual test apparatus may differ. Experimental studies attempt to replicate actual situations under controlled conditions. For example, during aircraft take-off and landing, debris flying from the runway can cause

Table 4.1. *Articles describing impact test apparatus*

Gas gun
Cantwell and Morton (1985b)
Dan-Jumbo et al. (1989)
Delfosse et al. (1993)
Hong and Liu (1989)
Hussman et al. (1975)
Jenq et al. (1992)
Jenq et al. (1994)
Liu (1988)
Malvern et al. (1987, 1989)
Qian and Swanson (1989)
Sharma (1981)
Takamatsu et al. (1986)
Dropweight
Ambur et al. (1995)
Cantwell et al. (1986)
Curtis et al. (1984)
Levin (1986)
Schoeppner (1993)
Tsai and Tang (1991)
Wu and Liau (1994)
Pendulum
Buynak et al. (1988)
Sjoblom and Hwang (1989)

damage; this situation, with small high-velocity projectiles, is best simulated using a gas gun. Another concern is the impact of a composite structure by a larger projectile at low velocity, which occurs when tools are accidentally dropped on a structure. This situation is best simulated using a dropweight tester.

Several variants in implementation are described in the literature (Table 4.1), but the main features of a gas gun apparatus are shown in Fig. 4.1a. High-pressure compressed air is drawn into an accumulator to a given pressure controlled by a regulator. The pressure is released by a solenoid valve, the breakage of a thin diaphragm, or other mechanisms. The projectile then travels through the gun barrel and passes a speed-sensing device while still in the barrel of the gun or right at the exit. A simple speed-sensing device consists of a single light-emitting diode (LED) and a photodetector. The projectile, which has a known length, interrupts the light beam, and the duration of that interruption in signal produced by the sensor is used to calculate the projectile velocity. Most experimental setups utilize two LED-photodetector pairs. The travel time between the two sensors is determined using a digital counter and is used to calculate the projectile velocity. Another system was used by Cantwell and Morton (1985c) who placed two thin wires across the barrel of a gun a given distance apart. The projectile velocity is determined knowing the time elapsed between the instants when the first and the second wires were broken.

Dropweight testers are used extensively (Table 4.1) and can be of differing designs. Heavy impactors are usually guided by a rail during their free fall from a given height. Usually, a sensor activates a mechanical device designed to prevent multiple impacts after the impactor bounces back up. Schoeppner (1993) describes a system in which the specimen is subjected to a controlled tensile preload. Small projectiles are simply dropped from a known distance (Fig. 4.1b), sometimes being guided by a tube.

Pendulum-type systems are also used to generate low-velocity impacts (Table 4.1). Pendulum-type testers (Fig. 4.1d) consist of a steel ball hanging from a string, or a heavier projectile equipped with force transducers or velocity sensors. The Hopkinson-type pressure bar technique is also used. Lal (1983a,b, 1984) used a cantilevered impactor (Fig. 4.1c), for which a 1-in.-diameter steel ball is mounted at the end of a flexible beam which is pulled back and then released to produce the impact.

During the impact by a spherical projectile, the stress distribution under the impactor is truly three-dimensional. As soon as the projectiles enters in contact with the target, a compressive wave, a shear wave, and surface waves propagate away from the impact point. For low-velocity impacts, no significant damage is introduced during the early stage of the impact. That is, stress levels remain low as these waves travel many times through the thickness of the laminate

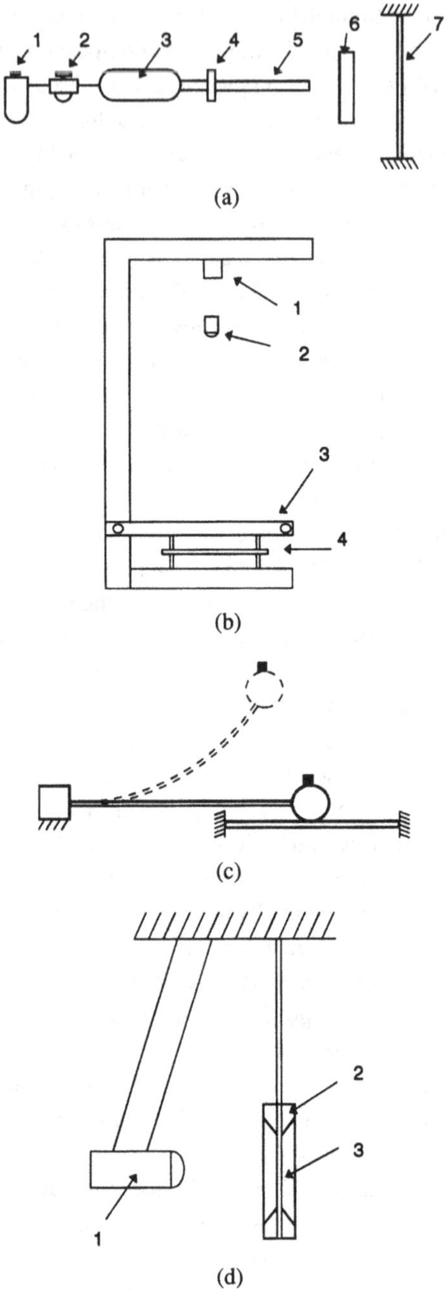

Figure 4.1. a) Gas gun apparatus: (1) air filter, (2) pressure regulator, (3) air tank, (4) valve, (5) tube, (6) speed sensing device, (7) specimen. b) Dropweight tester: (1) magnet, (2) impactor, (3) holder, (4) specimen. c) Cantilevered impactor. d) Pendulum-type tester: (1) impactor, (2) specimen holder, (3) specimen.

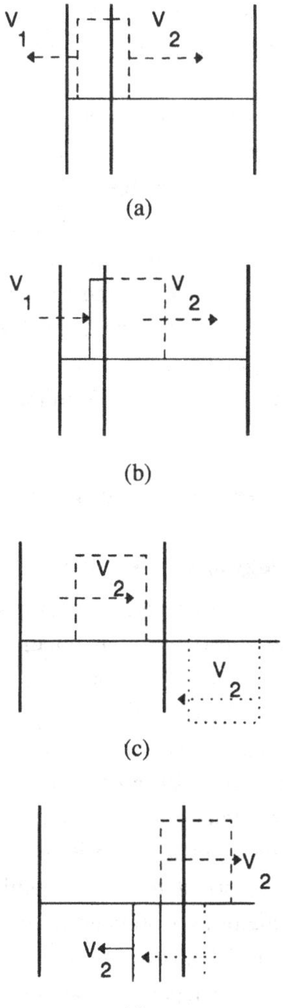

Figure 4.2. Stresses in flyer plate and target during flyer plate experiment.

and then the target deforms like a plate. Damage is introduced when overall bending motion is established. With higher-velocity impacts, the compressive wave, after reflection from the back surface, can generate tensile stresses of sufficient magnitudes to create failure near the back face. The thin flyer plate technique is used to study spall damage caused by one-dimensional stress pulses. In a typical arrangement, a thin flyer plate is accelerated by the explosion of a thin aluminum foil. As the flyer plate impacts the specimen, compressive stress waves propagate towards the left in the flyer plate with a velocity v_1 and towards the right in the specimen with a velocity v_2 (Fig. 4.2).

After the compressive wave in the flyer plate reaches the free boundary (i.e., after $t_1 = h_1/v_1$), the stress in the flyer plate is the sum of the compressive wave and a tensile wave of the same magnitude propagating to the right with velocity v_1 (Fig. 4.2b). At time $2t_1$, contact between the two solids ends, and therefore the duration of the compressive pulse in the specimen is $2t_1$. The compressive pulse reaches the back surface of the specimen at time $t_2 = h_2/v_2$. After t_2, the stresses in the specimen are the sum of the compressive pulse propagating to the right and a tensile pulse of the same magnitude starting to propagate to the left at time t_2. Tensile stresses develop in the specimen after time $t_2 + t_1$ and starting at a distance $d = v_2 t_1$ from the back face. If the tensile stress exceeds the strength of the laminate in that direction, failure occurs at that location, and sometimes a layer of thickness d is severed from the specimen and flies off.

4.2 Failure Modes in Low-Velocity Impact Damage

4.2.1 Morphology of Low-Velocity Impact Damage

For impacts that do not result in complete penetration of the target, experiments indicate that damage consists of delaminations, matrix cracking, and fiber failures. Delaminations, that is, the debonding between adjacent laminas, are of most concern since they significantly reduce the strength of the laminate. Experimental studies consistently report that delaminations occur only at interfaces between plies with different fiber orientations. If two adjacent plies have the same fiber orientation, no delamination will be introduced at the interface between them. For a laminate impacted on its top surface, at interfaces between plies with different fiber orientation, the delaminated area has an oblong or "peanut" shape with its major axis oriented in the direction of the fibers in the lower ply at that interface. This is illustrated schematically in Fig. 4.3. It must be noted that delamination shapes often are quite irregular and that their orientation becomes rather difficult to ascertain.

The delaminated area usually is plotted against the initial kinetic energy of the impactor (Table 4.2), and after a small threshold value is reached, the size of the delaminations increases linearly with the kinetic energy. Delamination size usually is defined as the damaged area measured from ultrasonic C-scans. In general, delaminations are introduced at several interfaces, and C-scans provide a projection of all these damaged surfaces on a single plane. The projected damage area is affected by the number of plies in the laminate so that for each layup, experimental results of area versus initial kinetic energy fall on different lines. Other experimental techniques are used to measure the area of delaminations at each interface (Table 4.3), and the total area varies

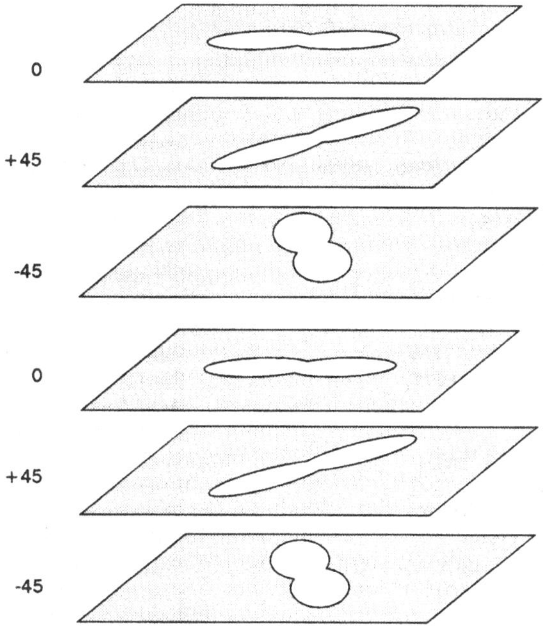

Figure 4.3. Orientation of delaminations.

linearly with the initial kinetic energy of the projectile. A single curve is obtained for all laminates regardless of the number of plies for that material system.

The threshold value of the kinetic energy is hard to determine experimentally because of experimental scatter from one specimen to another, and several tests are needed in order to determine the threshold level of the initial kinetic energy required to initiate delamination. Several studies indicate that damage is introduced when the contact force reaches a critical value which is the same for static tests and low-velocity impact tests (Sjoblom and Hwang 1989, Lindsay and Wilkins 1991). The force threshold value usually corresponds to the first discontinuity in the contact force history. The force threshold is very consistent from sample to sample and can be determined from a single test. The data presented by Sjoblom and Hwang (1989) for a $[-45,45,90,90,45,-45,0,\bar{0}]_S$ graphite-PEEK composite (AS4/APC2) shows that damage starts to develop after the impact force reaches a critical value, regardless of the initial kinetic energy of the projectile.

Sjoblom (1987) proposed a simple model to explain how the threshold force value varies with the several governing parameters and the laminate thickness in particular. With a spherical indentor of diameter D, a contact force P produces

Table 4.2. *Relationship between damage size and impact energy*

Avery and Grande (1990)	Leach et al. (1987)
Bouadi et al. (1992)	Liu (1987, 1988)
Cantwell (1988a,b)	Malvern et al. (1987, 1989)
Cantwell and Morton (1984b, 1989b,c, 1991)	Marshall and Bouadi (1983)
Cantwell et al. (1986, 1991)	Minguet (1993)
Caprino et al. (1984)	Moon and Shively (1990)
Choi and Chang (1992)	Morton and Godwin (1989)
Choi et al. (1990)	O'Kane and Benham (1986)
Curtis and Bishop (1984)	Peijs et al. (1990)
Davies et al. (1994)	Pelstring (1989)
Dempsey and Horton (1990)	Phillips et al. (1990)
Demuts and Sharpe (1987)	Poe (1990)
Dobyns (1980)	Ramkumar and Thakar (1987)
Dorey (1987)	Rhodes et al. (1981)
Dorey et al. (1978, 1985)	Sierakowski (1991)
Evans and Alderson (1992)	Sierakowski et al. (1976)
Gandhe and Griffin (1989)	Sjoblom and Hwang (1989)
Ghaffari et al. (1990)	Smith and Yamaki (1990)
Ghasemi Nejhad and Parvizi-Majidi (1990)	Spamer and Brink (1988)
Gong and Sankar (1991)	Srinivasan et al. (1991, 1992)
Gottesman et al. (1987)	Strait et al. (1992b)
Grady and Meyn (1989)	Swanson (1992, 1993)
Griffin (1987)	Takeda et al. (1981a)
Hong and Liu (1989)	Verpoest et al. (1987)
Hull and Shi (1993)	Wang and Khanh (1991)
Kumar and Narayanan (1990)	Williams and Rhodes (1982)
Kumar and Rai (1991)	Wu and Liau (1994)
Lal (1982)	Wu and Springer (1988a)

Table 4.3. *Detailed maps of delaminations*

Boll et al. (1986)
Buynak et al. (1988)
Clark (1989)
Davies et al. (1994)
Guynn and O'Brien (1985)
Hong and Liu (1989)
Joshi and Sun (1985, 1987)
Kumar and Rai (1991)
Liu (1988)
Liu et al. (1987a,b)
Preuss and Clark (1988)
Sierakowski et al. (1976)
Takeda et al. (1981a, 1982a)

a contact zone of radius $r = (D\alpha)^{1/2}$ where the indentation α is related to P and the contact stiffness k_c by Hertz contact law. The transverse shear stress is assumed to be uniformly distributed over a cylinder of radius r and height h, and failure occurs when this stress reaches a maximum allowable value. With these assumptions, the damage initiation force is given by

$$P = (2\pi h \tau_{\max})^{\frac{3}{2}} D^{\frac{3}{4}} k_c^{-\frac{1}{2}} \tag{4.1}$$

which indicates that P increases with $h^{3/2}$, and experimental results verify that trend.

After impact, there are many matrix cracks arranged in a complicated pattern that would be very difficult to predict, but it is not necessary to do so since matrix cracks do not significantly contribute to the reduction in residual properties of the laminate. However, the damage process is initiated by matrix cracks which then induce delaminations at ply interfaces. Two types of matrix cracks are observed: tensile cracks and shear cracks (Fig. 4.4). Tensile cracks are introduced when inplane normal stresses exceed the transverse tensile strength of the ply. Shear cracks are at an angle from the midsurface, which indicates that transverse shear stresses play a significant role in their formation. With thick laminates, matrix cracks are first induced in the first layer impacted by the projectile because of the high, localized contact stresses. Damage progresses from the top down, resulting in a pine tree pattern (Fig. 4.5a). For thin laminates, bending stresses in the back side of the laminate introduce matrix cracks in the lowest layer, which again starts a pattern of matrix cracks and delaminations and leads to a reversed pine tree pattern (Fig. 4.5b).

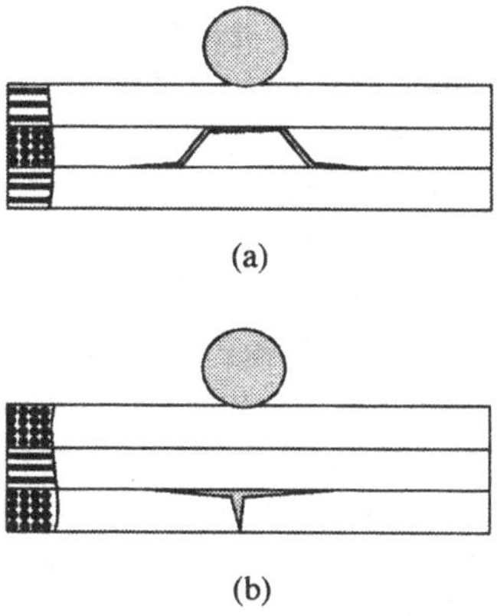

Figure 4.4. Two types of matrix cracks: a) tensile crack, b) shear crack.

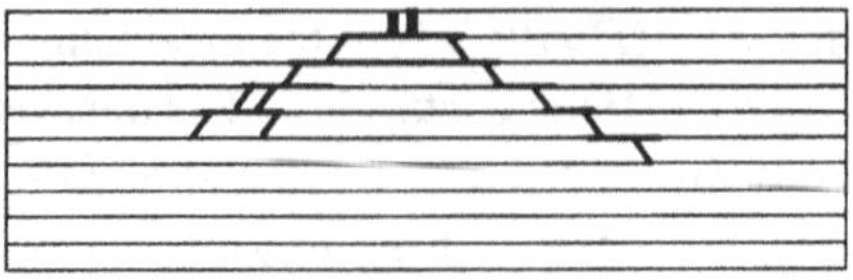

(a)

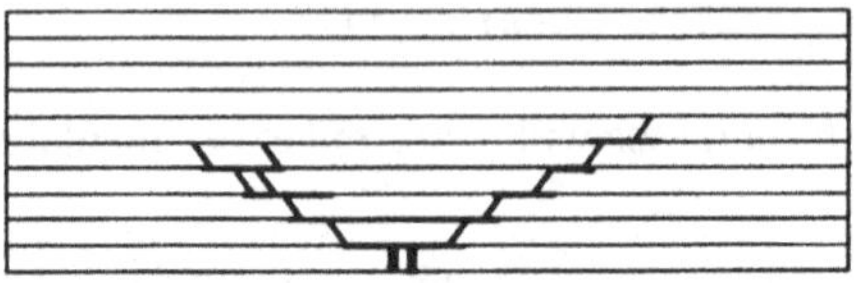

(b)

Figure 4.5. Pine tree (a) and reversed pine tree (b) damage patterns.

4.2.2 Damage Development

Several studies (e.g., Lee and Zahuta 1991) showed similarities between damage induced by impact loading and that produced by static loading. Curson et al. (1990) conducted tests on 16-ply carbon/PEEK composites with a $[-45,0,+45,90]_{2S}$ layup using a dropweight tester in which the specimen is supported on a ring with 50 mm internal diameter. For small levels of kinetic energy (1 J), the bottom ply (back face), when experiencing tension, exhibited an intralaminar crack. This crack did not propagate into subsequent plies or turn into a delamination crack. For an input energy of 2.5 J, the intralaminar crack has propagated through two plies linked by a small delaminated region. Up to this level, the force-time curves are relatively smooth with uniform rise and fall, implying an almost elastic response. The creation of matrix cracks does not require significant energy or cause a noticeable drop in stiffness. For higher energy levels (4 J), discontinuities in the force-time signals followed by oscillations indicate the onset of delaminations and noticeable changes in stiffness.

Very few investigators quantify the amount of fiber damage introduced during impact. Avery and Grande (1990) showed that for a $[45,0,-45,90_2,-45,0,45_2,0,-45,90]_S$ graphite-epoxy laminate, the total damage area determined by ultrasonic inspection increased almost linearly for impact energies up to about 30 J, reached a plateau until approximately 40 J, and then started to increase again. This plateau in the evolution of the damage area was shown to coincide with the introduction of significant fiber damage. Figures 4.6 and 4.7 show the fiber damage area in each ply for two specimens impacted with 34 J and 41 J, respectively.

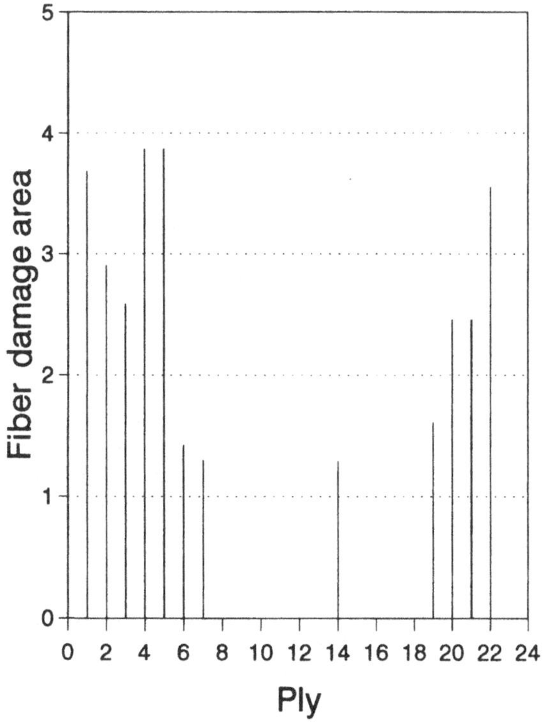

Figure 4.6. Fiber damage in laminate subjected to a 34 J impact.

4.2.3 Qualitative Models for Predicting Delamination Patterns

Two simple models have been put forward to explain why delaminations appear when laminates are subjected to localized loads. Both approaches are based on the fact that the laminate is made up of several orthotropic layers. Each layer tends to deform in a particular way, and transverse normal and shear stresses applied at the interfaces constrain the layup to behave as one plate. When these interlaminar stresses become too large under concentrated contact loads, delaminations are introduced. The orthotropic behavior of each ply and the mismatch in their bending stiffnesses is thought to be the basic cause of delaminations, and the study of this mismatch yields important information regarding the location, orientation, and size of delaminations in a laminate.

Liu (1988) studied the delamination of two-layer plates and proposed a "bending stiffness mismatch" model to predict the orientation, size, and shape of the delaminations based on the premise that delaminations occur because the sublaminates above and below a given interface have different bending rigidities.

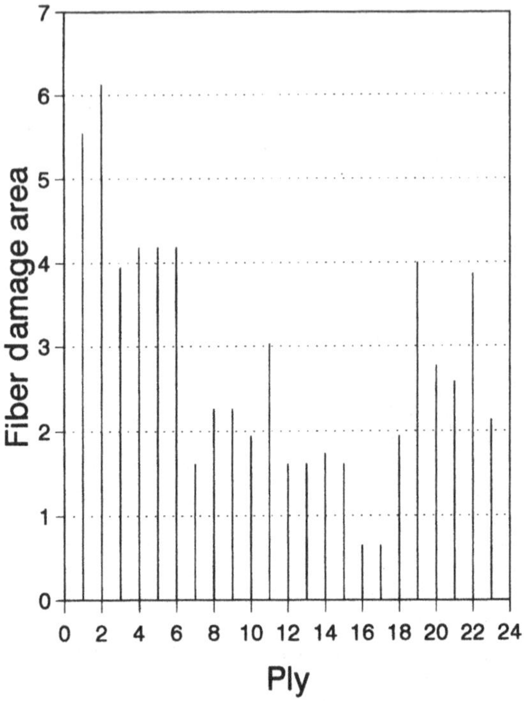

Figure 4.7. Fiber damage in laminate subjected to a 42 J impact.

Because of the anisotropy and of the different fiber orientations, this difference or "mismatch" in bending rigidities is different in different directions. In the experiments conducted to validate the model, the length and width of the specimens were kept the same, they were held in the same holder, and were subjected to the same impact. This way, the effect of difference in fiber orientation in the two plies on delamination at the interface could be isolated from other factors that could affect damage size. It is postulated that delaminations occur because of differences in bending rigidities between the two plies. Mismatch coefficients are defined as

$$M = \frac{[D_{ij}(\theta_b) - D_{ij}(\theta_t)]}{[D_{ij}(0°) - D_{ij}(90°)]} \tag{4.2}$$

where the D_{ij} are the components of the bending rigidity matrix D relating moment resultants to plate curvatures. Each ply is considered separately, so $D_{ij}(\theta_b)$ is the rigidity of the bottom layer acting alone, and the subscript t refers to the top layer. While a mismatch coefficient can be defined for each bending coefficient D_{ij}, usually only D_{11} is considered. The denominator is

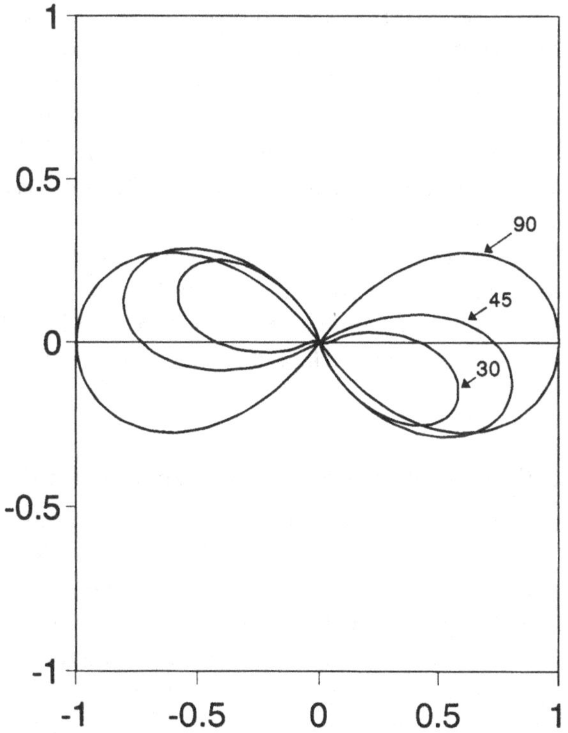

Figure 4.8. Bending stiffness mismatch coefficient for mismatch angles of 90, 45, and 30 degrees.

simply introduced to nondimensionalize M; for two-layer plates, $M = 1$ when the angle difference is 90°. As an example, consider $[\theta,\ 0]$ graphite-epoxy laminates with elastic properties

$$E_1 = 181\,\text{GPa}, \quad E_2 = 10.3\,\text{GPa}, \quad \nu_{12} = 0.30, \quad G_{12} = 7.17\,\text{GPa} \tag{4.3}$$

for $\theta = 90°, 45°, 30°$. Results are presented in Fig. 4.8, and for each direction defined by the angle α, the length OP is proportional to M. The three closed curves represent the variation of M with orientation for the three laminates considered. The shape of the curve generally resembles that of delaminations, but the orientation is not strictly in the direction of fibers in the lower ply except when $\theta = 90°$. The model also predicts smaller-sized delaminations as the difference in fiber orientations becomes smaller.

Another simple way to explain why two layers with different fiber orientations should delaminate when subjected to concentrated transverse loads was presented by Lesser and Filippov (1991). The transverse displacements of a simply supported rectangular plate consisting of a single composite layer with fibers oriented in the 0° direction subjected to a concentrated force applied in the center can be calculated using the Navier solution described in Chapter 3. The same problem was solved again for a fiber orientation of 90°. If two layers are stacked on top of each other but not bonded together, the two layers would separate under load because they deform differently. The difference between the displacements of the two layers has the same shape as the delaminations at the interface between the same two layers if they were bonded together. The idea behind this simple explanation is that when the two layers are bonded together, interlaminar stresses develop on the interface in order to force these layers to deform as a single plate. High interlaminar stresses are expected to cause delaminations.

4.3 Experimental Methods for Damage Assessment

Many techniques have been developed to determine the extent of impact-induced damage in composite structures. Impact damage is internal and generally consists of delaminations, matrix cracking, and fiber breakage, which usually cannot be detected by simply examining the surface of the specimen. The objectives for this section are to provide a brief description of the techniques used for damage assessment and to provide a guide to the existing literature for the interested reader. In applications where one seeks to detect the presence of impact-induced damage, nondestructive techniques providing whole field information are used. For research purposes when the location of impact is known, other, destructive methods provide more detailed information but also end up destroying the specimen in the process.

4.3.1 Nondestructive Techniques

Methods capable of detecting the presence of eventual impact damage over the whole structure are needed. It is necessary to determine if damage is present, where it is located, and its extent. With translucent material systems such as glass-epoxy or Kevlar-epoxy composites, impact damage can be observed using strong backlighting. The size and shape of delaminations and the presence of matrix cracks can be detected by visual observation. Several investigators have used this technique, and much of the early work on impact damage was carried out using this technique and brought about a great deal of understanding about the morphology of impact damage and impact damage development.

Other material systems such as graphite-epoxy are opaque, and thus this visual inspection approach cannot be used. Whole-field nondestructive methods such as ultrasonic imaging or radiography are used to visualize internal damage over large areas. C-scans and traditional x-rays provide a projected image of the damage zone and are useful in delineating the extent of the damage, but many of the features of the damage area are lost. It is important to understand how delaminations are distributed through the thickness, their size and orientation, and how they might be connected through intraply cracks. This knowledge provides a basis for developing a model for damage development during impact. Improved ultrasonic inspection techniques capable of resolving the distribution and size of delaminations through the thickness of the specimen have been developed (Buynak and Moran 1986, Frock et al. 1988, Ramkumar and Chen 1983a, Boll et al. 1986).

Many other nondestructive techniques for impact damage inspections have been reported in the literature, although ultrasonic and x-ray methods are by far the two most commonly used techniques. With infrared thermography, a heat flow directed in the direction normal to the plane of the laminate using a plastic-covered, heated water tank with a large thermal mass. The specimen is placed on the surface of the water, and the temperature of the upper surface of the specimen is analyzed using a thermovision system designed around an infrared camera (Hillman and Hillman 1985). The presence of delaminations causes detectable temperature gradients. Because heat is being conducted around delaminations, there is no sharp temperature difference at the delamination boundary, and temperature maps do not directly indicate the size of delaminations. With the vibrothermography technique (Potet et al. 1987), high-frequency (18 MHz) low-amplitude mechanical vibrations are used to induce localized heating in the specimen. Local defects, such as delaminations, introduce stress concentrations and act as heat sources by dissipating mechanical energy. The location of the delaminations can be determined from temperature maps generated by an infrared camera, but the size of the delaminations is more difficult to assert.

The "tapping" technique is also used empirically to detect the presence of internal defects by lightly tapping the specimen and listening to the change in radiated sound as the impact location is changed from an undamaged zone to a region containing a delamination (Kenner et al. 1985). A small sphere was dropped onto a small transducer placed on the surface of the specimen from a controlled height. The contact force measured by the transducer showed only minor differences between impacts on a damaged or an undamaged zone. After contact with the projectile ends, the frequency spectrum of the sound radiated by the specimen is significantly different depending on whether the impact occurred on or off the debonded area.

Spallation of a thin coat of brittle lacquer added to the back surface of the impacted specimen yields patterns that correspond to areas of internal damage determined from ultrasonic C-scans (Rhodes et al. 1981). With holographic interferometry, the passage of a laser beam through the holographic plate forms a reference beam that interferes with the light reflected from the object (Baird et al. 1993). When a delaminated specimen is loaded in compression, local buckling is detectable. Under tension, localized warping due to the asymmetric nature of the laminate can be detected.

In actual applications, it is often difficult to inspect a structure using ultrasonic, radiographic, or other nondestructive inspection technique because normal operations must be interrupted, accessibility for personnel and equipment must be provided, and sufficient time must be allowed for inspection. Using a widely spread array of embedded optical fibers (30–200 μm in diameter), damage can be continuously monitored (Hofer 1987). The structure is considered sound along the path of a given fiber if light is successfully transmitted from one end to the other. If damage is introduced and the fiber is broken, light transmission is interrupted. A network of optical fibers can be embedded in the structure and connected to a computer that continuously monitors damage. A review of the application of embedded optical fibers as damage detection sensors based on the failure of the sensor is given by LeBlanc and Measures (1992).

Electrical current can be injected along graphite fibers, and the resulting electric field can be measured. In areas with fiber damage, disruptions in current paths create detectable perturbations in measured magnetic fields (Rhodes et al. 1981). Eddy current methods can be used to detect fiber fractures (De Goeje and Wapenaar 1992, and Lane et al. 1991).

4.3.2 Destructive Techniques

Detailed maps of impact damage can be obtained by sectioning several strips of material at different locations and orientations throughout the impacted zone. After careful preparation, microscopic examinations of each section are used to construct detailed maps of delaminations at each interface and of matrix cracks in each ply. The use of micrographs in documenting impact damage is reported in many studies. Typically, slices are cut with a diamond lapidary saw using a water spray to minimize local heating, and then mounted in epoxy resin and ground on successively finer abrasive silicon carbide paper (240, 320, 400, 600, 1000, and 1200 grit).

With the often-used deply technique (Levin 1986), a gold chloride solution with an isopropyl carrier is used to infiltrate the damaged area. If the surface damage is not sufficient for the solution to penetrate, 1 mm holes can be drilled through the laminate. After drying, a precipitation covers the fracture surfaces. The matrix is pyrolysed in an oven at about 420°C, and afterwards the laminate can be separated into individual laminas. Delaminations and matrix cracks can be observed under an optical microscope.

4.3.3 Observation of Damage Development during Impact

The velocity of delamination crack propagation in semi-transparent glass-epoxy plates was measured by Takeda et al. (1982b) using high-speed photography. With strong backlighting, the entire impact event was filmed with a 16 mm camera at speeds of up to 40,000 frames per second. Shadow-Moire interferometry was used by several investigators to study delamination buckling during post-impact compression tests. A stationary master grid placed close to the specimen remains undeformed, while the grid formed by the shadow of that master grid on the specimen will deform with the specimen. The interference of the master grid and its shadow produce dark and bright fringes from which the normal displacements of the panel could be determined. Chai et al. (1983) used high-speed photography with the shadow-Moire technique to record failure propagation in composite panels subjected to compressive inplane loading and low-velocity impact. Pictures were taken every 50 μs to track the extent of the bulging corresponding to buckling of delaminated areas.

4.4 Parameters Affecting Impact Damage

The extensive experimental work performed to date produced an understanding of which parameters affect the initiation and growth of impact damage. Material properties affect the overall stiffness of the structure and the contact stiffness and therefore will have a significant effect on the dynamic response of the structure. We are also interested in which properties of the matrix, the fibers, and the fiber-matrix interface control the initiation and growth of impact damage. The thickness of the laminate, the size of the panel, and the boundary conditions are all factors that influence the impact dynamics, since they control the stiffness of the target. The characteristics of the projectile – including its density, elastic properties, shape, initial velocity, and incidence angle – are another set of parameters to be considered. The effect of layup, stitching, preload, and environmental conditions are important factors that have received various degrees of attention.

4.4.1 Material Properties

Identifying material properties that have a direct effect on impact damage is of great interest since it provides directions for the development of improved materials systems and the design of impact-resistant composite structures. Many material systems were considered in the open literature, but a large number of studies deal with carbon-epoxy laminates, reflecting the widespread use of that material in actual applications. Some material properties of the composite affect the impact dynamics or the strength of the laminate. Properties of the matrix material, the reinforcement, and the interface have been identified as having a distinct effect on impact resistance. Using high strain-to-failure fibers, tougher resin systems, compliant layers between certain plies, or woven or stitched laminate leads to improvements in impact resistance.

The elastic properties of the material (E_1, E_2, ν_{12}, G_{12}), along with the lamination scheme, define the overall rigidities of the plate which greatly influence the contact force history. As discussed earlier, the ratio E_1/E_2 has a major effect on the bending stiffness mismatch coefficient between plies with different fiber orientations. The transverse modulus E_2 has a major effect on the contact stiffness. Lowering the contact stiffness also lowers the contact forces and increases the contact area, which in turn significantly affects the stress distribution under the impactor. Anisotropy in elastic properties and coefficients of thermal expansion affect impact resistance because of the residual thermal stresses developed during the curing process. For correct prediction of impact damage initiation, thermal stresses must be accounted for in the analysis (e.g., Wang and Khanh 1991).

The threshold kinetic energy is strongly influenced by the properties of the matrix and is essentially independent of the properties of the fibers, the layup, and whether woven or unwoven layers are used. Experiments with the same matrix material and five different types of fiber reinforcements showed that the threshold damage energy was the same for these five composites, indicating that damage initiation is matrix-dominated (Griffin 1987). Experiments conducted by Strait et al. (1992) verify that incipient damage is matrix- and interface-dominated. The stacking sequence and the properties of the reinforcing fibers had no measurable effect on the energy required for damage initiation. Damage is initiated by matrix cracking; when a matrix crack reaches an interface between layers with different fiber orientations, delamination is initiated. Because the elastic modulus of the reinforcing fibers is usually much higher than that of the matrix, these fibers appear to be essentially rigid. Therefore, the type of fibers being used does not seem to affect the onset of matrix cracking and delaminations. For higher levels of impact energy, the properties of fibers and the stacking sequence become important.

Improved matrix properties are obtained using thermoplastic or rubber modified epoxy resins, and toughened epoxy matrices lead to increased damage resistance (Chen et al. 1993, Recker et al. 1989). But the development of higher strain capability for the matrix is limited by the need to maintain satisfactory performance at high temperatures and in difficult environmental conditions. Thermoplastics such as PEEK meet those requirements, and after impact, delaminations are less extensive with carbon-PEEK laminates; compression tests show that the residual strength of those laminates is superior to those of the carbon-epoxy laminates (Dorey et al. 1985, Boll et al. 1986, Demuts and Sharpe 1987, Griffin 1987). Tough BMI matrices can be developed in order to provide tough impact resistance without sacrifice in hot/wet performance. BMI resin systems used in composites have higher temperature resistance than epoxy and are easier to process than other high-temperature resins. For the particular BMI resin system tested by Ho (1989), the flexural strength of the neat resin at 450°F retains 48% of its room temperature value, and its modulus retain 63% of its room temperature value. The material properties of laminates made with this matrix and AS-6 carbon fiber prepregs show similar retention of stiffness and strength at high temperatures. The G_{Ic} of the laminate was measured by the double torsion method, and the G_{IIc} by the end-notched flexural test method. Experiments suggest that increasing both the mode I and mode II fracture toughness of the composite lead to improved damage resistance and damage tolerance.

Fibers with higher failure strains are expected to improve the impact resistance of the composite for high energy levels (Cantwell and Morton 1991, Curson et al. 1990). The introduction of a discrete layer of very-high-toughness, high-shear-strain resin at certain interfaces is called "interleaving" and has been shown experimentally to improve impact resistance (Hirshbuehler 1985, 1987; Evans and Masters 1987; Gandhe and Griffin 1989; Pintado et al. 1991a,b; Ishai and Shragai 1990; Wang et al. 1991). Typical thickness of the interleaf is 1/2 mil (Hirschbuehler 1985). Sun and Rechack (1988) showed that adhesive layers placed along interfaces of a laminate reduce interlaminar shear stresses and therefore lower delaminations. The extra resin used has the adverse effect of lowering the modulus of the finished laminate and of reducing its compressive strength.

Impact resistance of glass-epoxy composites can be significantly improved by introducing a rubbery interface between the glass fibers and the rigid epoxy matrix (Peiffer 1979). Tissington et al. (1992) reported that enhanced fiber/matrix adhesion resulting from oxygen plasma treatment reduced delamination during impact and also reduced the energy absorption capability of the laminate. The flyer plate experiments of Takeda et al. (1987) showed that the spall velocity for neat resin is significantly larger than that of the composites with the same resin material because of the stress concentration effect due to the presence of relatively stiff fibers. The effect of reinforcement type is found to be small, but the effects of interfacial strength on spall velocity are clearly observed.

4.4.2 Target Stiffness

Target stiffness depends on material properties, as already mentioned, but also on the thickness of the laminate, the layup, its size, and the boundary conditions. The stiffness of the thickness has a significant effect on the magnitude of the maximum contact force which, of course, will affect the extent of the damage induced. Cantwell and Morton (1985c) studied the influence of target bending stiffness on low-velocity impact damage in CFRP laminates using a dropweight tester. At low velocities, flexible targets responded primarily by bending, which caused high tensile stresses in the lowest ply. Matrix cracks then developed in the lowest ply, which in turn generated a delamination at the lowest interface. This matrix cracking-delamination repeats itself from ply to ply, resulting in an inverted pine tree appearance (Fig. 4.5b). For stiffer targets, damage is initiated by high contact stresses and propagates downwards through the same matrix cracking-delamination process, giving, a pine tree appearance (Fig. 4.5a). Hull and Shi (1993) determined that the overall stiffness of the laminate determined the overall geometry of the damaged area, but that the shape of the delamination at a given interface depends on the local stacking sequence near that interface.

4.4.3 Projectile Characteristics

While a lot of studies consider the effect of several parameters during impacts generated by a single impactor, the size and shape of the impactor, the material it is made of, and its angle of incidence relative to the surface of the specimen are all factors that will have a strong influence on the impact response of the specimen.

Kumar and Rai (1993) considered 32-ply graphite-epoxy plates impacted by aluminum and steel projectiles. Two layups used were $[0,45,-45,90]_{4S}$ and $[0_2,45,-45,0,90,0_2,-45,45,0,90,0_2,45,-45]_S$. The projectiles were 11 mm cylinders with hemispherical noses and a total length of 20 mm. Figure 4.9a shows the relationship between the total delamination area and the impact velocity. A significant difference is observed between the results obtained using steel and aluminum projectiles. This difference can be explained for the most part by the fact that the mass of the steel projectile is roughly three times that of the aluminum projectiles. For the same velocity, the kinetic energies of the steel and aluminum projectiles are therefore in the same ratio. When the overall damage area is plotted versus the kinetic energy of the projectile (Fig. 4.9b), the difference between the two sets of results is much smaller. However, the steel projectiles still produce slightly more damage. This can be explained by the fact that the higher modulus of elasticity of steel gives a higher contact stiffness, which induces higher contact forces and a smaller contact zone. Both of those factors potentially causing higher damage. It must be noted that for

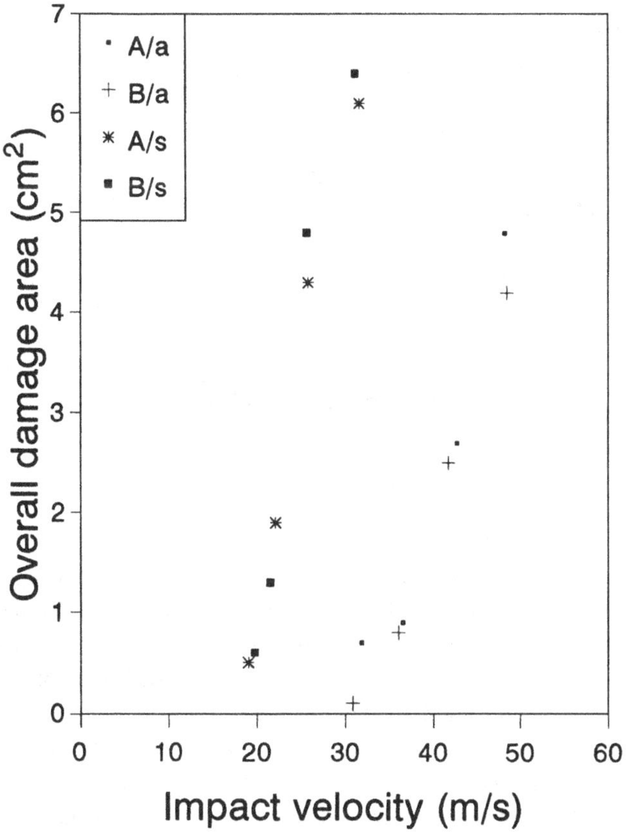

(a)

Figure 4.9. (a) Damage area versus impact velocity; (b) damage area versus kinetic energy for impacts with steel and aluminum projectiles.

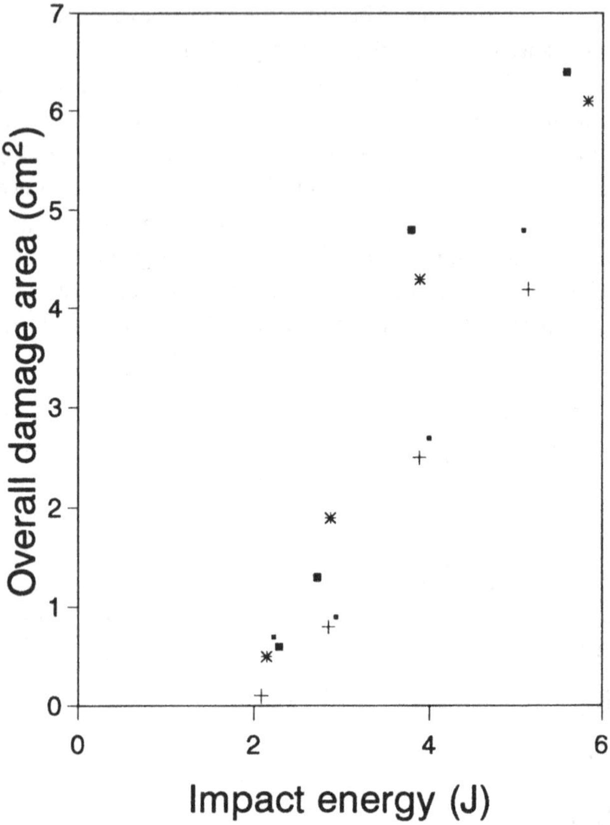

(b)

Figure 4.9. (*cont.*)

the same impactor, the total delamination area seems to be independent of the layup used in this case (Fig. 4.9). However, it must be noted that these two laminates have the same number of plies and that the lamination parameters are not very different. The bending rigidities of the plates will be similar, and the impact dynamics are not significantly affected.

Most of the experimental work on impact deals with normal impacts, that is, the initial velocity of the projectile is normal to the surface of the target. In practice, oblique impacts occur, and some studies address the problem of determining the effect of the angle of incidence on impact damage. For oblique impacts, the damage area depends on the angle of incidence, but when the damage area is plotted versus the normal component of the initial velocity, the data collapse into a single curve (Ghaffari et al. 1990, Jenq et al. 1991, Bouadi et al. 1992). Considering the oblique impact of a sphere with a composite laminate, both of which having smooth surfaces, it can be assumed that for low-velocity impacts, the coefficient of friction between the projectile and the target is small. This assumption is reasonable when no visible damage is introduced on the surface of the laminate. If the tangential component of the contact force is negligible, conservation of linear momentum in the tangential direction implies that the tangential component of the projectile velocity is unchanged by the impact. Therefore, under these conditions the contact force is induced by that fraction of the kinetic energy associated with the normal component of the initial velocity of the projectile.

4.4.4 Layup and Stitching

The importance of the stacking sequence on the impact resistance of laminates was first demonstrated by Ross and Sierakowski (1973). In a unidirectional laminate, since the reinforcing fibers are all oriented in the same direction, no delamination occurs. For two plates with the same thickness but with different stacking sequences, the one with the higher difference of angle between two adjacent plies will experience higher delamination areas. Increasing the thickness of each layer will also lead to increased delaminations. Increasing the difference between the longitudinal and transverse moduli of the material leads to higher bending stiffness mismatching and therefore increased delamination. However, damage initiation is matrix- and interface-dependent and therefore has little or no dependence on the stacking sequence. The peak load reached during impact, or the energy at peak load, is strongly dependent on the stacking sequence.

Stitching is used to introduce through-the-thickness reinforcement but in a different way than with weaving or braiding. The laminated structure is preserved, and stitching can be performed on either a prepreg or a preform. Stitching density and pattern and properties of the thread can be varied to improve delamination resistance. Stitching of laminates prior to curing limits the size of delamination when the composite is subjected to out-of-plane loading and improves its resistance to transverse fracture when subjected to inplane loading. Dry preform stitching improves the compression-after-impact strength for two reasons. First, during impact, stitching arrests delaminations and therefore limits the damage size. Second, during compression-after-impact (CAI) tests, stitching prevents the growth of delaminations. However, some drawbacks are also present. Fiber damage can be introduced by needle penetration during stitching, by waviness of the fibers, and by introduction of resin-rich pockets, which cause stress concentrations and can reduce the strength of the laminate.

Therefore, the extra manufacturing step of stitching the laminate to improve delamination resistance must be done carefully to minimize the reductions in inplane properties. Dransfield et al. (1994) presented a review of previous work dealing with the delamination resistance of graphite-epoxy laminates by stitching and discussed manufacturing issues at length.

4.4.5 Preload

Schoeppner (1993) conducted a series of experiments to determine the effect of a tensile preload on the damage resistance of graphite-epoxy laminates using a dropweight tester. The stiffening effect of the pre-tension is shown to decrease the time required to reach the maximum impact load and to increase the indentation depth. The maximum load was insensitive to the preload. It must be noted that in these experiments, the mass of the impactor was 13.95 kg and the kinetic energy of the impactor was 80 J. These impacts resulted in partial or complete penetration, which may explain the results concerning the independence of mass load on pre-tension whereas earlier studies of laminates with initial stresses showed a strong dependence. Phillips et al. (1990) conducted impact experiments on ceramic matrix composites under preload and showed that applied tensile loads drastically reduce the impact energy required to produce total fracture of the specimen.

4.4.6 Environmental Conditions

Changes in temperature and moisture content are known to affect both stiffness and strength of composites. It is logical to expect that impact resistance will also be affected by environmental factors. Strait et al. (1992) showed that the threshold kinetic energy is increased significantly when glass fiber–reinforced composites are subjected to long-term exposure to sea water prior to impact. However, peak load during impact and the total energy absorbed are substantially reduced. Pope and Kulkarny (1992) studied the effect of moisture absorption and freezing on impact resistance of resin transfer molded glass/vinyl-ester resin composites. Results were too limited to draw general conclusions.

4.5 Summary

Many types of impact tests have been proposed, but now two particular tests are being used: the dropweight tests and the airgun test, even though variants in implementation still exist. A lot of knowledge about impact damage has been gained through extensive testing. Impact damage consists of matrix cracking, delaminations, and fiber fracture. It is generally admitted that damage is initiated by matrix cracks that develop due to either excessive transverse shear stress or tensile bending stress. These matrix cracks will then induce delaminations at interfaces between plies with different fiber orientations. Delaminations do not occur between layers with the same fiber orientation. After a small threshold value, damage area generally increases linearly with the initial kinetic energy of the impactor. Delaminations development follows some precise rules governing their orientation and size. Simple heuristic models based on a bending stiffness mismatch between sublaminates above and below an interface explain how those patterns develop, even though the models cannot provide quantitative estimates of delamination size.

Many experimental techniques for detecting impact damage have been proposed, but ultrasonic methods and x-rays are most commonly used. Experiments have also brought a number of observations concerning the effects of material properties, target stiffness, layup, projectile characteristics, and other parameters. A clear understanding of these effects is necessary for interpreting test results and designing composite structures for impact resistance.

4.6 Review Questions

4.1 Can impacts induce delaminations in a unidirectional laminate?

4.2 Draw a figure showing the location of possible delaminations in a $[0_2,\pm45,90]_S$ laminate. Draw another figure showing the general orientation of possible delaminations at each interface.

4.3 Describe patterns for damage initiation and propagation through the thickness of the laminate.

4.4 What is the bending stiffness mismatch coefficient?

4.5 Explain the effect of obliquity on the low-velocity impact of a laminate by a sphere.

4.6 Why does stitching improve damage tolerance?

4.7 What is the difference between a shear crack and a bending crack?

4.8 What are the effects of target stiffness on the impact dynamics? What are the parameters controlling the stiffness of the target?

4.9 Explain what is meant by "damage resistance" and "damage tolerance."

4.10 List the most commonly used nondestructive inspection techniques used for damage detection.

4.11 What are the failure modes associated with impact damage?

4.12 Why is it important to detect delaminations and determine their size, whereas usually little attention is paid to the many matrix cracks also present in the damaged zone?

4.13 Why is it difficult to obtain direct experimental evidence of how damage develops in graphite-epoxy?

4.14 Explain the difference between a dropweight tester and a gas gun. Give reasons for choosing one system over the other.

5
Damage Prediction

5.1 Introduction

Impact damage usually follows some very complex distributions, and it may not be possible to reconstruct the entire sequence of events leading to a given damaged state. For low-velocity impacts, damage starts with the creation of a matrix crack. In some cases the target is flexible and the crack is created by tensile flexural stresses in the bottom ply of the laminate. This crack, which is usually perpendicular to the plane of the laminate, is called a *tensile crack*. For thick laminates, cracks appear near the top of the laminate and are created by the contact stresses. These cracks, called *shear cracks*, are inclined relative to the normal to the midplane. Matrix cracks induce delaminations at interfaces between adjacent plies and initiate a pattern of damage evolution either from the bottom up or from the top down. Therefore, while it is possible to predict the onset of damage, a detailed prediction of the final damage state cannot realistically be achieved.

Two types of approaches are used for predicting impact damage. The first type attempts to estimate the overall size of the damaged area based on the stress distribution around the impact point without considering individual failure modes. The general idea is that impact induces high stresses near the impact point and that these localized stresses initiate cracks, propagate delaminations, and eventually lead to the final damage state. Section 5.2 describes how this approach is used for predicting damage size for thick laminates, which behave essentially as semi-infinite bodies. In Section 5.3, the approach is applied to thin laminates, which are modeled as shear deformable plates.

The second approach used in the prediction of impact damage consists of determining damage initiation, usually the appearance of the first matrix crack, using a detailed three-dimensional stress analysis of the impact zone and appropriate failure criteria. Then, delamination sites are determined and the propagation of these delaminations are studied. This approach, involving extensive computations, has been applied to relatively simple layups, albeit with encouraging success. While providing useful insight into the damage process and an important understanding of which material properties govern the impact resistance, this approach is not practical for design purposes. Methods for predicting damage initiation are presented in Section 5.4, and the propagation of existing delaminations is discussed in Section 5.5.

5.2 Damage in Thick Laminates

When thick laminates are subjected to low-velocity impacts, bending deformations can generally be neglected and the laminates can be considered as semi-infinite bodies. The maximum impact force determined from an impact dynamic analysis is assumed to be distributed on the surface according to Hertz theory of contact. For isotropic materials, a solution due to Boussinesq is available for calculating the stresses created by a concentrated force acting on the boundary of a half-space (Timoshenko and Goodier 1970). Stresses in a semi-infinite body produced by the pressure distribution caused by contact can be calculated using this basic solution by integrating over the region of contact (Love 1929). This approach was used by Poe (1991a,b) to predict impact damage in 36-mm-thick graphite-epoxy cylindrical shells used as rocket motor cases. Since the laminate is made of orthotropic layers and is not a homogeneous isotropic body, the results are expected to be only qualitative. In addition, since damage is not modeled, the size of the damage region can only be approximated. A complex state of stress is present under the indentor; in order to predict failure, principal stresses and the maximum shear stress can be determined at each point, and the maximum compressive stress and the maximum shear stress distributions are predicted. The maximum shear stress occurs along the centerline of the contact zone below the surface. A maximum shear stress criterion and a maximum compressive stress criterion can be used to determine damage size. Damage initiated below the surface due to shear stresses, as well as the depth and width of damage, increases in proportion to impactor radius.

A closed-form solution for the stresses in a transversely isotropic half-space due to a point force has been obtained by Lekhnitskii (1981). Stresses due to Hertz contact loading can be determined by integrating the stress components due to a point force over the region of contact. This approach was used by Matemilola and Stronge (1995) to determine the effect of anisotropy on the stress distribution in the contact zone. As the ratio of the inplane to transverse Young's moduli E_r/E_z becomes larger, both the radial and the tangential stresses become larger on the surface of loading. As with isotropic material, the radial normal stress is tensile outside the loaded region, with the maximum value occurring at the periphery of the contact area. For brittle materials, fracture initiates where the radial tensile stress is maximum, which is why ring cracks are observed when glass is subjected to contact loading. The maximum value of the transverse shear stress occurs below the surface of the half-space towards the outside of the contact zone. As the anisotropy ratio E_r/E_z increases, the maximum shear stress also increases, and the depth below the surface at which it occurs becomes more shallow. The location where transverse shear stresses are maximum is highly susceptible to shear fracture.

To use available solutions for isotropic or transversely isotropic semi-infinite solids is that the laminate must be modeled as an equivalent homogeneous solid. This can be done for a laminate with many plies, and several authors have presented methods for estimating the elastic properties of the equivalent solid given the material properties and the layup of the laminate. A simple approach was used by Poe (1991b). Starting with the elastic properties of a lamina (E_1, E_2, ν_{12}, G_{12}), the lamination theory can be used to find the inplane elastic constants of the laminate (E_x, E_y, ν_{xy}, ν_{yx}, G_{xy}), and then the elastic constants of the equivalent transversely isotropic solid are calculated as

$$E_r = (E_x + E_y)/2, \quad E_z = E_{22}, \quad \nu_r = (\nu_{xy} + \nu_{yx})/2,$$
$$\nu_{rz} = \nu_{12}, \quad G_{zr} = G_{12}. \tag{5.1}$$

This approach allows the properties of the equivalent homogeneous solid to be determined quickly.

5.3 Damage in Thin Laminates

Often one is interested in determining the overall size of the damage created by a given impact, since damage size affects the residual properties of the structure. Damage is introduced only after the impact force reaches a minimum level. Therefore, it is also desirable to be able to predict this threshold impact force level. Two simple methods for performing those tasks will now be presented.

The first approach, proposed by Dobyns (1980) and Dobyns and Porter (1981), is aimed at predicting the overall damage size. It is based on the premise that delaminations, which are the critical component of impact damage, grow because of high transverse shear stresses in the vicinity of the impactor. The idea is to determine the distribution of the transverse shear force resultant around the point of impact and to use an appropriate failure criterion to estimate the size of the damaged zone. Transverse shear force resultants are determined using the first-order shear deformable plate theory. If the contact force is assumed to be a concentrated load, the transverse shear force tends to infinity near the

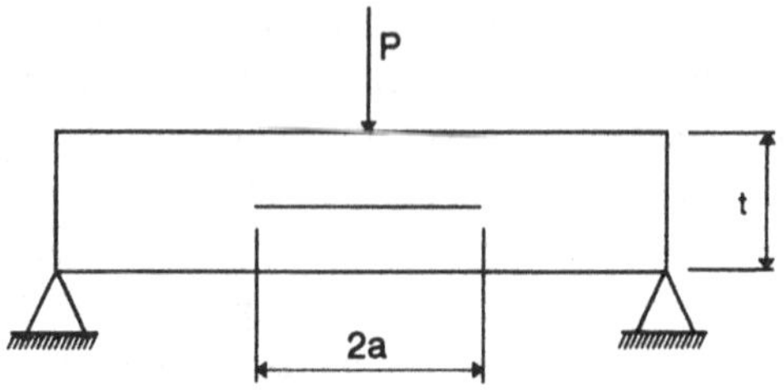

Figure 5.1. Model for estimating the critical load for damage initiation (Davies et al. 1994).

point of application of load. Since the load is applied through contact with the indentor, it is reasonable to assume that this load is distributed over a small area. Since the damage zone is usually significantly larger than the contact area, it is possible use a simplified distribution. Dobyns and Porter (1981) assumed that the contact force is uniformly distributed over a small square patch with each side equal to the radius of the impactor. No attempt is made to obtain accurate stress distributions through the thickness of the laminate. Delaminations are assumed to occur when the transverse shear force Q_n is such that the average transverse normal stress Q_n/h exceeds 34.5 MPa for graphite-epoxy. This approach allows to predict the damaged zone with good accuracy. Graves and Koontz (1988) followed the same approach and used a critical average transverse shear stress of 23.3 MPa for graphite-epoxy and 20.5 MPa for graphite-PEEK laminates.

The second approach to be discussed here deals with the prediction of the threshold value of the contact force that corresponds to damage initiation. When the damage area is plotted versus the maximum impact force, there is a clear sudden increase in damage size once the load reaches a critical value P_c (Davies et al. 1994). Below this value, the damage area is small due to Hertzian surface damage. P_c corresponds to the onset of delaminations. A very simple model for estimating this critical load is shown in Fig. 5.1. The quasi-isotropic laminate is treated as isotropic, and a single circular delamination of radius a is assumed to be located on the midplane of a circular plate on simple supports. The delamination is expected to extend when the energy release rate reaches the critical energy release rate for mode II fracture. This simple model yields the following expression for the energy release rate:

$$G_{\text{II}} = \frac{9P^2(1-\nu^2)}{8\pi^2 E h^3} \tag{5.2}$$

where E and ν are the equivalent inplane modulus and Poisson ratio for the laminate, and h is the laminate thickness. It is important to note that the strain energy release rate is independent of a, the radius of the delamination. As P is increased, a critical value G_{IIc} is reached at which point the delamination size is indeterminate. As a consequence, as soon as the critical value of the force is reached, the delamination area increases very rapidly. The critical force threshold is then

$$P_c^2 = \frac{8\pi^2 E h^3}{9(1-\nu^2)} G_{\text{IIc}}. \tag{5.3}$$

This equation indicates that the critical force threshold is proportional to $h^{3/2}$. The critical value of $G_{\text{IIc}} = 0.8$ N/mm was used. Equation (5.3) was used successfully to predict the onset of delamination damage for several quasi-isotropic graphite-epoxy laminates. With glass-polyester laminates, the existence of such a force threshold is evident from plots of the damage area versus contact force obtained from experiments (Davies et al. 1994). However, in this case a significant difference is observed between the results of static indentation tests and those of impact tests because material properties of glass-polyester are strain rate–dependent. In particular, (1) the modulus starts to increase as the strain rate becomes on the order of 10 s^{-1}, and (2) fracture toughness also is expected to be rate dependent, but contradicting reports in the literature indicate that it might increase or sometimes decrease with strain rate.

5.4 Damage Initiation

In this section, methods used for predicting the appearance of the first matrix crack are presented. As this crack reaches the interfaces between the current ply and the two adjacent plies, a significant stress redistribution takes place and four possible sites for delamination crack initiation are created. The next task is to determine if and where delaminations are created. The propagation of existing delaminations will be treated in the next section.

Interlaminar normal and shear stresses play an important role in impact damage initiation and must be determined accurately. Often two- or three-dimensional finite element models are employed for that purpose because of the versatility of the method. However, this results in a very large number of degrees of freedom, which may be prohibitive. Modeling the laminate using the first-order shear deformation plate theory usually provides a good estimate of the inplane stresses; using the approach described in Section 3.2.5, accurate estimates of the interlaminar stresses can be recovered very efficiently. When a laminate is subjected to a high-velocity impact, initially stress waves are confined to a small area near the impactor and propagate through the thickness with the speed of sound C. Compressive strains are on the order V/C, and failure is expected when V/C is around 1%. For graphite-epoxy, C is approximately 2000 m/s. Therefore, through-the-thickness stress waves must be considered for velocities in excess of 20 m/s. For lower velocities, engineering approximations for beams, plates, and shells are valid. When the transverse displacement exceeds the thickness of the laminate, it may be necessary to include the effect of geometrical nonlinearities in the impact dynamics analysis to obtain accurate contact force histories. A significant portion of the impact energy is used for membrane deformation in that case. Thermal residual stresses also must be included in the analysis because they often have a strong effect on damage threshold energy levels.

Once an accurate determination of the stress distribution is available, an appropriate failure criterion must be used to determine the location of the first matrix crack. One approach is to determine the maximum tensile stress transverse to the fibers for each ply. Failure is predicted using a maximum stress criterion. That is, tensile matrix failure occurs when this maximum stress exceeds the tensile strength in the transverse direction. Hashin's failure criterion was used by several authors to predict the appearance of matrix cracks:

$$\frac{1}{Y_t^2}(\sigma_{yy} + \sigma_{zz})^2 + \frac{1}{S^2}\left(\sigma_{yz}^2 - \sigma_{yy}\sigma_{zz}\right) + \frac{1}{S^2}\left(\sigma_{xy}^2 + \sigma_{xz}^2\right) = e_M^2 \qquad (5.4)$$

where Y_t is the transverse normal strength in tension and S is the transverse shear strength. The z-axis is normal to the laminate, and the x- and y-axes are local coordinates parallel and normal to the fiber direction in the layer under consideration. Failure occurs when e_M becomes larger than or equal to one. Since the transverse normal stress is usually small, σ_{zz} can be neglected in (5.4). For the purely two-dimensional problem of a beam subjected to a cylindrical impactor, σ_{xz} is also zero, and the criterion can be simplified further:

$$\left(\frac{\bar{\sigma}_{yy}}{Y}\right)^2 + \left(\frac{\bar{\sigma}_{yz}}{S_i}\right)^2 = e_M^2 \qquad (5.5)$$

where Y and S_i are the in-situ transverse normal and shear strengths (Choi et al. 1991b, Choi and Chang 1992). Overbars indicate that the stresses are averaged over the thickness of the ply, for example,

$$\bar{\sigma}_{yy} = \frac{1}{h_n}\int_{z_{n-1}}^{z_n} \sigma_{yy}\,dz. \qquad (5.6)$$

The strength in the transverse direction is taken as $Y = Y_t$ if $\sigma_{yy} > 0$, and $Y = Y_c$ if $\sigma_{yy} < 0$. The strength of a ply in the transverse direction when that ply is located between plies with different fiber orientation is different than when the same ply is part of a unidirectional laminate. Adjacent plies have the effect of increasing the apparent or "in-situ" strength of a ply. The in-situ strengths vary with $\Delta\theta$, the minimum ply angle change between the current ply and its neighbors, and M, the number of layers in the current ply. The in-situ strength can be estimated using

$$Y = Y^\circ\left(1 + A\frac{\sin(\Delta\theta)}{M^B}\right)$$
$$S = S^\circ\left(1 + C\frac{\sin(\Delta\theta)}{M^D}\right) \tag{5.7}$$

where A–D are constants determined from experiments.

Fiber failure can also be predicted using the following criteria. Compressive failure of the fibers occurs when $\sigma_{xx} < 0$ and

$$\|\sigma_{xx}\| > X' \tag{5.8}$$

where X' is the compressive strength of the lamina in the fiber direction. Tensile fiber failure occurs when $\sigma_{xx} > 0$ and

$$\left(\frac{\sigma_{xx}}{X}\right)^2 + \frac{\sigma_{xy}^2 + \sigma_{xz}^2}{S} \geq 1 \tag{5.9}$$

where X is the tensile strength in the fiber direction and S is the inplane shear strength.

To calculate the stress distributions in the neighborhood of the critical matrix crack, the elastic properties of the finite element in which the crack is predicted to occur are modified. The damaged element is assumed not to sustain any transverse tensile stress or out-of-plane shear stress. Results presented for a $[0_6, 90_2]_S$ laminate show that the shear matrix crack in the 90° layer leads to positive normal stresses along the interface away from the impacted area. Along the upper interface, the situation is reversed. Delaminations are initiated in mode I and lead to the damage pattern shown in Fig. 4.6a.

The introduction of a matrix crack due to transverse shear create both high interlaminar tension as well as shear stresses at the matrix crack tips. Sharp stress gradients suggest that matrix cracks result in interlaminar stress singularities that cause delaminations at interfaces. Wang and Yew (1990), Liu (1993), and Liu et al. (1993a,b) used a different criterion for predicting the onset delamination:

$$\left(\frac{\sigma_{zz}}{Y}\right)^2 + \left(\frac{\sigma_{xz}^2 + \sigma_{yz}^2}{S^2}\right) = e_D^2. \tag{5.10}$$

Failure occurs as e_D reaches one. Equation (5.10) is used to predict initial failure, and then fracture analysis was applied to simulate the growth of delaminations as the applied load continued to increase.

5.5 Propagation of Delaminations during Impact

The study of the propagation of delaminations during impact requires an accurate prediction of the stress distribution in the vicinity of the crack tip, the calculation of the strain energy relase rates for the various fracture modes, and the use of an accurate mixed-mode fracture criterion.

While in some cases beam or plate models can be used, the laminates containing delaminations are often analyzed as 2D or 3D solids. Because of the complex state of stress in the impact region with bending, transverse shear and contact stresses, and the presence of both matrix cracks and delaminations, mode I and II fractures are present at the delamination crack tip. Mode I and II strain energy release rates for a crack lying along the interface of two dissimilar elastic media are not well defined due to the oscillatory character of stresses in a small region near the crack tip (Sun and Jih 1987). However, the total strain energy release rate is always well defined. When a two-dimensional finite element model is used to obtain accurate stress distributions near the delamination crack tip, the total strain energy release rate G and its two components G_{I} and G_{II} can be determined using the virtual crack closure technique (Rybicki and Kanninen 1977). However, as expected, the values of G_{I} and G_{II} do not converge as the finite element mesh is refined (Sun and Jih 1987, Salpekar 1993). The region of oscillation is small, and the length of the crack increment used in the finite element analysis should be large enough to avoid those oscillations. On the other hand, small elements are needed to determine the total energy release rate accurately; and a balance must be struck between these conflicting requirements. The smallest element size recommended by Salpekar (1993) is one-fourth of the thickness of a single ply.

In addition to the difficulties related to the computation of accurate strain energy release rates mentioned above, two additional problems must also be addressed: accurate measurement of reliable critical strain energy release rates, and the adoption of an appropriate failure criterion. The development of reliable test procedures and the characterization of the critical energy release rates is the object of extensive efforts. A review of the literature (Jones et al. 1988) describes the various experimental procedures in details and shows that the values of G_{Ic} obtained in 24 different studies are remarkably close in spite of the differences in test methods. For mode II fracture, great variability in G_{IIc} is reported. A comprehensive review of mixed-mode fracture in composite laminates (Garg 1988) indicates that a reliable failure criteria is not yet available. A simple and commonly used failure criterion is written as

$$\left(\frac{G_{\mathrm{I}}}{G_{\mathrm{Ic}}}\right)^m + \left(\frac{G_{\mathrm{II}}}{G_{\mathrm{IIc}}}\right)^n = e_D \tag{5.11}$$

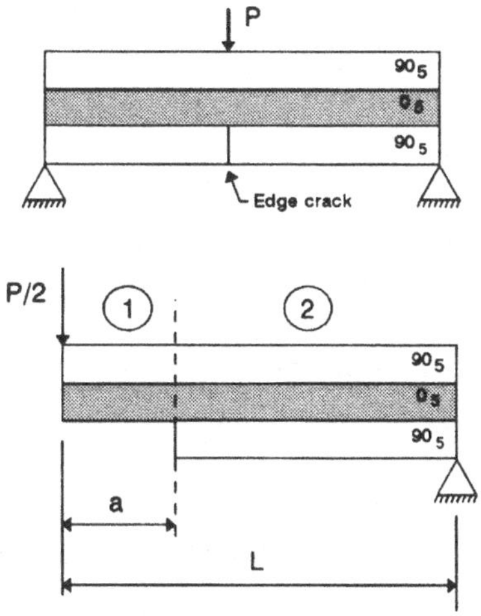

Figure 5.2. Analysis of delamination initiated by a bending crack in $[90_5,0_5,90_5]$ graphite-epoxy laminate (Sun and Manoharan 1989).

where m and n are constants used to fit experimental data. The delamination extends when $e_D \geq 1$. Liu et al. (1993a,b) and others used (5.11) with $m = n = 1$.

The growth of delaminations induced by matrix bending cracks is governed primarily by mode I fracture in the direction normal to the fiber direction of the bottom layer (Choi and Chang 1992, Liu et al. 1993a,b). The growth of such delaminations is quite stable. The quasi-static three-point bending tests of Liu et al. (1993a,b) on $[0_6,90_3]_S$ and $[0_4,90_4]_S$ carbon-epoxy laminates indicate that delaminations induced by shear cracks propagated in an unstable manner and led to catastrophic failures. Very good correlation between numerical predictions and experimental results was reported for the force versus indentor displacement. Similar results were also reported by Liu et al. (1993b) for $[0_4,90_3,0_2]_S$ and $[0_3,90_2,0]_S$ laminates. Salpekar (1993) analyzed $[0_2,90_8,0_2]$ glass-epoxy and graphite-epoxy laminates with inclined matrix cracks to show that large interlaminar tensile stresses are created at the crack tips and may give rise to delaminations, and that a significant mode I component of the strain energy release rate is present at the delamination initiation.

Sun and Manoharan (1989) studied $[90_5,0_5,90_5]$ graphite-epoxy laminates with an edge notch simulating a bending crack (Fig. 5.2) similar to those observed after low-velocity impacts of such laminates. A two-dimensional linear finite element analysis was performed in order to determine the energy release rates G_{I} and G_{II} corresponding to the fracture modes I and II and the total energy release rate G when the specimen is subjected to three-point bending.

A beam model including the effect of shear deformation was also developed to derive the following expression for the total energy release rate:

$$G = \frac{1}{2}\left[\left(\frac{A_{11}}{D}\right)_1 - \left(\frac{A_{11}}{D}\right)_2\right] \cdot \frac{P^2}{4} \cdot (a - L)^2 \tag{5.12}$$

G decreases as the crack length increases, which indicates that the crack growth is stable. In addition, G_{I} is much larger than G_{II}, indicating that mode I is the dominant mode of failure. Grady and Sun (1986), Sun and Grady (1988), and Sankar and Hu (1991) studied the propagation of an existing midplane delamination crack during the impact of a cantilever, 20-ply $[90,0]_s$ graphite-epoxy beam by a rubber ball. In this case, the mode I contribution to the strain energy release rate is negligible, and crack propagation is a mode II fracture process. These two simple examples show that depending on the particular case, the propagation of delaminations is governed by either mode I or mode II fracture. This has implications for the selection and development of materials. Under one type of tests, one material system may prove to have a superior damage resistance, but the situation might be reversed if a different type of test is performed.

Razi and Kobayashi (1993) conducted tests on $[0_4,90_4]_s$ and $[0_4,90_4]_s$ cross-ply laminated beams and plates to study the development of delaminations. Damage observed during low-velocity impact was similar to that induced during quasi-static tests. Detailed finite element analyses coupled with a failure criterion based on the energy release rate were used to predict the size of delaminations in beam and plate specimens. Mode II was shown to be the dominant failure mode, and the contribution of G_{I} was negligible. The dynamic initiation strain energy release rate was estimated to be 15% lower than G_{IIc}, and delaminations were arrested when the strain energy release rate reached a value G_{IIa} that was approximately 75% lower than G_{IIc}. Very good comparison was shown between the shape of delaminations predicted using this approach and experimental results.

5.6 Conclusion

Impact damage with its complicated patterns of delaminations, matrix cracks, and fiber failures cannot be predicted in full detail. Predicting the location and size of delaminations is desirable since this type of damage accounts for most of the degradation of the properties of the structure. The overall size of the damage zone usually can be estimated using simplified approaches. The onset of matrix cracking, which is thought to initiate the entire damage process, can also be predicted accurately. Difficulties arise when attempting to proceed to determine at which interfaces delaminations will develop and to predict the growth of these delaminations. While such calculations have been successfully performed, they are complex and costly and therefore cannot be considered for routine calculations.

5.7 Exercises

5.1 Given the elastic constants for a graphite-epoxy lamina ($E_1 = 111$ GPa, $E_2 = 1.92$ GPa, $G_{12} = 4.28$ GPa, $\nu_{12} = 0.267$), find the properties of the equivalent transversely isotropic solid for a $[0, \pm 56.5]_{2S}$ laminate using (5.1).

5.2 Use the Boussinesq solution (Timoshenko and Goodier 1970) and plot the radial stress, the transverse shear stress, and the transverse normal stress near the point of application of a unit force on the surface of a half-space. Find the principal stresses and the maximum shear stresses and plot their distributions.

5.3 Starting with the Boussinesq solution, use numerical integration to find the stress distributions when the unit load is (a) uniformly distributed over a patch of width $2a$, and (b) parabolically distributed over a patch of similar width.

5.4 A 3 in. × 3 in. graphite-epoxy plate with a $[0_4, 90_4]_S$ layup is subjected to a quasi-static loading by a $\frac{1}{4}$-in.-diameter indentor. The elastic constants for a graphite-epoxy lamina are $E_1 = 178$ GPa, $E_2 = 12.4$ GPa, $G_{12} = 4.62$ GPa, $\nu_{12} = 0.39$. Plot contours of the average transverse shear stress, and determine the size of the damage zone if the critical value of the transverse shear stress is 30 MPa. Estimate the critical value of the critical force threshold using (5.3).

6
Residual Properties

6.1 Introduction

Having studied the dynamics of impact, damage development, and damage prediction methods, the next area of interest is the effect of damage on the mechanical properties of laminated composite structures. This is often called the study of *damage tolerance* since it refers to the experimental determination or the numerical prediction of the residual mechanical properties of the damaged structure. Many organizations have developed guidelines or requirements for residual strength after impact. For example, U.S. Air Force draft requirements for damage tolerance for low-velocity impacts are that laminates should maintain a minimum design strength after impacts with 100 ft-lb kinetic energy by a 1-in.-diameter hemispherical indenter or after impacts resulting in a 0.10 in. dent, whichever is less severe (Schoeppner 1993).

An understanding of damage tolerance can be gained through experiments and available models for predicting residual properties. The general trend for the residual strength of laminated composites with impact damage is that, for low initial kinetic energy levels, the strength is not affected since little or no damage is introduced. As damage size increases, the strength drops rapidly and then levels off. The effects of impact damage on the residual strength in tension, compression, shear, and bending have been investigated at length and follow the same general trend. Experimental techniques, general results obtained from experiments, and models for predicting residual properties are presented in this chapter.

6.2 Compressive Strength

The large number of articles devoted to understanding the compression after impact behavior of composite materials (Table 6.1) is evidence of the importance of the topic and the level of effort focused on it. Several new test fixtures

Table 6.1. *Articles on residual compressive strength of impact-damaged laminated composites*

Altus and Ishai (1992)	Griffin (1987)
Avery and Grande (1990)	Guynn and O'Brien (1985)
Avva (1993)	Hirschbuehler (1985, 1987)
Basehore (1987)	Ishai and Shragai (1990)
Bishop (1985)	Ishikawa et al. (1996)
Boyd et al. (1989)	Leach et al. (1987)
Cantwell et al. (1983, 1986, 1991)	Levin (1986)
Caprino (1984)	Manders and Harris (1986)
Chen et al. (1993)	Moon and Shively (1990)
Chuanchao and Kaida (1991)	Morton and Godwin (1989)
Clerico et al. (1989)	Mousley (1984)
Curtis and Bishop (1984)	Nettles and Hodge (1991)
Dempsey and Horton (1990)	Olesen et al. (1992)
Demuts (1993)	Ong et al. (1991)
Demuts and Sharpe (1987)	Pelstring and Madan (1989)
Demuts et al. (1985)	Pintado et al. (1991a,b)
Dorey et al. (1985)	Prandy et al. (1991)
Dost et al. (1991)	Prichard and Hogg (1990)
Dow and Smith (1989)	Recker et al. (1989)
Evans and Masters (1987)	Sjoblom and Hwang (1989)
Finn et al. (1992)	Spamer and Brink (1988)
Ghasemi Nejhad and Parvizi-Majidi (1990)	Srinivasan et al. (1991)
Gong and Sankar (1991)	Teh and Morton (1993)
Gottesman et al. (1994)	Tsai and Tang (1991)
Greszczuk (1982)	Xiong et al. (1995)

have been developed in order to use smaller specimens and to reduce costs while producing reliable data. Extensive testing brought about better insight into the compressive failure of impact-damaged laminates and the influence of material properties and factors such as stitching. Significant efforts to develop mathematical models capable of predicting the onset and propagation of delaminations also brought about a much better understanding to this topic.

6.2.1 Test Procedures

Compression is critical for impact-damaged specimens because under this type of loading, strength reductions are the largest. Major experimental efforts have been directed toward understanding the behavior of impact-damaged specimens under compressive loading. The first objective is to develop test procedures to measure the Compression After Impact (CAI) strength and study the failure mechanisms involved. Procedures must specify both how the impact test is to be performed and how the compression test is to be conducted. Test methods

Table 6.2. *Summary of common CAI test methods*

	NASA	Boeing	Pritchard Hogg	CRAG
Material				
Specimen thickness	6.35 mm	4 to 5 mm	2 mm	3 mm
Layup	(45,0,−45,90)	(−45,0,45,90)	$(-45,0,45,90)_{2S}$	(45,−45,0,90)
Impact				
Tup diameter	12.7 mm	15.75 mm	20 mm	10 mm
Mass	4.5 kg	4.6 to 6.8 kg	3.96 kg	As required
Drop height	28 J	As needed	As needed	1 m
Support	127 mm square	127 × 76 mm	40 mm diameter	140 mm diameter
	Clamped	Clamped at 4 points	Clamped	Clamped
Compression tests				
Specimen size	$h = 254$–317 mm	$h = 152$ mm	$h = 53$ mm	$h = 180$ mm minimum
	$w = 178$ mm	$w = 102$ mm	$w = 45$ mm	$w = 50$ mm
Loading	End loading	End loading	End loading	End tabs
Loading rate	1.27 mm/min	0.5 mm/min	0.3 mm/min	Adjusted to achieve Failure in 30–90 s

described in NASA Reference Publication 1142 (1985) or Boeing Standard Specification BSS 7260 (1982) are commonly used to measure CAI strength. In the United Kingdom, the CRAG method is often used, and many more methods have been reported. A summary of some of these methods is provided in Table 6.2. Differences between the NASA and Boeing methods have to do with specimen size and the exact manner in which the impact tests are performed prior to compression testing. During CAI tests, the specimens were clamped along the top and bottom edges and supported along the two sides in a fixture similar to that shown in Fig. 6.1. The lateral support is designed to prevent overall buckling of the specimen.

CAI strength depends on the energy absorbed by the specimen, which can be strongly affected by the particular design of the test fixture. For the same impact energy level, damage is less extensive if the holding fixture is more flexible (Prandy et al. 1991) or if some of the other parameters affecting the impact dynamics are different. Therefore, the method for performing the impact test must be specified.

The cost of performing CAI tests is high because relatively large and thick coupons requiring large quantities of material, expensive machining, and high-

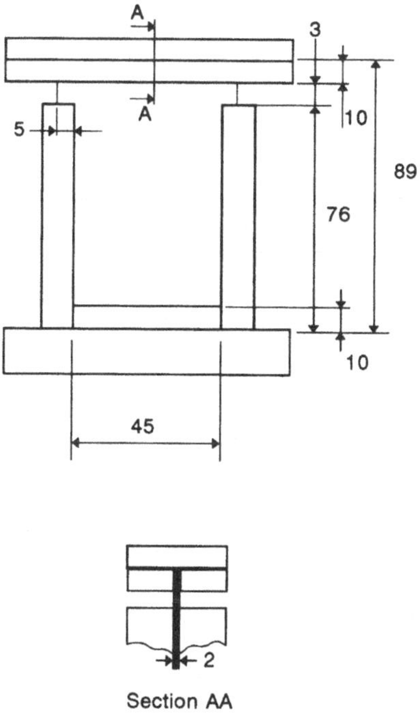

Figure 6.1. Pritchard and Hogg's fixture for compression-after-impact test (dimensions are in mm).

capacity test equipment are required. For example, if the undamaged strength of the material is 400 MPa, testing of a NASA-type specimen requires the application of a 452 kN force, and for a Boeing-size specimen, 163 to 204 kN are needed. New procedures using smaller specimens with fewer plies were proposed by several investigators. With the method proposed by Sjoblom and Hwang 1989, the 76.2 mm × 177.8 mm (3 in. × 7 in.) specimen is equipped with 1.59 mm (1/16 in.) thick glass-epoxy end tabs that leave a 3 in. × 4 in. test section (Fig. 6.2a). To prevent overall buckling (or macro-buckling) of these thin specimens (2 mm), 12.7 mm (1/2 in.) thick anti-buckling plates (Fig. 6.2b) placed on either side of the specimen are held in place by four bolts (Fig. 6.2c). These bolts are hand-tightened so as to allow the specimen to compress freely. A 31.75 mm (1.25 in.) diameter hole is made in the center of the anti-buckling plates because impact damage creates surface deformation and damage. The specimen equipped with the anti-buckling plates is then gripped in an MTS machine and tested. Similar fixtures were designed by Sarma Avva and Padmanabha (1986), Nettles and Hodge (1991), Breivik et al. (1992).

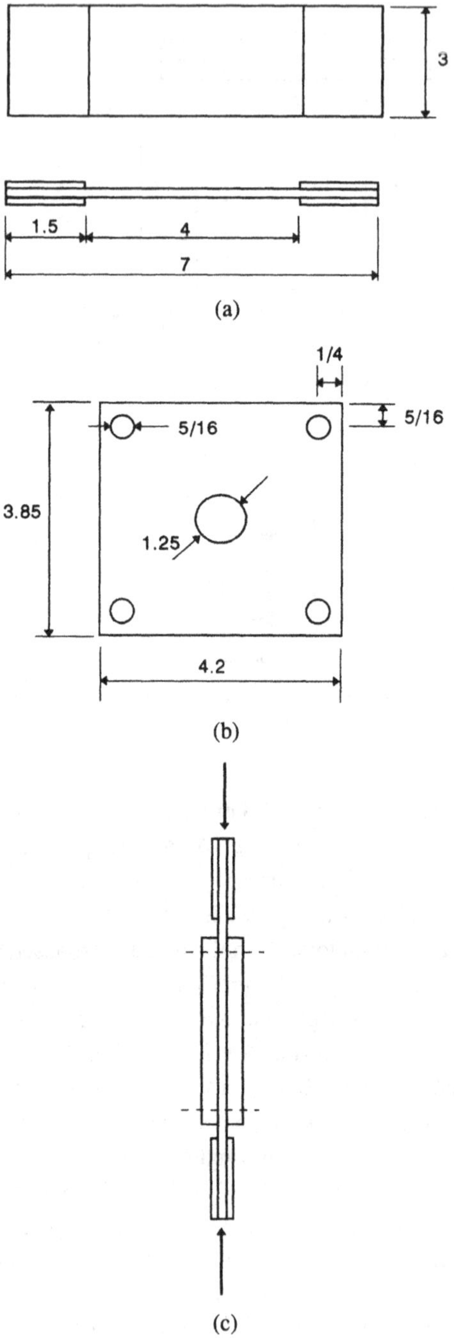

Figure 6.2. Sjoblom's fixture for compression-after-impact test: a) specimen; b) antibuckling guide; c) specimen with antibuckling guides.

Test results are used to screen materials, but caution is advocated when attempting to extrapolate to actual components (Sjoblom and Hwang 1989). CAI strength depends on the size of impact-induced delaminations. Smaller damage areas lead to smaller reductions in residual strength (Ghasemi Nejhad and Parvizi-Majidi 1990). In CAI tests, Srinivasan et al. (1992) observed localized buckling of the sublaminates formed by impact damage, delamination growth, and final failure by buckling.

6.2.2 Experimental Results

Extensive experimental investigations have brought out general trends concerning the effect of impact damage on the compressive strength of laminated composites. In this section, selected results are presented to illustrate the effects of impact energy, projected delamination area, dent depth, fiber and matrix properties, and stitching on the residual compressive strength are also examined.

Cantwell et al. (1986) conducted an experimental study to determine the effect of fiber properties on the residual tensile and compressive strengths of laminated composites. This example shows that the compressive strength is reduced more than the tensile strength and that these reductions can be substantial. The use of higher strain to failure leads to higher undamaged strength, which translates into a higher damaged strength. Material system A consists of a 3501-6 epoxy resin and high strain AS4 carbon fibers with a mean tensile strength of 3.59 GPa, an elastic modulus of 235 GPa, a 1.53% strain to failure, and a 7.0 μm diameter. Material system B consists of a similar but not completely identical matrix material and carbon fibers with a mean tensile strength of 2.70 GPa, an elastic modulus of 235 GPa, a 1.14% strain to failure and a 7.7 μm diameter. With the first material systems, a $[(0_2,\pm45)_2]_S$ layup with a nominal ply thickness of 0.125 mm was used, while for the second material system, the layup used was $[0_2,\pm45,0_2,\pm45]_S$. Since the elastic modulus of the fiber is 235 GPa in both cases and since the matrix material and the layups are nearly identical, the bending rigidities of the two laminates will be very close. Therefore the impact dynamics will not be affected, and the only difference will be the strain to failure of the fiber.

For the range of impact velocities considered, the damage area obtained from C-scans was always smaller for the laminate with the high-strain-to-failure carbon fibers (Fig. 6.3). The compressive strength experiences a much more drastic reduction than the tensile strength for a given impact energy level (Fig. 6.4). The residual strengths in both tension and compression were always significantly higher for material system A (Fig. 6.4), as pointed out by Cantwell et al. (1986), who mentioned that for a 6 J impact the residual tensile strength of

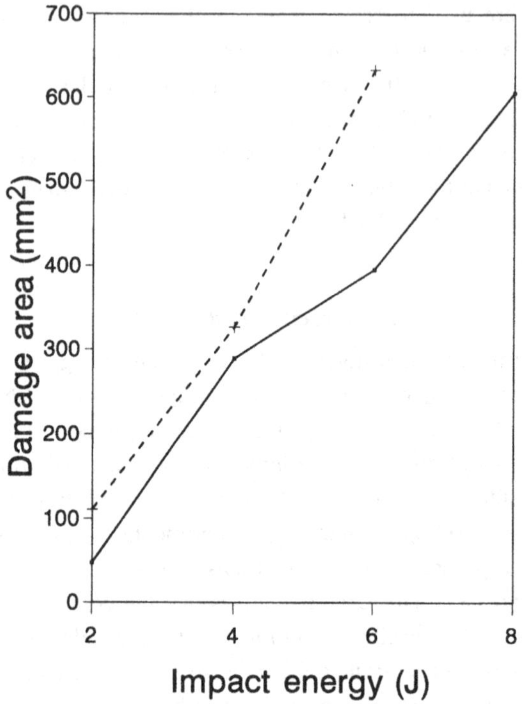

Figure 6.3. Damage versus impact energy with two types of reinforcing fibers (Cantwell et al. 1986). Solid line: material system A; dashed line: material system B.

specimen A was 100% greater than that of specimen B. However, most of the difference comes from the fact that the undamaged strength of material A is significantly larger than that of system B. The undamaged tensile strength of system A is 38% higher than that of system B, which is directly related to the 39% increase in strain to failure of the fibers. In compression, however, the undamaged strength of system A is only 18% higher than that of system B. Therefore, plotting the retention factor (that is, the ratio of the damaged to undamaged strengths) versus the impact energy (Fig. 6.5) gives a better indication of the impact tolerance of the two systems. Figure 6.5 shows that impact damage has a more drastic effect on the compressive strength of the laminate than on its tensile strength. However, the difference in strength retention factors between the two material systems is relatively modest. The damage area is larger for system B, so the use of the high-strain carbon fiber improves damage resistance was also improved. The significant increase in residual strength resulting from the use of the high-strain-to-failure fibers is due mainly to the increase in undamaged strength and to a smaller extent to a better resistance

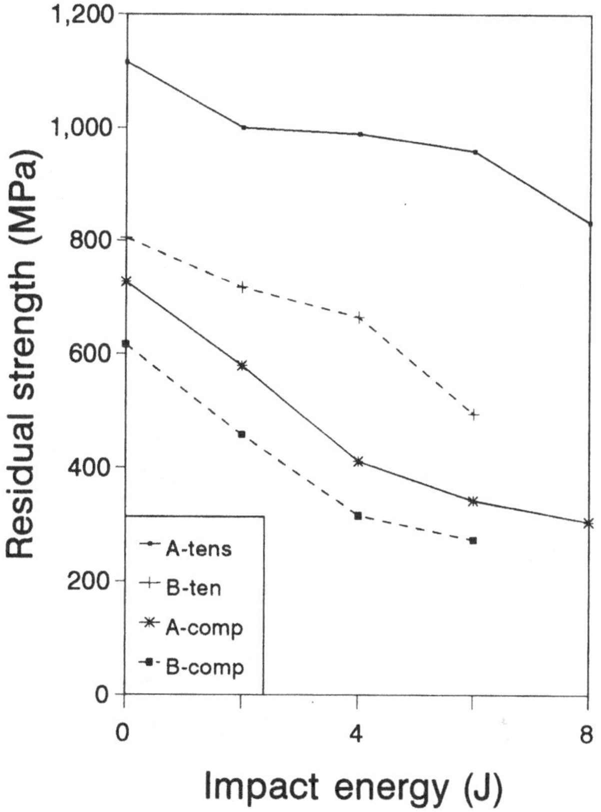

Figure 6.4. Residual tensile and compressive strengths versus impact energy with two types of reinforcing fibers (Cantwell et al. 1986).

to impact damage. These experiments do not provide evidence of improved impact damage tolerance.

Pritchard and Hogg (1990) compared the residual compression strength of carbon fiber–reinforced composites with toughened epoxy and PEEK matrices. This study was designed to examine the influence of matrix properties. The two material systems were Ciba-Geigy Fibredux 924C, a toughened epoxy with Torayca T800H (12 K) carbon fibers, and ICI APC-2 (PEEK with AS4 carbon fibers). Quasi-isotropic $[-45,0,45,90]_{2S}$ with a nominal thickness of 2 mm were clamped between two plates with a circular 40 mm opening and impacted by a 20-mm-diameter hemispherical tup weighing 3.96 kg. Post-impact compression tests were carried out on 89 × 55 mm specimens. The results shown in Fig. 6.6 indicate that the residual compressive strength of the thermoplastic composite is far superior to that of the thermoset composite used in this

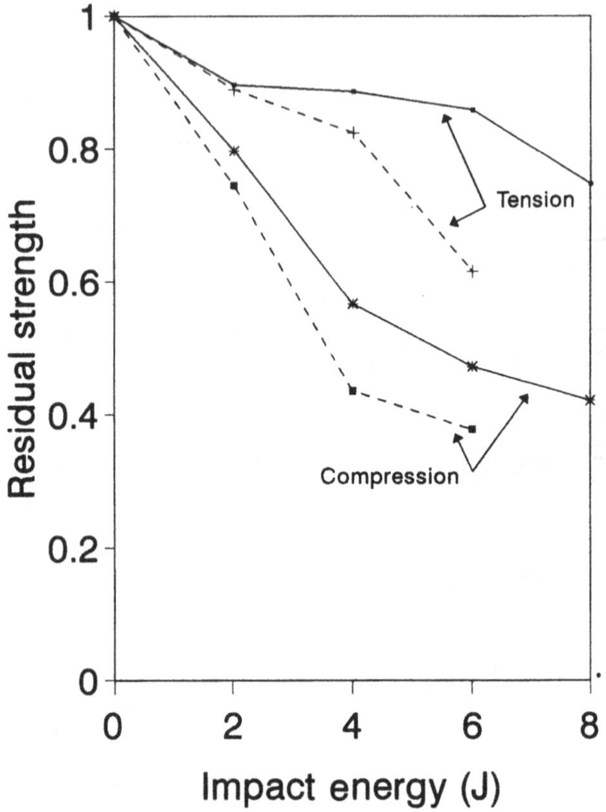

Figure 6.5. Tensile and compressive strength retention factors versus impact energy with two types of reinforcing fibers (Cantwell et al. 1986). Solid line: material system A; dashed line: material system B.

study. The width of the damage zone was also measured, and, as observed by other researchers, more damage is introduced into the thermoset composites under the same test conditions. Plotting the residual compressive strength versus the width of the damage area (Fig. 6.7) shows that for the same damage size, the residual strength of the thermoplastic composite is only slightly higher than that of the thermoset composite. This example highlights the points that thermoplastic composites are more impact damage–resistant and that, because less damage is introduced during impact, the residual strength will be higher.

Understanding the influence of matrix and fiber–matrix interface properties on the compressive strength after impact is of particular interest for the development of new material systems. Experimental results show a strong correlation

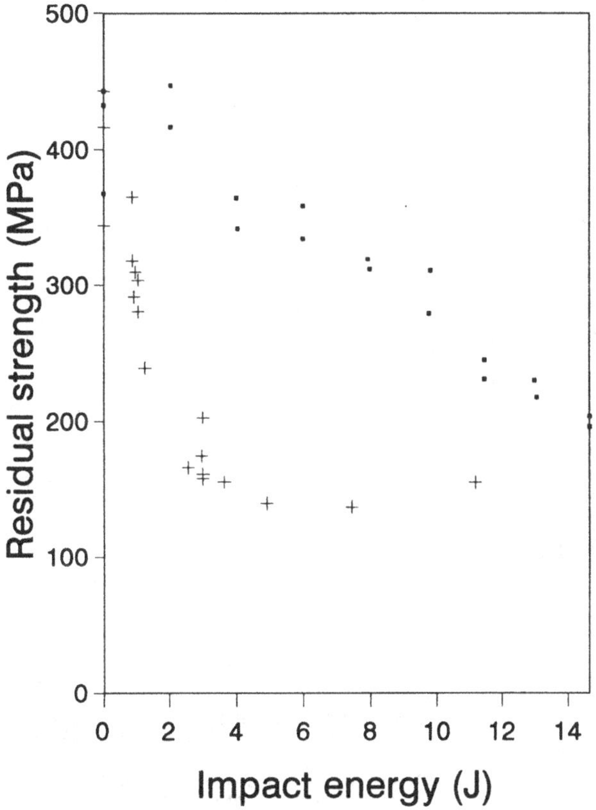

Figure 6.6. Compressive strength of thermoset and thermoplastics (Pritchard and Hogg 1990). Squares: thermoplastics; crosses: thermosets.

between the residual compressive strength of impact-damaged composites and the flexural strain to failure of the neat matrix resin (Evans and Masters 1987, Hirshbuehler 1987). This is consistent with the idea that impact damage is initiated by matrix cracks which then initiate interlaminar or delamination cracks. As discussed in Chapter 4, damage initiation is governed by the elastic properties of the matrix; if for the same impact less damage is introduced because of improved material properties of the matrix, the residual strength should be improved. For example, impact damage reduced the compressive strength of thin-walled graphite-epoxy struts by a maximum of 45–55% (Chen et al. 1993). With a toughened epoxy matrix, the maximum strength reduction was approximately 10%. In addition to the use of toughened matrices, interleaved systems can be used to improve the residual strength of impact-damaged laminates in compression and in bending (Pintado et al. 1991a,b). Fiber surface

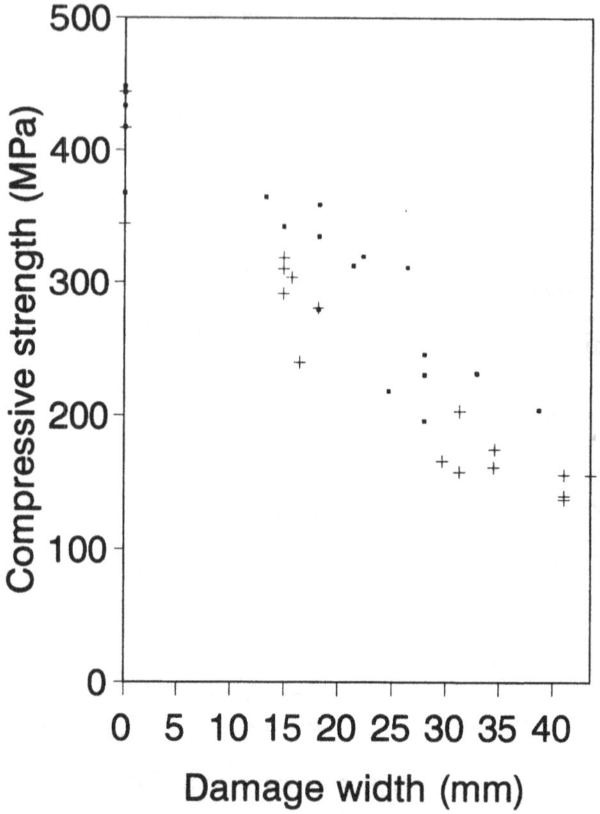

Figure 6.7. Compressive strength versus width of damage zone (Pritchard and Hogg 1990). Squares: thermoplastics; crosses: thermosets.

functionality, which promotes adhesion between fiber and matrix, is also a key requirement for damage tolerance (Manders and Harris 1986). Fiber tensile strength has relatively little influence on damage tolerance (Dempsey and Horton 1990).

Many experiments were performed in order to determine how the CAI strength of composites is affected by the stacking sequence, the use of fabric reinforcement, stitching, and multidimensional reinforcement. Dost et al. (1991) presented CAI test results for 24-ply graphite-epoxy laminates (Fig. 6.8). In general, for 24-ply quasi-isotropic layups, laminates with multiple-ply layers have lower compressive strength than layers with single plies. Larger changes in orientation angles from layer to layer tend to result in lower CAI strength. The same observations apply to the variation in damage size as a function of impact energy. Manders and Harris (1986) reported earlier that for the layups considered in their tests, variations in layups and orientations for quasi-isotropic laminates have no significant effects. This is in apparent contradiction to the results of Dost et al. (1991), but closer examination of their results indicates that there is no discrepancy because the stacking sequences considered do not show drastic differences in layer thicknesses or angle changes from layer to layer.

Delaminations occur because of the low resistance of laminated composite materials to transverse normal stresses and transverse shear stresses, particularly at the interfaces between plies with different orientations when localized

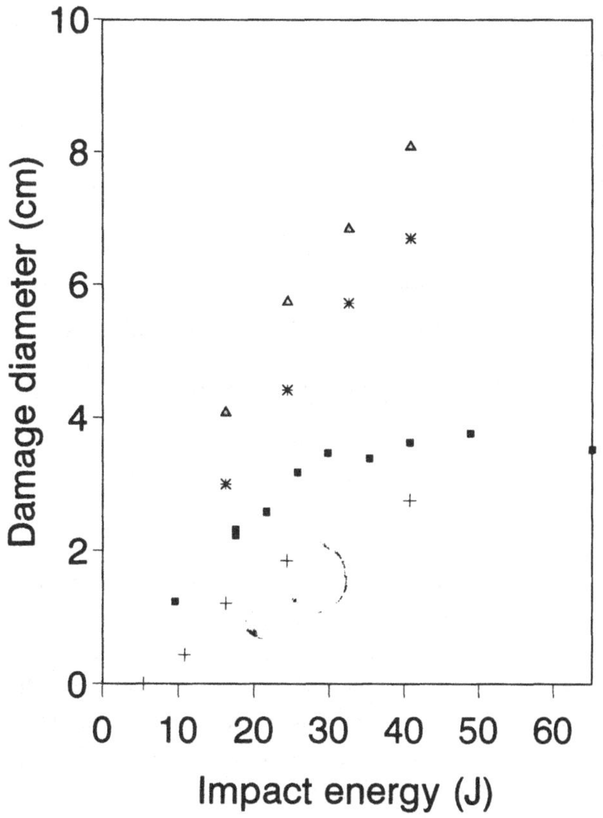

(a)

Figure 6.8. Influence of the stacking sequence on a) the damage size and b) the residual compressive strength of graphite-epoxy laminates (Dost et al. 1991). Crosses: $[45,90,-45,0]_{3S}$; asterisks: $[45_2,90_2,-45_2,0_2]_{2S}$; triangles: $[45_3,90_3,-45_3,0_3]_S$; squares: $[30,60,90,-60,-30,0]_{2S}$.

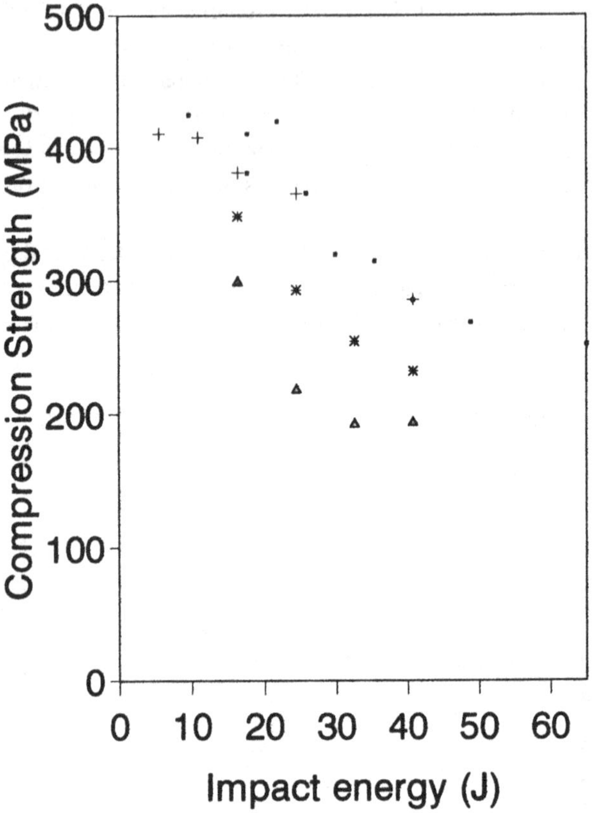

(b)

Figure 6.8. (*cont.*)

loads are applied. Reinforcements in the transverse direction introduced by stitching or the use of 2D or 3D composites can improve damage resistance and damage tolerance. Dry preform stitching improves the CAI strength, first by limiting the impact damage area by arresting delaminations, and then by preventing delamination growth during compression tests (Pelstring and Madan 1989, Dow and Smith 1989). Multidimensional laminates show improved impact resistance (Cantwell and Morton 1991, Chou et al. 1992, Gong and Sankar 1991). Improved damage tolerance compared to quasi-isotropic laminates was reported.

While most impact testing is performed at room temperature, composite materials are often used in areas in which they are exposed to either elevated temperatures or cold temperatures. Material properties of composite materials being temperature dependent, the residual strength is also expected to be affected by temperature. Bishop (1985) reported that the compressive failure stress of undamaged carbon fiber–reinforced composites with epoxy or PEEK matrices and $[\pm 45, 0_3, \pm 45, 0_2]_S$ layups decreases as the temperature increases from 20 to 120°C.

6.2.3 Prediction of Residual Compressive Strength

Models capable of predicting the residual strength of impact-damaged laminates in compression are needed in order to help interpret experimental results and to reduce the need for testing. Several models assume that impact damage provides the same stress concentration effect as either crack, a clean hole, or a softer inclusion of the same size. While not based on a rigorous stress analysis of the actual damage in the part, these models are quite useful for data reduction and design purposes. One such model is presented in this section.

Caprino (1983) developed a simple model based on linear elastic fracture mechanics. The critical stress in a wide sheet containing a notch of length $2L$ is

$$\sigma_c = K_{1c} \, (\pi L)^{\frac{1}{2}} \tag{6.1}$$

where K_{1c} is the fracture toughness of the material. With composite materials, a pseudo-plastic zone is formed at the crack tip and, to account for that, the $\frac{1}{2}$ exponent is replaced by m in (6.1). The strength of an unnotched material is written as

$$\sigma_o = K_{1c} \, (\pi L_o)^{-m} \tag{6.2}$$

where L_o is the size of an intrinsic flaw. Combining (6.1) and (6.2) gives

$$\sigma_c / \sigma_o = (L_o / L)^m \tag{6.3}$$

which can be used to predict the residual strength of a notched laminate. Equation (6.3) successfully fits experimental results for various laminates with notches or circular holes and loaded in tension (Caprino 1983). For the cases investigated, the best value for the exponent m was 0.31.

For composites with impact-induced damage, the damaged area is modeled as a notch of the same size (Caprino 1983, 1984). Considering that the size of the damage area is related to the kinetic energy of the projectile, the size of the equivalent notch is taken as

$$L = kU^n. \tag{6.4}$$

This assumption is reasonable since many investigators report that the projected damage area increases linearly with the initial kinetic energy of the projectile

Table 6.3. *Parameters α and C in least square curve fitting of experimental results by Dost et al. (1991)*

Laminate	Layup	α	C
A	$[45_2,90_2,-45_2,0_2]_{2S}$	.45004	1225.5
B	$[45_3,90_3,-45_3,0_3]_S$	.49227	1126.8
C	$[45,(90,-45)_3,(0,45)_2,0]_S$	.46962	1248.8
D	$[45,(0,-45)_3,(90,45)_2,0]_S$	.29649	644.6
E	$[45,90,-45,0]_{3S}$	.25923	779.9
F	$[30,60,90,-60,-30,0]_{2S}$	.31785	966.8
G	$[30,60,90,-30,-60,0]_{2S}$	.25541	685.6

(Table 4.2). Defining U_o, the maximum energy level the material can withstand without strength reduction, as

$$L_o = kU_o^n, \tag{6.5}$$

the ratio of the residual strength to the undamaged strength is given by

$$\sigma_r/\sigma_o = (U_o/U)^\alpha \tag{6.6}$$

where $\alpha = mn$. Since the undamaged strength is often unknown, it is more convenient to rewrite (6.6) as

$$\sigma_r = CU^{-\alpha} \tag{6.7}$$

and determine the two constants C and α using a least square fit of the experimental results. Dost et al. (1991) presented extensive experimental results on the compressive strength of carbon-epoxy laminates with the same material system but different stacking sequences. Experimental results show the effect of layup on damage size (Fig. 6.8a) and the reduction in compressive strength (Fig. 6.8b). Table 6.3 gives the values of C and α that provide the best fit of experimental results for the 7 layups tested. Equation (6.6) adequately fits the experimental results for residual tensile and compressive strength of impact damaged laminates. However, in this case the exponent α does not appear to remain constant but assumes different values for different stacking sequences.

6.3 Buckling

Impact-induced delaminations can significantly reduce the compressive strength of the structure. A number of investigators studied the stability of laminated plates with impact-induced delaminations (Table 6.4). Buckling and delamination growth are thought to be the first steps in the compressive failure process. The question is how much load the damaged structure can withstand. Because

Table 6.4. *References on stability of impact damaged laminates*

Analytical studies	Experimental studies
Adan et al. (1994)	Horban and Palazotto (1987)
Avery (1989)	Ilcewicz et al. (1989)
Bottega and Maewal (1983)	Jones et al. (1985)
Chai et al. (1981, 1983)	Kassapoglou and Abbott (1988)
Davidson (1989, 1991)	Mousley (1984)
Dost et al. (1988)	Palazotto et al. (1989)
Grady et al. (1989)	Romeo and Gaetani (1990)
Jones et al. (1985)	Seifert and Palazotto (1987)
Kapania and Wolfe (1987, 1989)	Zheng and Sun (1995)
Kardomateas (1989)	
Kardomateas and Schmueser (1987)	
Kassapoglou (1988)	
Kassapoglou and Abbott (1988)	
Moon and Kennedy (1994)	
Naganarayana and Atluri (1995)	
Palazotto et al. (1991)	
Peck and Springer (1991)	
Romeo and Gaetani (1990)	
Simitses et al. (1985)	
Tracy and Pardoen (1989)	
Vizzini and Lagace (1987)	
Wilder and Palazotto (1988)	
Williams et al. (1986)	
Yeh and Tan (1994)	
Yin (1985, 1986, 1987)	
Yin and Fei (1984, 1985)	
Yin and Wang (1984)	
Yin et al. (1986)	

of the presence of delaminations, the load-carrying capacity of the damaged structure will be lowered and, in addition, once buckling occurs, delaminations might extend and further decrease the load-carrying capacity of the structure. An understanding of the stability of a delaminated structure can be gained by considering an homogeneous, axially loaded beam-plate with a symmetrically located delamination (Fig. 6.9). Since the delamination extends across the entire width of the specimen, a one-dimensional analysis can be used. A particular case that has received considerable attention is that of thin delaminations. The delamination is located near the surface of the specimen and will buckle while the rest of the laminate remains straight (Fig. 6.10a) in what is called a local buckling mode, as opposed to global or mixed modes (Fig. 6.10). When a short delamination is located near the midplane of the laminate, its effect on the stability of the laminate is small, the laminate buckles as if undamaged, and in this

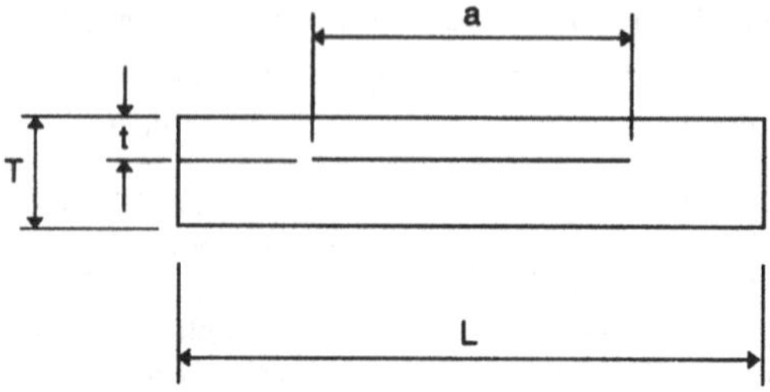

Figure 6.9. Delaminated beam-plate.

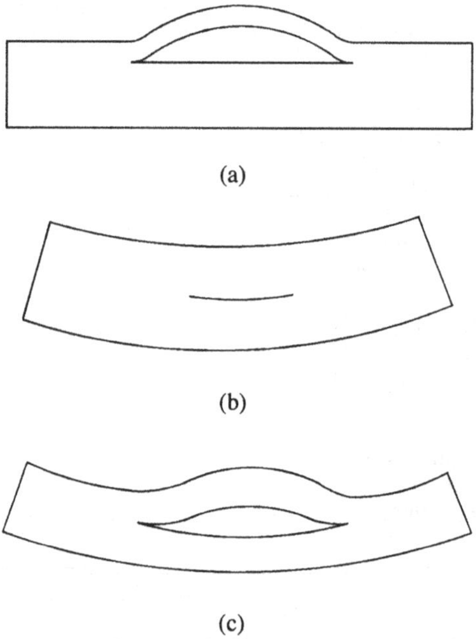

Figure 6.10. Local, global, and local-global buckling modes.

case we have a global buckling mode (Fig. 6.10b). For longer delaminations that are not located near the surface, buckling of the delamination reduces the overall rigidity of the laminate, and the remaining portion is no longer symmetric and will buckle at a much lower load in a different mode (Fig. 6.10c). This is what is called a mixed mode.

In this section, the analysis of local buckling modes for delaminated beams is presented first. Then, a more general model capable of predicting local, global, and mixed modes for delaminated beams is described. Finally, mathematical models used for analyzing the stability of delaminated composite plates are discussed.

6.3.1 Thin-Film Delaminations

A laminate with a through-the-width delamination is usually referred to as a beam-plate. In order to simplify the analysis, it is assumed that the laminate is symmetric and that both sublaminates created by the delamination are symmetrically laminated. In this case, extension-bending coupling due to material anisotropy is not present. When the delamination is located near the surface of the specimen, it can be assumed that as compressive stresses are applied, the delaminated region will buckle while the rest of the beam will remain straight. The objective is to determine the failure load of such a delaminated beam. It can be assumed that as the load is gradually increased, delamination buckling will be observed first, followed by a stable growth of the delamination, and finally unstable growth. The condition when the delamination begins to grow is generally taken as the maximum load-carrying capacity of the laminate. While some have studied dynamic delamination growth, this is a complicated problem needing more attention.

If the delaminated region is a symmetric laminate, there will be no extension-bending coupling, and the transverse displacement is governed by

$$\frac{d^4 w}{dx^4} + \frac{P'}{D}\frac{d^2 w}{dx^2} = 0 \tag{6.8}$$

where D is the bending rigidity of the delaminated region and P' is the axial force, which is taken to be positive in tension. The general solution to (6.8) is of the form

$$w = A + Bx + C\sin(\beta x) + D\cos(\beta x). \tag{6.9}$$

The buckling load is related to the parameter β by $P' = -\beta^2 D$, and the constants A–D are determined from the boundary conditions. Assuming that the displacements and the rotations are zero at both ends of the delaminated zone, buckling occurs when the axial force in the delaminated region reaches the critical value

$$P^* = -4\pi^2 \frac{D}{a^2} = -\frac{\pi^2}{3}\frac{Ebt^3}{a^2} \tag{6.10}$$

where a is the length of the delamination, b is the width, t is the thickness, and E is the modulus of elasticity. The compressive stress at this stage is

$$\sigma = \frac{\pi^2}{3}\frac{Et^2}{a^2}, \tag{6.11}$$

and the length of the specimen is shortened by

$$u = \frac{\sigma L}{E} = \frac{PL}{bTE} \tag{6.12}$$

where L and T are the total length and the total thickness of the specimen respectively. P^* is the compressive force acting on the entire laminate. The system compliance is defined as $\lambda = \frac{\partial u}{\partial P}$, and the energy release rate is given by

$$G = \frac{P^2}{2b}\left(\frac{\partial \lambda}{\partial a}\right). \tag{6.13}$$

From (6.12), it is clear that, prior to buckling, $G = 0$ regardless of the delamination length.

Subsequent to local buckling, the delaminated portion is assumed to withstand only the force P' and the end shortening is found to be

$$u_P = P\left[\frac{at + L(T-t)}{bTE(T-t)}\right] - \frac{\pi^2 t^3}{3a(T-t)} \tag{6.14}$$

where $P > P^*$. Substituting into (6.13) and using the nondimensional quantities

$$\alpha = a/t, \quad \beta = \frac{t}{T-t} \tag{6.15}$$

we find that, above the buckling load, the energy release rate is related to the size of the delamination by

$$G = \frac{\beta T \sigma^2}{2E}\left[1 + \left(\frac{\alpha^*}{\alpha}\right)^2\right] \tag{6.16}$$

where the value of α at which buckling occurs for a given applied stress σ is

$$\alpha^* = \left(\frac{\pi^2 E}{3\sigma}\right)^{\frac{1}{2}}. \tag{6.17}$$

Figure 6.11 shows that, for a given stress σ, this model predicts a zero energy release rate until the critical value of the delamination length is reached. At the critical length, the nondimensional energy release rate assumes its maximum value of 2, and as the delamination length increases, an asymptotic value of 1 is reached.

This analysis suggests three possible scenarios: (1) no delamination extension occurs if initially the energy release rate is lower than the critical energy release rate; (2) initially the delamination extends, but eventually the energy release becomes lower than the critical value and the growth stops; (3) continuous and perhaps unstable growth if G always exceeds the critical value.

6.3.2 Delaminated Beam-Plates

The preceding analysis (Section 6.3.1) deals with thin delaminations. In this section the more general case of a beam-plate with a single delamination is

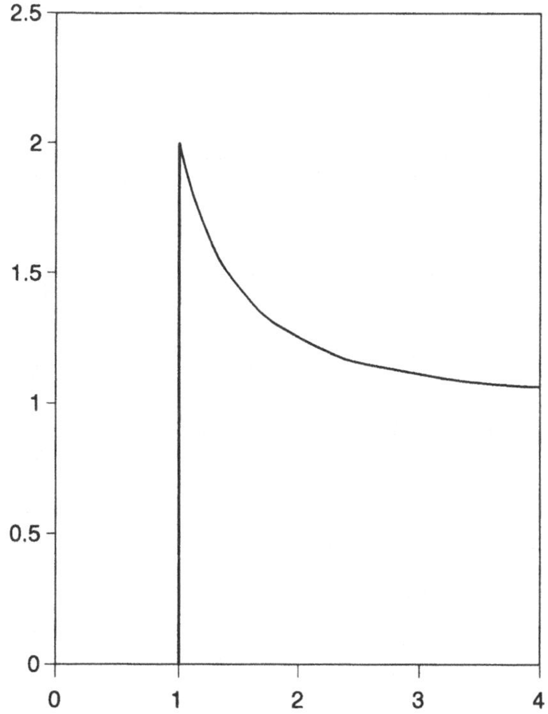

Figure 6.11. Nondimensional energy release $\frac{2EG}{\beta T\sigma^2}$ as a function of the nondimensional crack length α^*/α.

considered without restrictions on the size or location of the delamination through-the-thickness. Other modes of instability are observed, and limits of applicability of the simpler model can be determined.

Following Tracy and Pardoen (1989), the delaminated beam can be divided into four segments numbered 1–4 (Fig. 6.12a). The governing differential equations for each segment are

$$\frac{d^4 w_i}{dx^4} + \frac{P_i}{D_i}\frac{d^2 w_i}{dx^2} = 0 \tag{6.18}$$

and

$$\frac{dP_i}{dx} = -A_i \frac{d^2 u_i}{dx^2} = 0 \tag{6.19}$$

where $i = 1$–4, P is the axial force, D the bending rigidity, A the cross-sectional

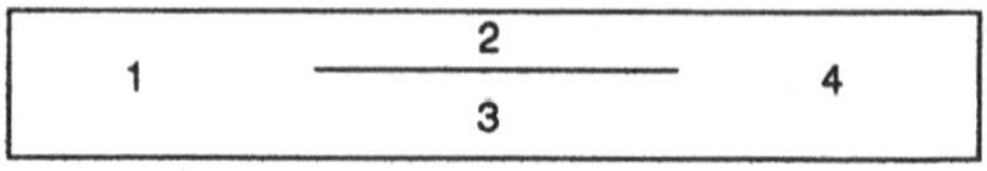

(a)

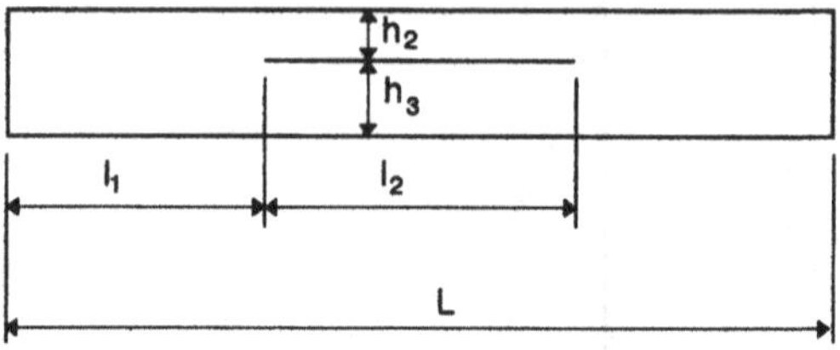

(b)

Figure 6.12. Beam-plate with central delamination.

area, and u and w are the axial and transverse displacements, respectively. The solutions to these differential equations are

$$\begin{aligned} w_i &= C_{i1}\cos(\alpha_i x) + C_{i2}\sin(\alpha_i x) + C_{i3}x + C_{i4} \\ u_i &= B_{i0} + B_{i1}x \end{aligned} \tag{6.20}$$

where $\alpha_i^2 = P_i/D_i$ and B_{ij} and C_{ij} are unknown constants. The boundary conditions for simply supported ends are

$$\begin{aligned} w_1(0) = 0, &\quad \frac{d^2 w_1(0)}{dx^2} = 0 \\ w_4(L) = 0, &\quad \frac{d^2 w_4(L)}{dx^2} = 0. \end{aligned} \tag{6.21}$$

Continuity conditions at $x = l_1 = x_1$ can be written as

$$\begin{aligned} &w_1(x_1) = w_2(x_1), \quad w_1(x_1) = w_3(x_1), \\ &\frac{dw_1(x_1)}{dx} = \frac{dw_2(x_1)}{dx}, \quad \frac{dw_1(x_1)}{dx} = \frac{dw_3(x_1)}{dx}, \\ &V_1(x_1) = V_2(x_1) + V_3(x_1), \\ &M_1(x_1) = M_2(x_1) + M_3(x_1) - P_3\left(\frac{t}{2} - \frac{h_3}{2}\right) + P_2\left(\frac{t}{2} - \frac{h_2}{2}\right) \\ &u_2(x_1) = u_1(x_1) - \left(\frac{t}{2} - \frac{h_2}{2}\right)\frac{dw_1(x_1)}{dx} \\ &u_3(x_1) = u_1(x_1) - \left(\frac{t}{2} - \frac{h_3}{2}\right)\frac{dw_1(x_1)}{dx}. \end{aligned} \tag{6.22}$$

While at $x = l_1 + l_2 = x_2$, the continuity conditions are

$$
\begin{aligned}
& w_2(x_2) = w_4(x_2), \quad w_3(x_2) = w_4(x_2), \\
& \frac{dw_2(x_2)}{dx} = \frac{dw_4(x_2)}{dx}, \quad \frac{dw_3(x_2)}{dx} = \frac{dw_4(x_2)}{dx}, \\
& V_4(x_2) = V_2(x_2) + V_3(x_2), \\
& M_4(x_2) = M_2(x_2) + M_3(x_2) - P_3\left(\frac{t}{2} - \frac{h_3}{2}\right) + P_2\left(\frac{t}{2} - \frac{h_2}{2}\right) \\
& u_2(x_2) = u_4(x) - \left(\frac{t}{2} - \frac{h_2}{2}\right)\frac{dw_4(x_2)}{dx} \\
& u_3(x_2) = u_4(x) + \left(\frac{t}{2} - \frac{h_3}{2}\right)\frac{dw_4(x_2)}{dx}.
\end{aligned}
\tag{6.23}
$$

Sallam and Yin (1985) gave the results for clamped-clamped and simply supported beams shown in Fig. 6.13 with the geometry defined in Fig. 6.9. Buckling loads are not affected by the presence of very small delaminations, but they drop rapidly to a very small value as damage size increases. The buckling load depends very strongly on the location of the delamination through-the-thickness of the laminate and the boundary conditions.

6.3.3 Buckling of Beam-Plates with Multiple Delaminations

Analyses of beams with a single delaminations are useful in that they provide insight into the behavior of composite structures with delaminations. Because of the relative simplicity of the model, most analyses follow an analytical approach. In practice, delaminations can be found at several interfaces through the thickness; in that case, analytical approaches are not practical. Finite element analyses usually are performed for beam-plates with multiple delaminations.

Consider a beam with two delaminations (Fig. 6.14a) of equal size, perfectly aligned with each other but arbitrarily located along the length of the beam and through the thickness. The two delaminations create three sublaminates in the damaged zone. This case can be modeled using beam elements with three degrees of freedom per node: axial displacement, transverse displacement, and rotation of the cross section. The beam is divided into 5 regions (Fig. 6.14b). Regions A and E are outside of the delaminated zone, and the delaminated zone is modeled as three beams in parallel (regions B, C, and D). Regions A–E are modeled as beams 1–3, 2–6, 4–8, 5–9, and 7–10, respectively (Fig. 6.14b). Each beam can be divided into as many elements as necessary to minimize discretization error. The delaminated zone consisting of regions B–D

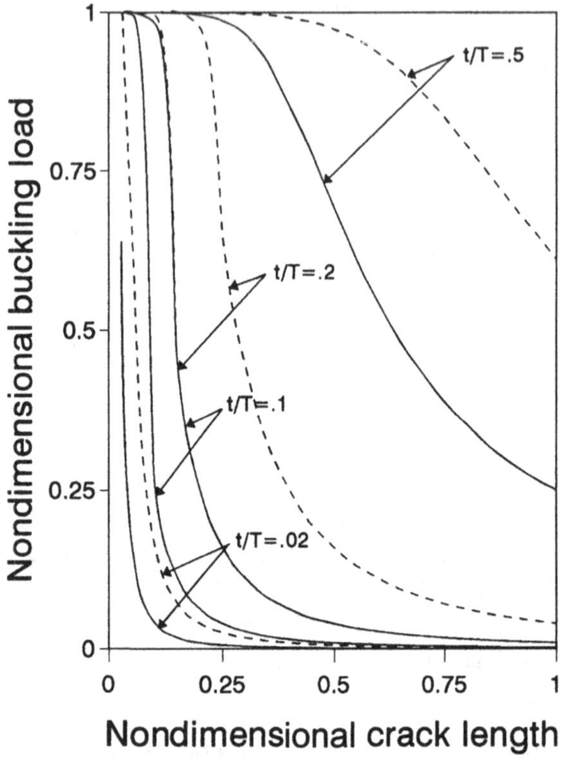

Figure 6.13. Nondimensional buckling load P/P_{perf} of delaminated beam-plates as a function of nondimensional delamination length a/L. Solid lines: clamped; dashed lines: simply supported.

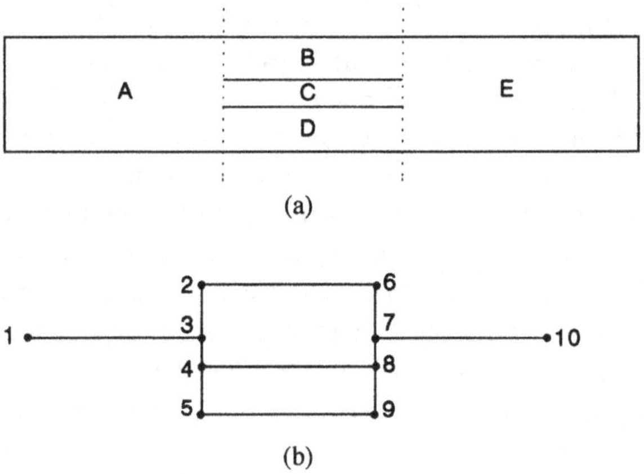

Figure 6.14. Beam with multiple delaminations.

is connected to the undamaged region A on the left and E on the right. Compatibility conditions can be written using the "rigid connector idea" (Kapania and Wolfe 1989), which means that at the junction between the undamaged region and the three sublaminates, all transverse displacements and all rotations are the same. The axial displacements can also be related knowing the rotation angle and the vertical distances between the respective neutral axes. This approach can be extended to deal with for any number of delaminations through-the-thickness. The finite element model can then be used to studying the buckling and post-buckling behavior of beams with multiple delaminations.

6.3.4 Buckling of Delaminated Plates

In practice, impact on composite plates introduces delaminations with approximately elliptical shapes. As a result, the stability of composite elliptical plates has been studied extensively, even though delaminations are sometimes modeled as rectangular orthotropic plates (e.g., Gottesman et al. 1994). This is the extension of the work on thin-film delamination where, assuming a near-surface delamination, the substrate is assumed to remain straight and a local delamination mode is produced. In that case, only the delaminated area needs to be considered in the analysis. As with delaminated beams, we determine the buckling load first and then the load required for delamination growth. To predict the CAI strength of a damaged laminate, one must check the stability and the compressive failure of each sublaminate. If the sublaminate buckles, it cannot sustain any additional load and stresses are redistributed. If a sublaminate fails, then its load carrying capacity is reduced. The process involved in the prediction of the CAI strength is complicated to implement, although conceptually simple.

The stability of a composite plate subjected to inplane loading can be studied using the Rayleigh-Ritz method. The strain energy in the plate is given by (3.194) for a generally laminated plate, or by (3.195) for a symmetrically laminated plate. The potential energy is given by

$$V = \frac{1}{2} \int_{\Omega} \left[N_x \left(\frac{\partial w}{\partial x} \right)^2 + N_y \left(\frac{\partial w}{\partial y} \right)^2 + 2N_{xy} \frac{\partial w}{\partial x} \cdot \frac{\partial w}{\partial y} \right] dx\, dy \qquad (6.24)$$

where N_x, N_y, and N_{xy} are the inplane forces applied on the delaminated plate. The total potential energy of the plate $\pi = U + V$ is to be minimized. For a symmetric laminate, inplane and transverse motion are uncoupled. Using the Rayleigh-Ritz method, the transverse displacements w are expressed as in (3.198) in terms of displacement approximation functions that must satisfy the essential boundary conditions of the problem. For a clamped elliptical plate, these displacement approximation functions can be taken as

$$\phi_i = x^\alpha y^\beta \left[\left(\frac{x}{a}\right)^2 + \left(\frac{y}{b}\right)^2 - 1 \right]^2. \tag{6.25}$$

The exponents α and β take the values 0, 1, 2, 3, ..., and the quantity inside the brackets vanishes on the boundary of the plate. Having this quantity squared in the approximation functions ensures that both ϕ_i and its derivative in the direction normal to the boundary vanish along the boundary.

The Rayleigh-Ritz method leads to the eigenvalue problem

$$([K] - \lambda[K_G])\, X = 0 \tag{6.26}$$

where X is the pseudo-displacement vector, K is the elastic stiffness matrix given by (3.203b), and K_G is the geometric stiffness matrix defined by

$$K_{ij}^G = N_x \int_\Omega \varphi_{i,x}\varphi_{j,x} dx\, dy + N_y \int_\Omega \varphi_{i,y}\varphi_{j,y}\, dx\, dy$$
$$+ N_{xy} \int_\Omega (\varphi_{i,x}\varphi_{j,y} + \varphi_{i,x}\varphi_{j,y})\, dx\, dy \tag{6.27}$$

and λ is the load factor. Solving this eigenvalue problem yields the buckling load of the sublaminate.

A three-term approximation was used by Davidson (1991). Since the first mode is expected to be symmetrical with respect to the x- and y-axes, the three approximation functions retained in the model were such that (α, β) = (0, 0), (2, 0), and (0, 2). Results showed good agreement with experimental values for small-size delaminations. For larger delaminations, the model overestimates the buckling load, and the discrepancy is caused by inability of the three-term displacement approximation to represent the lowest buckling mode adequately.

Avery (1989) also used the Rayleigh-Ritz method and the classical plate theory to study the stability of elliptical composite plates. However, it is assumed that the sublaminate being analyzed is not symmetric and, therefore, anisotropy causes some extension-bending coupling which is accounted for in the analysis. The transverse displacements and both inplane displacements are expressed in terms of series, with approximation functions vanishing along the edge of the plate. The strain energy of the plate is determined using (3.194), neglecting the last term dealing with shear deformation. Good agreement with experimental results is reported. Xiong et al. (1995) considered the same problem using a six-term polynomial expansion but the effect of extension-bending coupling was accounted for by using the reduced bending moduli approach. While approximate, this approach allows to consider only the transverse displacements in the analysis.

Peck and Springer (1991) investigated this problem using the Rayleigh-Ritz method also but with the higher-order shear deformation plate theory discussed in Chapter 3. Geometrical nonlinearities were included in order to investigate the post-buckling behavior. As the sublaminate deforms, contact forces may exist between the sublaminate and the balance of the plate. This contact is modeled by considering the sublaminate as a plate resting on a unilateral, linear elastic foundation. Thermal residual stresses affect the buckling load when there is mismatch between the sublaminate layup and the plate layup.

More realistic modeling of impact damage involves accounting for multiple delaminations with different sizes and orientations. Buckling for laminates with multiple elliptical delaminations through-the-thickness was studied by Davidson (1989) using a variational model. The finite element method lends itself well to detailed modeling of intricate details such as those encountered with impact damaged laminates. For example, Guedra-Degeorges et al. (1991) used a finite element model; based on experimental evidence showing that damage size increased from the front to the back of the laminate, the size of delaminations was taken to vary parabolically through-the-thickness.

Initial local buckling usually does not coincide with catastrophic failure. In order to predict the CAI strength of impact-damaged laminates, several authors (e.g., Ilcewicz et al. 1989, Xiong et al. 1995) consider stress redistributions around impact damage due to fiber failures and loss of stability. In the post-buckling regime a sublaminate is assumed to carry constant loads, which effectively corresponds to a loss of stiffness. Failure of the composite plate with a softer inclusion is predicted using a maximum strain criterion or other appropriate failure criteria such as those discussed in Section 6.4.2. Failure can also be defined as delamination growth under load. Peck and Springer (1991) investigated the onset of delamination growth using a fracture mechanics approach based on a detailed stress analysis of the post-buckling regime. The load required for delamination growth was calculated assuming that the total potential energy released exceeds the critical strain energy release rate of the material. The load–strain relations and the loads for the onset of delamination growth predicted by the model agreed reasonably well with experimental data. The loads required for delamination growth were significantly larger than the buckling loads for the examples presented.

To determine the strain energy release rate, delaminations often are assumed to grow in a self-similar mode, that is, the aspect ratio of the elliptical delamination is assumed constant. In fact, the strain energy release rate varies significantly along the delamination front, and a detailed finite element analysis is necessary to capture this effect (Naganarayana and Atluri 1995). An analysis of the simple double cantilever beam specimen shows that, even in that simple situation, the SERR distribution is not uniform along the delamination front (Zheng and Sun 1995). For such analyses, finite element models can be built by modeling each sublaminate in the delaminated region as a separate plate and connecting them to the undamaged portion of the plate (Naganarayana and Atluri 1995). Another approach is to consider the whole laminate as two plates, one above the delamination and one below, and to impose both constraints tying the top and bottom along the interface outside the delamination and contact constraints inside the delaminated region (Zheng and Sun 1995).

Buckling and growth of through-width delamination in thermoset and thermoplastic composites was analyzed by Gillepsie and Carlsson (1991). The onset of delamination growth was predicted using the mixed mode failure criterion

$$\frac{G_{\mathrm{I}}}{G_{\mathrm{Ic}}} + \frac{G_{\mathrm{II}}}{G_{\mathrm{IIc}}} = 1. \tag{6.28}$$

It was demonstrated that long and thin delaminations grow in mode II fracture, while short and thick delamination grow under mode I. This is consistent with the observations made by Hull and Shi (1993) concerning the experimentally found correlation between the CAI strength and sometimes G_{Ic} and at other times G_{IIc}. This result is interesting from a material development viewpoint, and indeed there are many articles announcing the development of new materials claiming improved damage tolerance based on higher values of the mode I or mode II critical strain energy rates. However, there are difficulties with this approach. As discussed in Chapter 5, it is difficult to determine G_{I} and G_{II}, the critical values of the strain energy rates are difficult to determine experimentally, and many failure criteria have been proposed besides the one given by (6.28) (Garg 1988).

6.4 Residual Tensile Strength

A number of publications deal with the effects of impact damage on the residual tensile strength of laminated composite materials (Table 6.5). Experimental studies show that fiber strength and strength to failure have a significant effect on the tensile strength of damaged as well as undamaged laminates, while the material properties of the matrix have a relatively insignificant effect. Models with different levels of sophistication have been presented to account for the effect of impact damage on the residual strength of the laminate.

Table 6.5. *References on residual tensile strength of impact-damaged laminated composites*

Adsit and Wazczak (1979)	Gandhe and Griffin (1989)
Avery et al. (1975, 1981)	Husman et al. (1975)
Avva et al. (1986)	Jenq et al. (1991, 1992a,b)
Awerbuch and Hahn (1976)	Lal (1983b, 1984)
Bishop (1985)	Llorente (1989)
Butcher (1979)	Llorente and Mar (1989)
Butcher and Fernback (1981)	Ma et al. (1991a)
Cairns and Lagace (1990, 1992)	Madaras et al. (1986)
Cantwell and Morton (1984, 1985b, 1989a,b)	Malvern et al. (1987)
Cantwell et al. (1983, 1984, 1986, 1991)	O'Kane and Benham (1986)
Caprino et al. (1983b)	Poe (1990, 1991a)
Clerico et al. (1989)	Poe et al. (1986)
Crivelli-Visconti et al. (1983)	Sankar and Sun (1986)
Curtis and Bishop (1984)	Sharma (1981)
Demuts et al. (1985)	Sun et al. (1993)
Dorey et al. (1985)	Takamatsu et al. (1986)
El-Zein and Reifsnider (1990a,b)	Verpoest et al. (1987)

Table 6.6. *Material data for the experimental study of Husman et al. (1975)*

		Ultimate tensile strength (ksi)	Tensile modulus (msi)	Ultimate strain (%)
Fiber				
Type I	HMS	300	55	0.545
Type II	HTS	400	38	1.05
	MODMOR II	380	40	0.95
Type III	AS	400	30	1.33
Matrix				
	ERL 2256	15.2	15.2	6.5
	ERLA	14.8	0.78	2.2
	Hercules 3004	10.2	0.36	50
	3M 1002	6.14	0.474	...
	PPQ 401	13.1	0.46	...

6.4.1 Experimental Results

Husman et al. (1975) studied the effect of impact damage on the tensile strength of cross-ply laminates to identify which material properties have significant effects on the residual tensile strength. Several combinations of fibers and matrix materials were used (Table 6.6). Figure 6.15 shows the effect of fiber

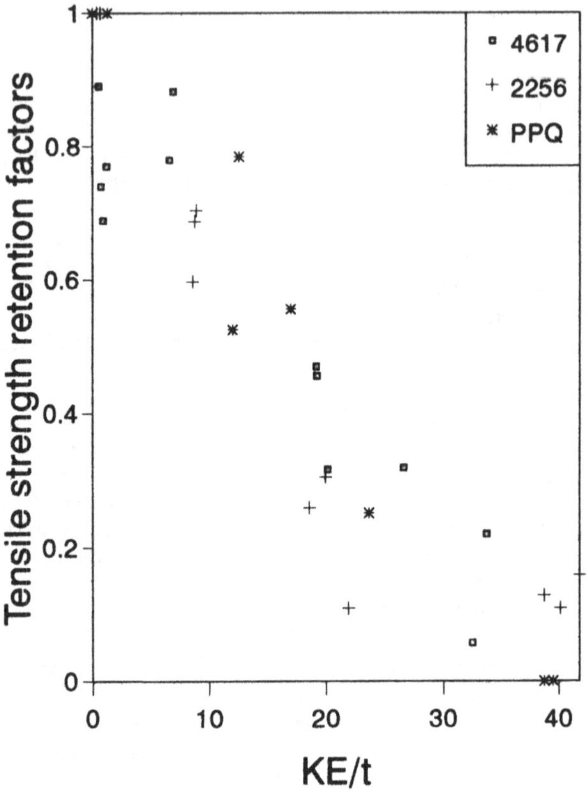

Figure 6.15. Effect of fiber properties on residual tensile strength (Husman et al. 1975).

properties on the residual tensile strength for laminates with the same matrix and three different types of fibers. As with the experiments of Cantwell et al. (1986), these results imply improvements in tensile strength using fibers with higher strength and higher strain to failure. Experiments on laminates with the same fiber reinforcement (MODMOR II) and several types of matrix materials (4617, 2256, PPQ) indicated that, considering the scatter in the experimental results, matrix properties have little effect on the residual strength of the laminate (Fig. 6.16). Because narrow specimens were tested (0.5 in.), the residual tensile strength essentially dropped to zero for high values of the impact energy.

The residual tensile strength of graphite-epoxy laminates with $[0,\pm45,90]_S$, $[0,\pm45,0]_{2S}$, and $[(0,90)_4]_S$ laminates was studied by El-Zein and Reifsnider (1990b). Initially, the projected damage area for the $[(0,90)_4]_S$ laminate is larger than for the other two laminates because of the larger angle between fiber orientations in adjacent plies (Fig. 6.17a). The undamaged strength of the

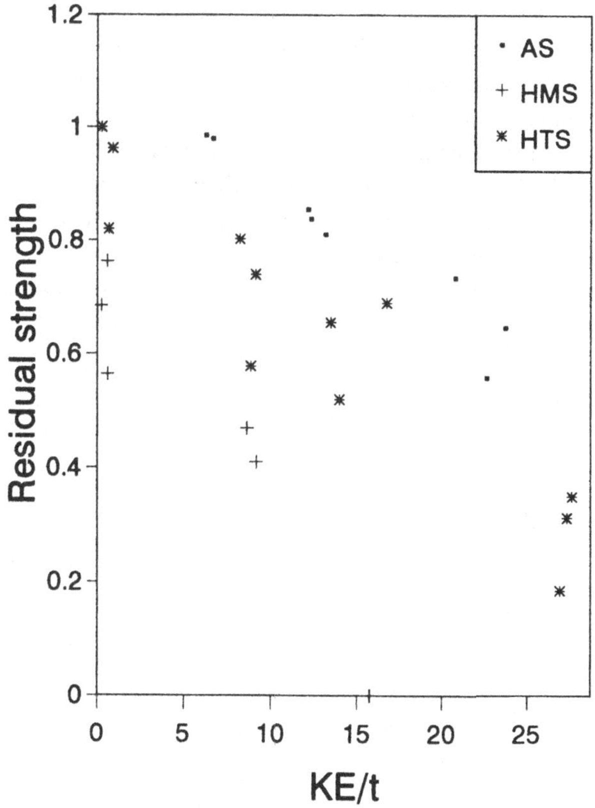

Figure 6.16. Effect of matrix properties on residual tensile strength (Husman et al. 1975).

$[0,\pm45,0]_{2S}$ laminate was 952 MPa, and that of the $[(0,90)_4]_S$ laminate was 876 MPa. Figure 6.17b indicates that low-energy impacts reduce the tensile strength because of fiber damage to the first 0° ply. Afterwards, the tensile strength is not significantly affected by the introduction of damage, which consists mostly of matrix cracks and delaminations, until the impact energy is large enough to create fiber damage. The specimens are 6 in. long and 3.5 in. wide, and Fig. 6.17b indicates that even when significant impact damage has been introduced, the laminate retains a significant fraction of its undamaged residual tensile strength.

6.4.2 Prediction of Residual Tensile Strength

Impact damage with delaminations, matrix cracks, and fiber failures is usually too complex to be modeled in details. Therefore, simplified models that capture the effect of impact damage on the residual tensile strength usually are adopted. The projected damage area obtained from ultrasonic inspection provides an overall measure of damage size. Several investigators have suggested that the behavior of impact damaged laminates is similar to that of a laminate with a clean hole or a crack of same size. The model first presented by Caprino (1983) and discussed in Section 6.2.2 has been used by many authors for the analysis of experimental results on the effect of impact damage on the residual strength of laminated composite materials.

The projected damage area determined from C-scans is an area in which interply delaminations, matrix cracks, and some fiber damage are present but the

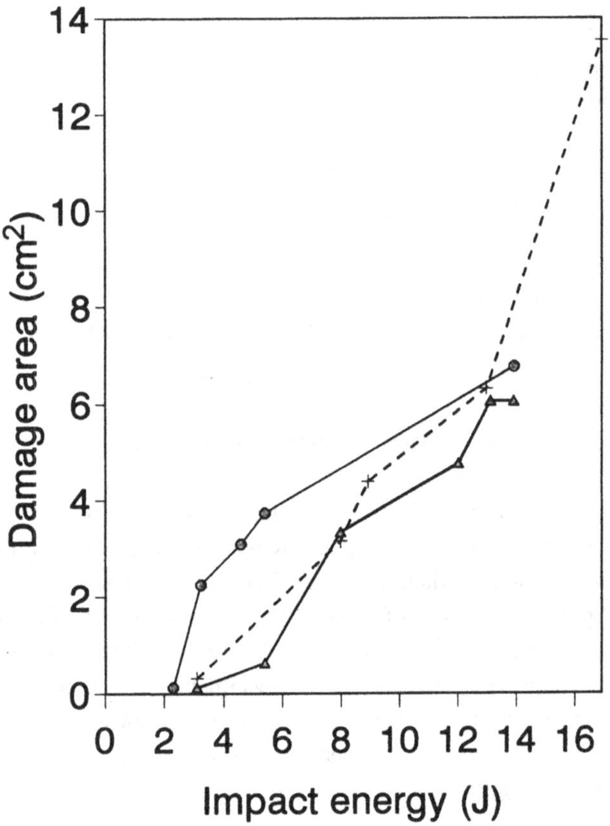

(a)

Figure 6.17. Impacts on graphite-epoxy laminates (El-Zein and Reifsnider 1990b). a) Damage area versus impact energy; b) residual tensile strength versus impact energy. Triangles: $[0,\pm45,90]_S$; crosses: $[0,\pm45,0]_{2S}$; circles: $[(0,90)_4]_S$.

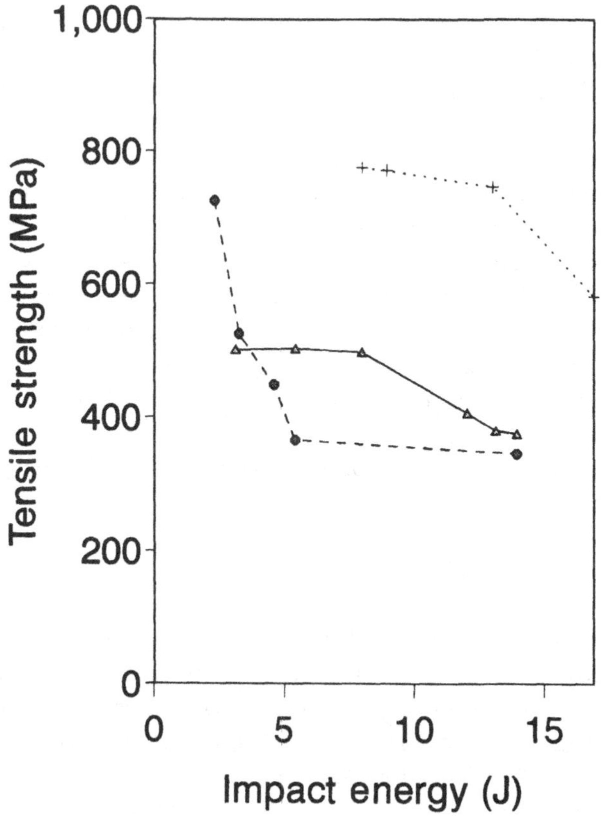

(b)

Figure 6.17. *(cont.)*

load-carrying capacity of the laminate in this area is reduced to a certain extent. It is then logical to model the damage area as a circular hole, in order to develop a model for predicting the residual strength. For an infinite plate with a circular hole of radius R subjected to a stress applied at infinity in the y-direction, the normal stress along the x-axis in front of the hole is approximated by

$$\sigma_y(x,0) = \frac{\sigma}{2}\left\{2 + \left(\frac{R}{x}\right)^2 + 3\left(\frac{R}{x}\right)^4 - (K_T^{\infty} - 3)\left[5\left(\frac{R}{x}\right)^6 - 7\left(\frac{R}{x}\right)^8\right]\right\} \tag{6.29}$$

where the stress concentration factor is given by

$$K_T^{\infty} = 1 + \left[\frac{2}{A_{22}}\left((A_{11}A_{22})^{\frac{1}{2}} - A_{12} + \frac{A_{11}A_{22} - A_{12}^2}{2A_{66}}\right)\right]^{\frac{1}{2}} \tag{6.30}$$

where the A_{ij} are the inplane stiffnesses of the laminate. The presence of a circular hole introduces stress concentration as indicated by (6.29)–(6.30). However, it is also known that larger holes cause greater strength reductions than smaller holes. This hole size effect cannot be explained by (6.19)–(6.20). For the same value of the stress concentration factor, the perturbation from the uniform stress distribution ahead of the hole is much more concentrated near the boundary for a small hole than for a larger one. Statistical distribution of initial flaws which, when subjected to stress, will grow and coalesce to initiate fracture. Therefore, since a smaller volume of material is subjected to higher stress levels, the strength of a plate with a smaller hole should see a smaller reduction. Nuismer and Whitney (1975) proposed two failure criteria to account for that stress distribution in some manner. The average stress criterion assumes that failure occurs when the average value of the stress over some characteristic distance a_o reaches the unnotched tensile strength of the material. That is,

$$\frac{1}{a_o}\int_R^{R+a_o} \sigma_y(x,0)dx = \sigma_o. \tag{6.31}$$

The point stress criterion assumes that failure occurs when the stress at some distance d_o from the hole reaches the unnotched tensile strength of the material. Nuismer and Whitney (1975) showed that, for graphite-epoxy laminates, the characteristic distances remain constant with $a_o = 0.15$ in. and $d_o = 0.004$ in. Liu et al. (1993c) found that, for SMC composites with short glass fiber reinforcement and a polyester matrix, $a_o = 11.49$ mm and $d_o = 3.7948$ mm.

For an infinite anisotropic plate with a crack of length $2c$, the exact anisotropic elasticity solution for the normal stress in a panel with a crack of length $2c$ in an infinite anisotropic plate under uniform uniaxial tension is

$$\sigma_y(x,0) = \frac{\sigma x}{(x^2-c^2)^{\frac{1}{2}}} = \frac{K_1 x}{\langle \pi c(x^2-c^2)\rangle^{\frac{1}{2}}}. \tag{6.32}$$

With the point stress failure criterion, the notched strength is given by

$$\frac{\sigma_N^\infty}{\sigma_o} = \langle 1-\xi_3^2\rangle^{\frac{1}{2}} \tag{6.33}$$

where $\xi_3 = c/(c+d_o)$. With the average stress failure criterion, the notched strength is given by

$$\frac{\sigma_N^\infty}{\sigma_o} = \left(\frac{1-\xi_4}{1+\xi_4}\right)^{\frac{1}{2}} \tag{6.34}$$

where $\xi_4 = c/(c+a_o)$. For a finite-width specimen, approximate correction factors are

$$K_T/K_T^{\infty} = \frac{2 + (1 - 2R/W)^3}{3(1 - 2R/W)} \tag{6.35}$$

$$K_T/K_T^{\infty} = [(W/\pi c)\tan(\pi c/W)]^{\frac{1}{2}} \tag{6.36}$$

where W is the finite specimen width.

New approaches for addressing this problem have been presented. Cairns and Lagace (1990) and El-Zein and Reifsnider (1990) modeled damage as an elliptical inclusion in an anisotropic layer to predict residual tensile strength. A three-dimensional finite element analysis was presented by Tian and Swanson (1992) to predict the residual tensile strength of laminates after impact damage. The analysis accounted for the detailed description of fiber breakage in each ply, and delaminations at each interface. Mode I and II energy release rates were calculated using the crack closure method, and the failure criterion given by Eq. (6.28) was used to estimate the extent of delamination propagation. Fiber failure was modeled as a crack through the ply and failure was predicted if the fiber strain ahead of the crack tip reached the ultimate fiber strain at a distance d_o from the crack tip. The characteristic distance d_o is assumed to be a material property independent of the layup. Delaminations reduce the ability of interlaminar shear stresses to transfer load from plies with broken fibers to adjacent plies and lower the laminate residual strength.

6.5 Residual Flexural Strength

Even though composite materials are often used in applications where they are subjected to bending, the residual flexural strength has received relatively little attention (Table 6.7). Rotem (1988) studied the effect of flexural damage on the bending rigidity and flexural strength of laminated beams. $[0,90]_{2S}$ graphite-epoxy laminates and $[45,-45]_{2S}$ glass-epoxy laminates were tested. Specimen size was 60 × 19 mm with thicknesses of approximately 1 mm for graphite-epoxy and 2 mm for glass-epoxy specimens. Ductile specimens experienced significant reductions in both stiffness and strength, whereas for brittle specimens, no losses were observed until complete failure occurred. The retention of flexural stiffness and flexural strength of graphite-epoxy, glass-epoxy, and Kevlar-epoxy plates after impact was studied in details by Malvern et al. (1989). Three symmetric cross-ply laminates each with a total of 30 plies were used. The first laminate, with a $[0_5,90_5,90_5]$ layup, has two interfaces where delamination can develop. The 5-layer $[0_3,90_3,0_3,90_3,0_3]$ laminate has four such interfaces, and the 15-layer [0,90,0,90,0,90,0,90,0,90,0,90,0,90,0] layup has 14 interfaces. Both graphite-epoxy and Kevlar-epoxy laminates with

Table 6.7. *References on residual flexural strength of impact-damaged laminated composites*

Cantwell et al. (1991)
Johnson and Sun (1988)
Kim et al. (1993)
Ma et al. (1991a,b)
Peijs et al. (1990, 1993)
Pintado et al. (1991a,b)
Wang et al. (1991)

those layups were tested. The projected delaminated area increased linearly with the initial kinetic energy of the impactor (Fig. 6.18b), but the area is smaller for laminates with more interfaces. More scatter in the results was observed for graphite-epoxy laminates than for Kevlar-epoxy laminates. The strength retention factor for the three Kevlar-epoxy laminates fall on a single curve when plotted versus the projected delamination area (Fig. 6.19). This indicates that the strength reduction is directly related to the damage size measured from C-scans. The stiffness retention factor appears to decrease linearly with the projected delamination area (Fig. 6.20). The fact that the experimental results for the three layups follow the same curve again supports the use of models in which damage is modeled as a clean hole with the size of damage measured from C-scans.

6.6 Fatigue

Generally, the degradation of composite laminates under cyclic loading occurs by general growth of cracks through the matrix or at the fiber–matrix interface (Challenger 1986). With existing, fiber-dominated, carbon-epoxy matrix material, the fatigue SN curve is almost flat (Curtis et al. 1984, Demuts 1990), and it is usually considered that impact damage has more effect on static strength than fatigue life. Relatively few studies are concerned with the fatigue of impact-damaged composites or the effect of previous fatigue loading on the impact resistance (Table 6.8).

Impact damage grows inconsistently under cyclic loading; in some cases the residual tensile strength of the laminate increases slightly after fatigue loading, whereas for other cases the reverse is true (Curtis et al. 1884). Microcracks that appear around the damage area are thought to reduce stress concentrations and increase the residual strength (Ong et al. 1991, Swanson et al. 1993).

With graphite-thermoplastic laminates, the main failure mechanism is matrix cracking, so that fatigue-induced microcracking extends existing damage, and progressive reduction in compressive strength is observed (Ong et al. 1991). Griffin and Betcht (1991) compared the fatigue behavior of graphite-reinforced composites with thermoplastic (APC-2) and thermoset (BMI) matrices; they concluded that thermoplastic composites sustained less impact damage and that damage growth during fatigue was negligible while for thermosets damage grew during fatigue. Basehore (1987) showed that the endurance limit for graphite-epoxy laminates after being subjected to a 20 ft-lb low-velocity impact was of the order of 70% of the static stress. Lauder et al. (1993) compared

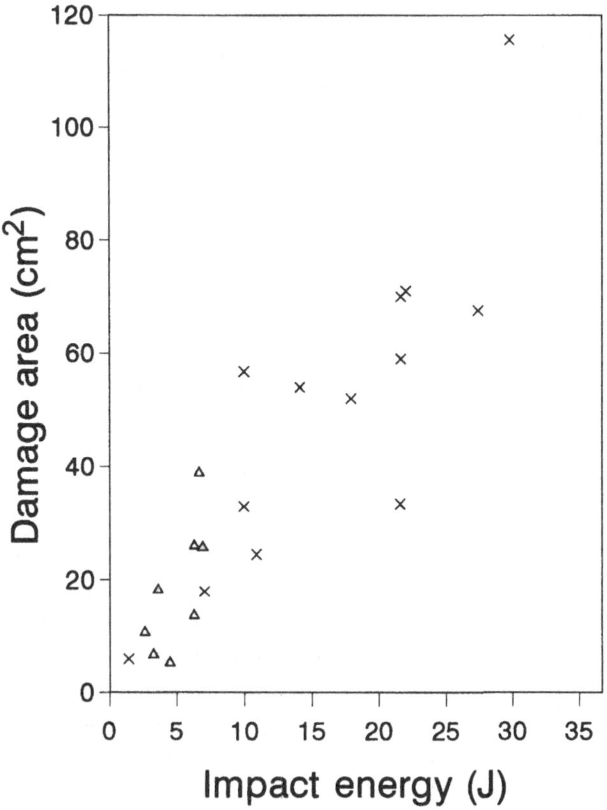

(a)

Figure 6.18. Delaminated area versus impact energy for impacts on graphite-epoxy cross-ply laminates with 3 and 5 layers (Malvern et al. 1989). a) Total delaminated area; b) projected delaminated area. Triangles: $[0_5, 90_5, 0_5]$ layup; crosses: $[0_3, 90_3, 0_3, 90_3, 0_3]$ layup.

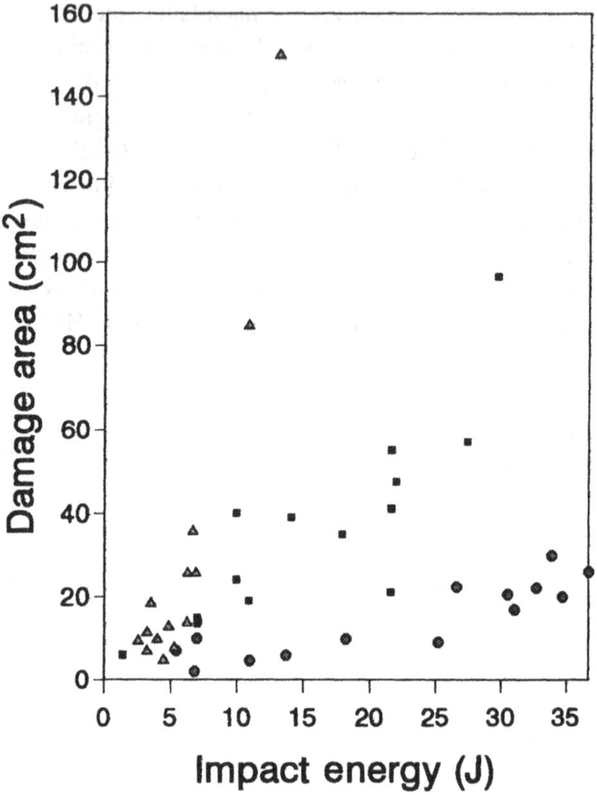

(b)

Figure 6.18. (*cont.*)

the effect of low-velocity impact damage and that of through holes on the fatigue life of glass-epoxy composites in four-point bending. Since impact damage consists mainly of delaminations, smaller increases in specimen compliance are observed compared to specimens with drilled holes of the same size. The endurance limit is reduced by one or two orders of magnitude, and as long as the same displacements are applied to the specimens, endurance is reduced to the same extent by impact damage and by drilled holes of the same size.

Experiments (Curtis et al. 1984) indicated that prior cyclic loading did not affect the impact damage resistance and residual strength of carbon-epoxy laminates. A clear distinction between delamination surfaces created during impact and those created during subsequent fatigue loading is often difficult to observe (Clark and Saunders 1991), adding to the difficulty in studying the fatigue behavior of impact-damaged specimens.

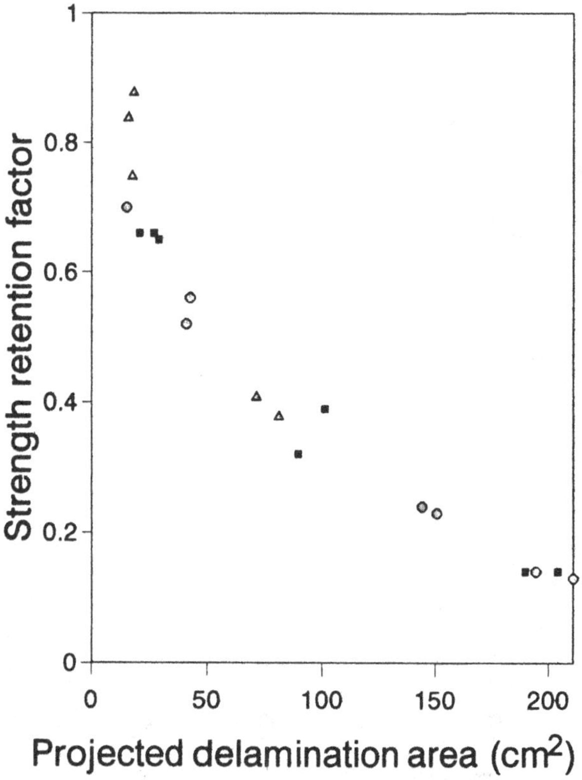

Figure 6.19. Strength retention factor as a function of the delaminated area (Malvern et al. 1989). Triangles: $[0_5,90_5,0_5]$ layup; crosses: $[0_3,90_3,0_3,90_3,0_3]$ layup; circles: [0,90,0,90,0,90,0,90,0,90,0,90,0,90,0] layup.

While it is difficult to generalize observations made from a limited number of studies, it appears that no drastic effect is to be expected when impact-damaged laminates are subjected to fatigue loading. Impact damage usually does not grow under subsequent fatigue loading, and prefatigued specimens do not experience any reduction in impact damage resistance.

6.7 Effect of Impact Damage on Structural Dynamics Behavior

Several investigators have studied the effect of damage on the natural frequencies and mode shapes of the structure both analytically and experimentally (Table 6.9). One type of studies determines how a given damage state affects the dynamics of the structure. In general, if the damage is small enough, changes in natural frequencies and mode shapes are negligible. Therefore, they

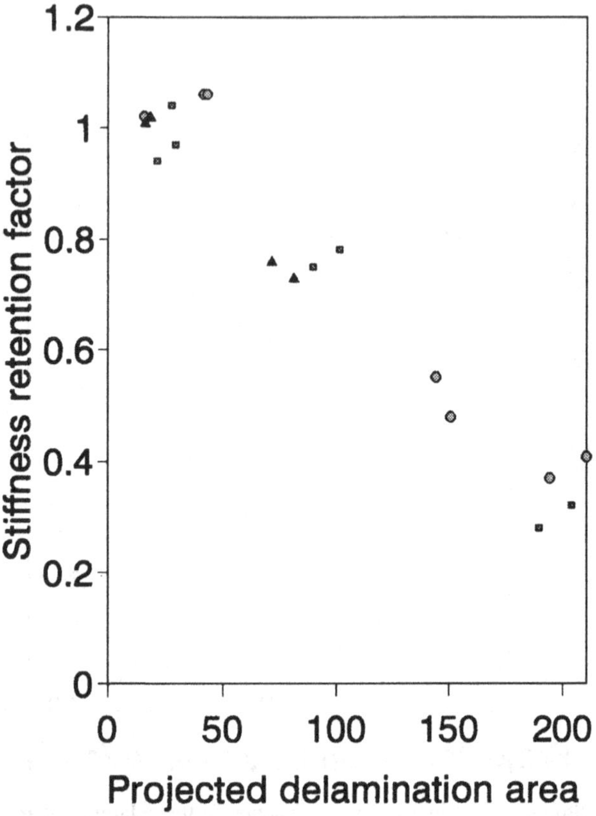

Figure 6.20. Stiffness retention factor decreases linearly with projected delamination area for Kevlar-epoxy laminates with 3, 5, and 15 layers (Malvern et al. 1989). Triangles: $[0_5,90_5,0_5]$ layup; crosses: $[0_3,90_3,0_3,90_3,0_3]$; circles: [0,90,0,90,0,90,0,90, 0,90,0,90,0,90,0] layup.

do not have to be included in the dynamic analysis of the impact. Another type of investigation is concerned with using measured changes in free vibration characteristics to detect the presence and determine the location of the damage.

Wang et al. (1982) presented an analytical approach for studying the vibrations of split beams: a model problem for understanding the effect of delaminations on the natural frequencies and mode shapes of laminated composite beams. Delaminations create two sublaminates that are modeled as beams subjected to bending and axial deformations. The nondimensionalized natural frequencies defined as

$$\Omega_i = \omega_i \left(\rho A L^4 / E I\right)^{\frac{1}{2}} \tag{6.37}$$

Table 6.8. *Articles on post-impact fatigue of laminated composite materials*

Avva et al. (1986)
Basehore (1987)
Cantwell et al. (1983, 1984)
Clark and Saunders (1991)
Curtis et al. (1984)
Davidson (1989)
Griffin and Becht (1991)
Krafchak et al. (1993)
Lauder et al. (1993)
Ma et al. (1996)
Moon and Kennedy (1994)
Ong et al. (1991)
Ramkumar (1983)
Spamer and Brink (1988)
Swanson et al. (1993)

Table 6.9. *References on free vibration of impact-damaged laminated structures*

Analytical studies	Experimental studies
Campanelli and Engblom (1995)	Balis-Crema et al. (1985)
Cawley and Adams (1979a,b)	Grady and Meyn (1989)
Chen H.P. (1991, 1993, 1994)	Peroni et al. (1989)
Chen H.P. et al. (1995)	Ramkumar et al. (1979)
Mujumbar and Suryanarayan (1988)	Tracy et al. (1985)
Ramkumar et al. (1979)	
Wang et al. (1982)	
Yin and Jane (1988)	

are calculated as a function of the nondimensional delamination length $S = a/L$ where a and L are defined in Fig. 6.9. For fixed-fixed beams with a delamination along the midplane, the effects of delamination length on the first three natural frequencies are shown in Fig. 6.21. When the delamination is small compared to the length of the beam, changes in natural frequencies remain small. The change in the first natural frequency is less than 1% for delaminations up to 30% of the total length of the beam. The effects of the delamination are more pronounced for the higher modes than for the fundamental mode. Because damage size is usually small, changes in natural frequencies and mode shapes caused by low-velocity impact damage are so small that they are difficult to

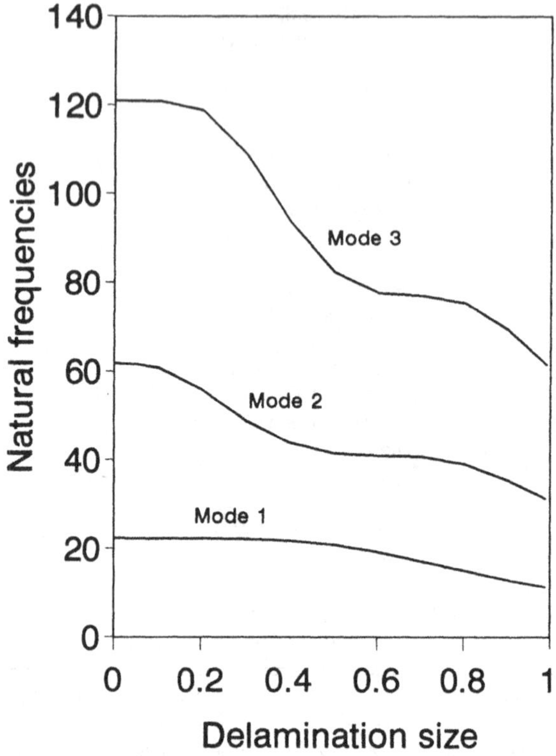

Figure 6.21. Effect of centrally located midplane delamination on natural frequencies (Wang et al. 1982).

detect. This result is important because it justifies not including the effect of damage in the model for impact dynamics analysis, as is commonly done.

The location of structural defects can be determined from shifts in natural frequencies induced by damage. Measured changes in natural frequencies can be compared with the predicted changes for various assumed damaged locations selected on a regular grid on the structure. Since a rather fine mesh is usually required to accurately study the free vibration, generally a sensitivity analysis method is used to predict the natural frequencies of the damaged structure, with the eigenvectors of the undamaged structure known. Therefore, a single eigenvalue problem needs to be solved. The estimated change in the natural frequency for a given mode is obtained as follows. The natural frequencies and mode shapes of the undamaged structure are solutions of an eigenvalue problem of the form

$$(K - \lambda M)x = 0 \tag{6.38}$$

where K is the stiffness matrix, M is the mass matrix, λ is the eigenvalue (frequency squared), and x is the eigenvector. A small change in the stiffness and mass matrices result in a corresponding change in the eigenvalues and eigenvectors and the preceding equation becomes

$$\{(K + \delta K) - (\lambda + \delta\lambda)(M + \delta M)\}(x + \delta x) = 0. \tag{6.39}$$

For low-velocity impacts, changes in stiffness results mainly from delaminations and fiber fracture without any change in mass. Neglecting second-order terms in the expansion of (6.39) yields

$$\delta K x - \delta\lambda\, M x + K\delta x - \lambda M \delta x = 0. \tag{6.40}$$

Recalling (6.40), the sum of the last two terms on the left-hand side of this equation vanish. Premultiplying by x^T, (6.42) gives

$$\delta\lambda = \frac{x^T \delta K x}{x^T M x}. \tag{6.41}$$

This expression provides an efficient way to estimate the change in natural frequencies caused by a change in the stiffness matrix once the free vibration behavior of the undamaged structure is known. With a finite element model of a composite plate, damage can be assumed to occur inside any one element. The stiffness of the damaged element is then set to zero. So, if K is the global stiffness matrix for the undamaged structure, $-\delta K$ is the contribution of the damaged element to the global stiffness matrix. Changes in natural frequencies for several modes have to be examined in order to pinpoint the correct damage location. This approach was successful in prediction different types of damage in composite plates (Cawley and Adams 1979a,b, Balis-Crema et al. 1985). The difficulty with this approach is that, for small-size localized damage, the frequency shifts are small, particularly if damage occurs in areas of small deformations for the undamaged structure in that particular mode.

6.8 Exercise Problems

6.1 The equation of equilibrium for an orthotropic plate subjected to initial inplane forces N_x^o, N_y^o, and N_{xy}^o is

$$\begin{aligned} &D_{11}w_{,xxxx} + 2(D_{12} + 2D_{66})w_{,xxyy} + D_{22}w_{,yyyy} \\ &= N_x^o w_{,xx} + N_y^o + w_{,yy} + 2N_{xy}^o w_{,xy}. \end{aligned}$$

(a) Show that for a simply supported plate, functions of the form

$$W_{mn} = \sin\left(\frac{m\pi x}{a}\right) \sin\left(\frac{n\pi y}{b}\right)$$

satisfy both the equation of equilibrium and the boundary conditions when $N^o_{xy} = 0$. A and b are the lengths of the plate in the x- and y-directions, respectively.

(b) Find an expression for the buckling loads of such a plate under biaxial loading.

6.2 Derive an approximate solution for the first buckling load of a simply supported, orthotropic, rectangular plate using a one-term polynomial approximation with the Rayleigh-Ritz method.

6.3 Find a set of polynomial approximation functions for analyzing the stability of fully clamped rectangular shapes using the Rayleigh-Ritz method.

6.4 Show that the approximation functions

$$\phi_{ij} = x_i y_j \left[\left(\frac{x}{a}\right)^2 + \left(\frac{y}{b}\right)^2 \right]^2$$

satisfy the two essential boundary conditions of zero displacement and zero normal slope along the boundary of an ellipse with semi-axes of length a and b.

6.5 Use the approximation functions found in Exercise 6.3 to develop a variational approximation for studying the stability of symmetrically laminated rectangular plates using the Rayleigh-Ritz method.

6.6 Write a computer program to study the stability of delaminated beam-plates using the approach presented in Section 6.3.2.

6.7 Write a computer program for predicting the tensile strength of a laminated plate with a circular hole using either the average stress criterion or the point stress criterion (Section 6.4.2).

7
Ballistic Impact

7.1 Introduction

Several definitions of ballistic impacts are used in the literature. Impacts resulting in complete penetration of the laminate are often called *ballistic impacts*, whereas nonpenetrating impacts are called low-velocity impacts. Although nonpenetrating impacts were studied extensively, impact penetration in composite materials has received considerably less attention. A different classification consists of calling low-velocity impacts those for which stress wave propagation through the thickness of the specimen plays no significant role. As soon as the projectiles enter in contact with the target, a compressive wave, a shear wave, and Rayleigh waves propagate outward from the impact point. Compressive and shear waves reach the back face and reflect back. After many reflections through the thickness of the laminate, the plate motion is established. Impacts for which damage is introduced after plate motion is established are called low-velocity impacts.

A simple method can be used to evaluate a transition velocity beyond which stress wave effects dominate. A cylindrical zone immediately under the impactor undergoes a uniform strain accross each cross section as the wave progresses from the front to the back face. With this simplifying assumption, the problem is reduced to that of the impact of a rigid mass on a cylindrical rod. Then, the initial compressive strain on the impacted surface is given by

$$\epsilon = V/c \tag{7.1}$$

where V is the impact velocity and c is the speed of sound in the transverse direction. Typically, critical strains between .5 and 1.0% are used to calculate the transition velocity. For common epoxy matrix composites, the transition to a stress wave–dominated impact occurs at impact velocities between 10 and 20 m/s. Dropweight testers generally induce low-velocity impacts since a drop height of 5 m will produce an impact velocity of 9.9 m/s and most testers have a shorter drop height.

In studying ballistic impacts, it is important to measure the residual velocity of the projectile accurately. This is a difficult task because many small particles, fibers, and shear plugs are pushed out by the projectile during penetration. This material can trigger the speed-sensing device being used and yield erroneous values. Special sensors designed to provide accurate measurements of the projectile have been designed in light of those difficulties and will be discussed in this Section 7.2.

For a given projectile and a given target, it is important to know the minimum impact velocity that will result in complete penetration. That velocity is usually called the *ballistic limit*. Since several complex failure modes are involved in the penetration process and since some degree of variability is always present, the ballistic limit is often defined as the initial velocity of the projectile that will result in complete penetration for 50% of the specimens. As the initial velocity increases above the ballistic limit, the residual velocity of the projectile becomes of interest since it may pose a threat to the equipment or the occupants inside the structure. Considering the balance of energy reveals important features of ballistic impact, including the effects of laminate thickness, projectile size, shape, and initial velocity (Section 7.3).

Failure modes include shear plugging, tensile fiber failure, and delaminations (Section 7.4). Knowing the failure modes involved in a particular impact empirical approach can be developed to determine the energy required for penetration, which can then be used to estimate the ballistic limit (Section 7.5). For applications where weight is a critical factor, fiber-reinforced composites may be considered as the primary ballistic protection, as with, for example, high-mobility land vehicles or combat aircraft. For very high velocities (greater than 2 km/s), combinations of a hard ceramic outer layer and a composite material backing can be used for light armor application (Section 7.6). Ceramic-composite armor systems are starting to be introduced for aircraft applications and are replacing metallic armor in ground vehicles achieving substantial weight savings (Arndt and Coltman 1990). The hard ceramic layer is used to blunt the projectile, fracture, and distribute the contact load over a larger area. The composite backing contains the projectile and the fragments that may have formed during the impact. In Section 7.6, we examine the performance of ceramic-composite armors and the models used to predict their ballistic limit.

7.2 Experimental Techniques

With high-velocity impacts, it is very important to measure the incident and residual velocities of the projectile in order to calculate the energy absorbed during the penetration process. Measuring the residual velocity of the projectile using optical sensors is difficult, since spalled material, shear plugs, and small particles can move ahead of the projectile as it exits the other side of the laminate. Zee et al. (1991) developed a microvelocity sensor to measure the velocity of a projectile during the penetration process. The basic principle behind the sensor design consists of placing a small magnet on the 6.3-mm-diameter, 50-mm-long projectile. The magnet triggers a succession of small coils as it passes through. Detailed information about the slowing down or energy loss of a projectile during ballistic impact is essential in understanding the failure process. For all the cases considered, the velocity dropped rapidly due to the fracture of the material, and then a much slower rate was observed as the projectile was being slowed by frictional forces. Another approach to gain insight into the penetration process was proposed by Azzi et al. (1991). The force was measured as a flyer composite disk impacted an instrumented stationary steel bar. The force history is strongly affected by the shape of the end of the bar.

7.3 Ballistic Limit and Residual Velocity of the Projectile

For a given target-projectile combination, the ballistic limit is defined as the lowest initial velocity of the projectile that will result in complete penetration. At that impact velocity, the residual velocity of the projectile is zero. Recognizing that a certain amount of variability is always present, the ballistic limit is often defined as the velocity that will result in penetration of 50% of the samples when a large number of tests are performed. It is necessary to determine the ballistic limit and to predict the residual velocity of projectile when the initial velocity exceeds the ballistic limit.

A first observation made from tests resulting in complete penetration of the laminate is that the residual kinetic energy of the projectile varies linearly with its initial kinetic energy. Therefore, the energy required for penetration of the laminate is constant, and the conservation of energy can be written as

$$\frac{1}{2} M v_i^2 = U_p + \frac{1}{2} M v_r^2 \tag{7.2}$$

where U_p is the perforation energy and v_i and v_r are the incident and residual velocities respectively. This equation indicates that the energy required for perforating the laminate is independent of the projectile velocity and that the penetration process is rate-independent. Plots of residual versus initial kinetic energy are straight lines with unit slopes (Fig. 7.1).

In (7.2), the kinetic energy of the material in the path of the projectile that is ejected during penetration is neglected. This assumption is reasonable near the ballistic limit since the residual velocity is small and the velocity of those

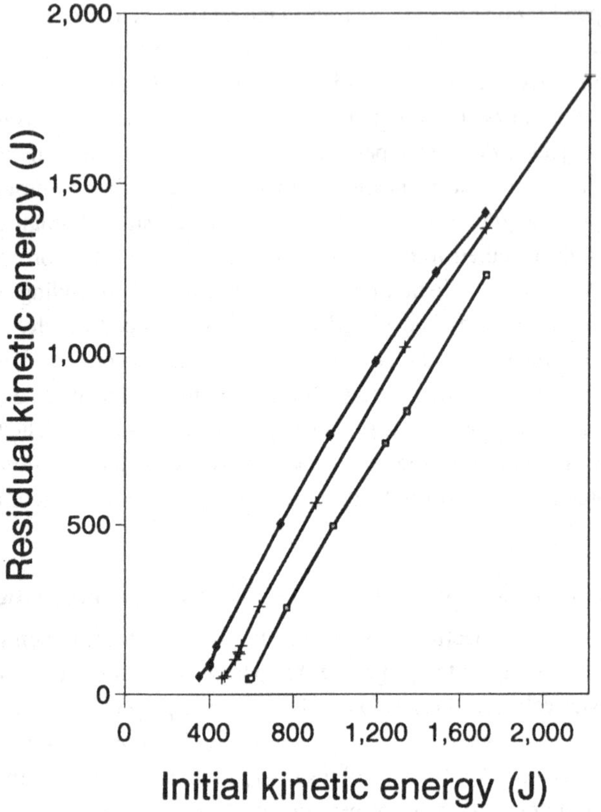

Figure 7.1. Residual versus initial kinetic energy of the projectile (Lin and Bhatnagar 1991).

small particles is also expected to be small. As the impact velocity becomes significantly larger than the ballistic limit, the material removed by the projectile during penetration is ejected at increasingly higher velocities. The kinetic energy stored in the ejecta is no longer negligible, and it effectively increases the apparent mass of the projectile. For this reason, the slope of E_r versus E_i is lower than that for high-impact velocities. Lin and Bhatnagar (1991) conducted an extensive experimental study on ballistic impacts on composite with polyethylene fabric and vinylester matrix and showed that the slope of the line representing the variation of the residual versus initial kinetic energy of the projectile remains between .98 and 1.08.

Knowing the general trend given by (7.2), the effect of laminate thickness, and projectile size and shape, material properties need to be examined. Several studies show that increasing the thickness of the laminate increases the ballistic

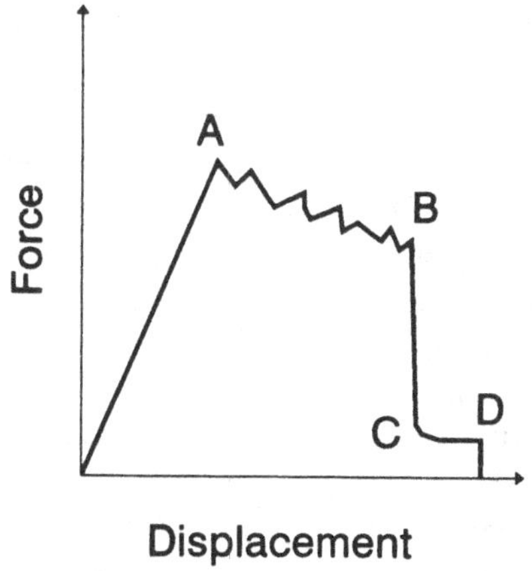

Figure 7.2. Load versus indentor displacement in static tests (Zhu et al. 1992).

limit (e.g., Vasudev and Mehlman 1987, Hsieh et al. 1990, Zhu et al. 1992a, Goldsmith et al. 1995). The ballistic limit often is said to be proportional to the aerial density of the composite (e.g., Segal 1991). This statement can be misleading because, besides the density, the mechanical properties of the material are also very important. In fact, what is meant is that with everything else remaining constant, the ballistic limit increases with the thickness of the laminate. Perforation energy increases with projectile diameter. Energy loss during penetration is strongly affected by projectile nose shape (Bless et al. 1990) because failure modes can be significantly different.

Zhu et al. (1992a,b) studied the quasi-static and high-velocity impact perforation of Kevlar-polyester laminates by conical projectiles. Plotting the load applied to the projectile during quasi-static tests versus its displacement (Fig. 7.2) shows three distinct phases: 1) the load increases due to both indentation and global deflection of the plate, 2) a plateau is reached corresponding to fiber failure, and 3) as penetration is completed, the load suddenly drops to a much lower level as friction against the side of the hole provided the only resistance to the motion. Similar results were found for quasi-static penetration of woven graphite-epoxy laminates by conical indentors (Goldsmith 1995). During dynamic tests, global deformation of the target was much smaller, and the size of delaminations was also much smaller. Local deformation (bulging) and fiber failure constituted the major energy absorption mechanism.

7.4 Failure Modes

Failure during laminate penetration depends in a large part on the shape of the projectile, which strongly affects the perforation energy. In general, shear plugging occurs near the impacted side, followed by a region in which failure occurs by tensile fiber fracture, and near the exit, delaminations occur. Depending on target rigidity, impactor mass, and velocity, only a shear failure mode is observed. In other cases, the first two modes are present, and for thicker targets all three modes are present. The penetration energy can be estimated by adding the energies required to produce each type of failure involved in the penetration process: fiber failure, matrix cracking, delamination, and friction between the projectile and the target. Good agreement with experiments is often obtained when failure modes are known from experiments. Which failure modes occur in a particular case depend on the laminate thickness and impact velocity and cannot yet be predicted.

Zee and Hsieh (1993) designed experiments to determine how the energy lost during penetration of a laminate is partitioned into energy of penetration, and frictional energy. The contribution of fiber failure, matrix cracking, delaminations, and friction between the projectile and the laminate are evaluated for several material systems. Delaminations were a major factor for graphite-epoxy laminates but had only a minimum effect for laminates with Kevlar and polyethylene fibers. This was explained by the fact that G_{IIc} was .15 J/cm^2 for graphite-epoxy, .09 J/cm^2 for Kevlar, and .014 J/cm^2 for PE. Energy losses by friction accounted from 10 to 20% of the total loss.

Sykes and Stoakley (1980) showed that with graphite-epoxy laminates, cure conditions had a significant effect on the force–displacement relations during perforation. Initial fracture occurred at the same force level, but for specimens cured at 450°K, the force decreased linearly down to zero, and for a specimen cured at 394°K, the force continued to increase to about twice the value required for damage initiation and then decreased. The penetration energy, which is the area under the force–displacement curve, was obviously very different in those two cases. This example shows the significant effect of residual thermal stresses on the penetration resistance.

7.5 Prediction of Ballistic Limit

Predicting the ballistic limit from first principles is a complex task that some researchers have undertaken. Failure modes involved during the penetration of the target by the projectile can be different at different locations through the thickness. For simple cases where failure modes are known, simple energy considerations allow estimation of the ballistic limit. When perforation occurs in thin laminates, a conical-shaped perforation zone is observed (Cantwell 1989b), starting with a diameter equal to that of the ball. The cone makes a 45° angle with respect to the impact direction, and extensive delaminations are observed.

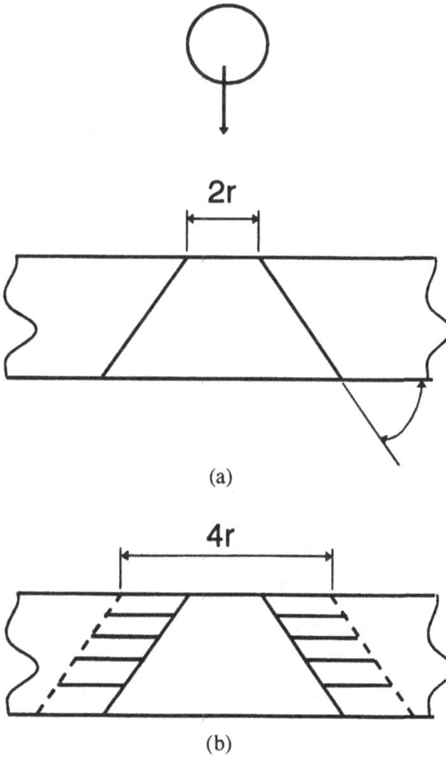

Figure 7.3. Failure mode: a) shear, b) delamination.

Cantwell and Morton (1985a) calculated the ballistic perforation energy as the sum of the energy the target absorbs by flexure E_f, contact deformation E_c, delamination E_d, and shear-out E_s. For thicker targets, two distinct failure processes are observed for the upper and lower portions of the specimens (Bless and Hartman, 1989, Bless et al. 1990, Lin et al. 1990, Cantwell and Morton 1990). A lower-bound estimate for the penetration energy for a thin laminate (4–16 plies) was obtained by Cantwell and Morton (1989c). The hole produced by shearing of the fibers during perforation was shaped as a truncated cone starting with the diameter of the projectile and with a 45° half angle (Fig. 7.3a). Tests determined that the transverse fracture energy required for shearing this carbon epoxy in a direction normal to the fibers was 37.5 kJ/m^2. Neglecting the strain energy stored in the structure and the energy required to produce delaminations, the energy required to produce the hole was estimated by multiplying the fracture energy per unit area by the area of the frustrum

$$A = \sqrt{2}\pi h(h + 2r) \tag{7.3}$$

where h is the laminate thickness and r is the projectile radius. This expression gives the trend observed during experiments. Sizable delaminated areas were also observed near the point of impact. The energy used to create these delaminations is estimated assuming that the delamination zone was also conical (Fig. 7.3b), starting with a radius $R = 9.1$ mm as determined from experiments. The fracture energy for delamination was taken as 500 J/m^2. The energy absorbed by bending of the composite beam is the only term that depends on the length of the beam. For simply supported beams subjected to a central force P, the strain energy is given by

$$U = \frac{P^2 L^3}{96EI}. \tag{7.4}$$

The energy required for shearing U_S and the energy E_d required to produce delaminations are independent of the length of the beam, so the total energy required for penetration is the sum of a constant term and a term varying with the cube of the beam length if failure occurs for a constant value of the contact force P. The study of 16-ply $[0_2, \pm 45]_{2S}$ carbon-epoxy beams 2 mm thick, 25 mm wide, and of varying length presented by Cantwell and Morton (1990) indicates that this assumption is not supported by experimental results. Instead, the energy required for penetration varies linearly with the beam length. If failure occurs when the lower surface strains exceed ϵ_c, the strain to failure of the fibers, then it can be shown that U_b varies linearly with both the length and the thickness of the beam.

Wang and Jang (1991a) proposed a simple approach based on the assumption that the failure mode was either shear at the periphery of the hole or compressive flow under the impactor. They showed that the ballistic limit should proportional to $(h/r)^{1/2}$ and obtained good agreement with experiments. This can be derived from (7.3), since U is proportional to hr for $h \ll r$ and the kinetic energy of the impactor is proportional to $r^3 v^2$.

Zhu et al. (1992b) presented simple models to estimate the resistance to striker motion and to predict the ballistic limit of Kevlar-polyester plates. The event was divided into three consecutive phases: indentation, perforation, and exit. The global deflections were determined using laminated plate theory, and dissipative mechanisms included the indentation of the striker tip, bulging at the back surface, fiber failure, delamination, and friction. Good agreement between predictions and experimental results was obtained. An interesting aspect in the development of the model was the assumption that "the reflection of the pressure wave created by the impact generates a tensile wave, which produces incipient mode I delamination, that is extended by the penetration of the projectile." This is confirmed by the work of Czarnecki (1992a,b) that shows the possibility of such a delamination mechanism.

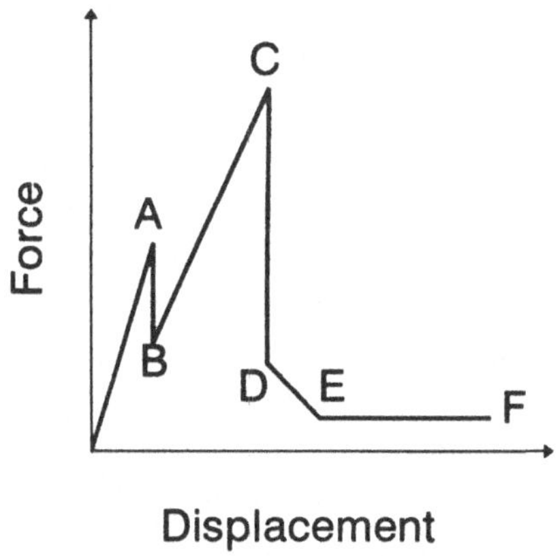

Figure 7.4. Load versus displacement of flat cylindrical indentor (Lee and Sun 1991).

Lee and Sun (1991, 1993b) studied the static penetration process in $([0,90,45,-45]_S)_2$ carbon-epoxy laminates subjected to blunt-ended punch. Experiments were conducted on an MTS machine using a flat cylindrical indentor with a diameter of 14.5 mm acting on circular plates with diameters equal to 3, 4, and 5 times that of the indentor. Typically, the force applied by the indentor increased until matrix cracks induced delaminations started (Fig. 7.4, point A), and a sudden drop occurred (point B). The stiffness of the plate was reduced, and as the load increased to a maximum (point C), a plug was formed, and the force dropped immediately (point D). Afterwards, friction between the indentor and the hole provided the only resistance to the motion. Delaminations were initiated by matrix cracks, were located in the double-ply layers, and extended fully throughout the entire plate. An axisymmetric 2D finite element analysis of quasi-isotropic laminates under a rigid flat-ended punch was performed. The material was modeled using effective moduli so that the details of the stacking sequence did not have to be included in the model. The first damage mode was delamination initiated by matrix cracks. The appearance of matrix cracks on the back face of the specimen under the indentor and on the front face near the edge of the plate was predicted using the maximum stress criterion. Residual thermal stresses were found to be significant in predicting the onset of matrix cracking since the residual thermal stress was 38 MPa, while the in-situ strength is 77 MPa for a double-ply layer and 84 MPa for a single-ply layer. The load-displacement curves predicted by the analysis were in good agreement with those obtained experimentally. The load at plug initiation was not predicted by the analysis, but experiments showed that it was independent of the diameter of the specimen. The analysis indicated that plug formation is an unstable fracture process and that the entire plug is formed immediately after initiation. Statically determined load-projectile displacement relations were used by Lee and Sun (1993a) to predict the ballistic limit of graphite-epoxy laminates.

Czarnecki (1992a,b) measured the transverse normal stress during high-velocity impact of graphite-epoxy laminates at different locations through the thickness. These results indicated the presence of two pulses propagating at different velocities. Experiments by Zhu et al. (1993) and others indicated that in the through-the-thickness direction, laminated composites have a nonlinear stress-strain behavior for the strain levels obtained during impact. Abrate (1993) used a bilinear stress strain relationship to model the material behavior and the numerical results obtained from a nonlinear finite element analysis captured the main features of the measured signals.

Jenq et al. (1992a,b) studied the strength degradation of glass-epoxy laminates subjected to ballistic impact. The residual strength can be predicted accurately using Caprino's model provided that the energy absorbed by the laminate during impact is used instead of the initial kinetic energy of the projectile. The residual tensile strength is predicted by

$$\left(\frac{\sigma_r}{\sigma_o}\right) = \left(U_o \Big/ \left[\frac{1}{2} m\left(v_i^2 - v_r^2\right)\right]\right)^{\alpha} \tag{7.5}$$

where U_o and α are coefficients to be determined experimentally. Jenq et al. (1994) showed that the material properties are strongly rate-dependent by comparing the energy required for static and dynamic perforation.

Prevorsek et al. (1993) used a finite element model for simulating the deformation of a composite plate during ballistic impact, and a finite difference model for determining the temperature rise during that event. The analysis showed that a significant temperature rise occurs at the projectile–composite interface, but because of the short duration of the impact and the low thermal conductivity of the composite, this temperature rise is confined of a very small region around the interface. The volume of material affected is too small to have any effect on performance.

7.6 Ceramic-Composite Armor

Ceramic-composite armor is used to defeat small-caliber projectiles while reducing weight, as indicated by Savrun et al. (1991), Arndt and Coltman (1990), and Navarro et al. (1993). Armor piercing projectiles are first shattered or blunted by the hard ceramic, and the load is then spread over a larger area. The composite backing deforms to absorb the remaining kinetic energy of the projectile. The experiments of Savrun et al. (1991) confirmed that the ceramic

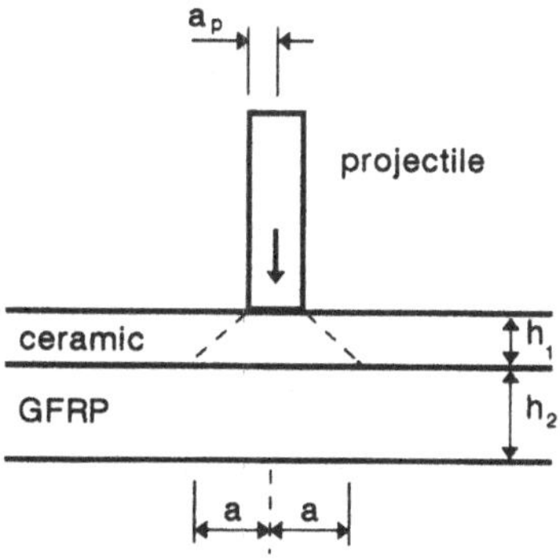

Figure 7.5. Ceramic-composite armor (Hetherington and Rajagopal 1991).

facing is very effective in blunting the projectile and that the backing absorbs the residual kinetic energy.

Ceramic targets backed by a thin metallic plate were analyzed using a finite difference model of both the target and the projectile (Cortes et al. 1992). Elastoplastic material behavior was assumed for both steel and aluminum, and a special material model was used for the ceramic component. Navarro et al. (1993) described experiments on ceramic faced armors with backing made out of aramid fibers in a vinylester matrix or polyethylene fibers in a polyethylene matrix (Fig. 7.5). The design of the armor was based on an energy absorption model proposed by Hetherington and Rajagopalan (1991) for composite armor. The ballistic limit was estimated using

$$V = \left(\frac{\epsilon_c S}{.91 f(a) M} \right)^{\frac{1}{2}} \tag{7.6}$$

where M is the mass of the projectile, S the ultimate strength of the backing plate, ϵ_c the breaking strain of the backing plate, and

$$f(a) = M/[M + (k_1 d_1 + k_2 d_2)A]A \tag{7.7}$$

where a is the radius of the affected zone at the ceramic-composite interface. In terms of the radius of the projectile a_p and the thickness of the ceramic layer h_1, this zone radius is given by

$$a = a_p + 2h_1. \tag{7.8}$$

In (7.7), A is the surface area of that zone ($A = \pi a^2$). For oblique impacts (Fig. 7.6), the area over which the composite is load is elliptical and

$$A = \pi C D \tag{7.9}$$

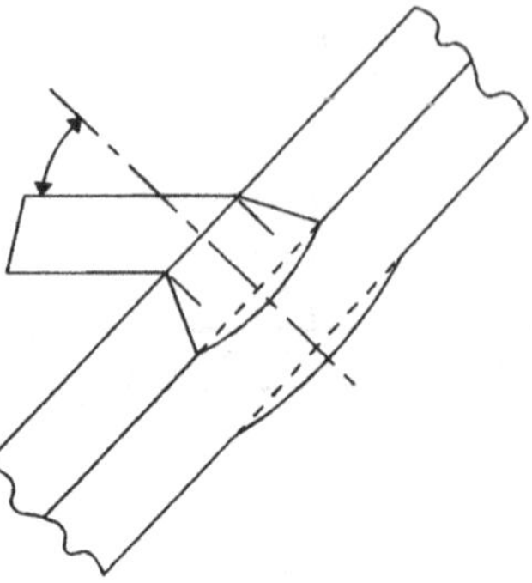

Figure 7.6. Oblique ceramic-composite armor.

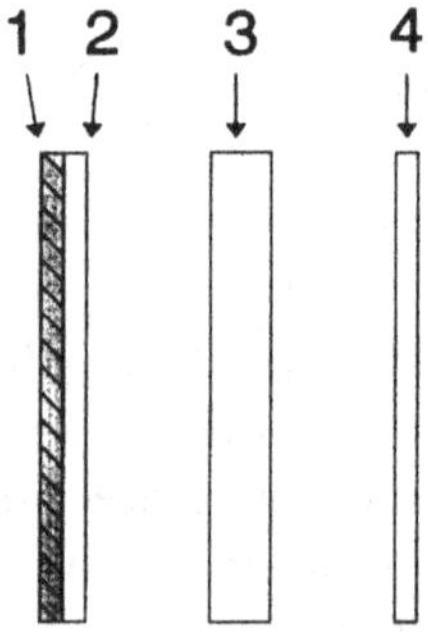

Figure 7.7. Example of retrofit armor augmentation (Meffert 1988).

where

$$C = \frac{a_p}{\cos\theta} + h_1 \tan\alpha, \quad D = a_p + h_1 \tan\alpha.$$

The model was shown to predict the ballistic limit fairly well for both normal and oblique impacts. Experimental evidence indicated that an inclined composite armor is more effective for a given thickness than one normal to the impact. However, ceramic-composite armor provides protection at a minimum weight when arranged perpendicular to the threat.

Meffert et al. (1988) investigated the use of composite materials as armor augmentation for armored personal carrier applications. A typical arrangement is shown in Fig. 7.7 with a ceramic-backed applique armor placed in front of the base aluminum armor and a spall suppressor placed behind it. The applique armor consists of a rigid panel of ceramic backed by a number of plies of glass, nylon, or aramid fabrics. Suitable ceramics are aluminum oxide, silicon nitride, silicon carbide, and boron carbide. The ceramic facing fractures and deforms the projectile core, and much of the residual kinetic energy is absorbed by the deformation and breaking of the backing fibers. The applique armor system reduces the probability of penetration by as much as 80%. Projectiles penetrating the aluminum core generate spall or vaporize the aluminum, which might explode when mixed with air. The aramid liner, called spall suppressor, is designed to prevent these two effects from occurring to prevent injury to personnel inside the vehicle. In that particular application, a Kevlar laminate was selected for the spall protector.

8
Repairs

8.1 Introduction

The expression *composite repairs* usually refers to one of three areas of extensive research activity: (1) repair of a damaged composite structure, (2) repair of aging (metallic) aircraft with composite patches, (3) and repair of civil engineering structures using composite reinforcement. Each one of those areas has received considerable attention. In this chapter we focus on the repair of impact-damaged composite structures.

Once damage is detected and the effects on the residual properties of the structure have been estimated, a decision must be made as to whether this composite part should be repaired or replaced. There are cases where damage cannot be repaired. For example, highly stressed members may not have sufficient strength after repair. Three types of repairs are possible:

(1) Large damages that reduce the load-carrying capability of the component below the ultimate load must be repaired immediately. In time, temporary field repairs using bolted metal patches must be replaced by permanent shop repairs. Permanent repairs can be either bolted or bonded repairs with precured composite parts. Another option, called laminate repair or cocured repair, consists of simultaneously curing the reinforcing patches and the bonded joints.
(2) Minor damages such that the part can sustain the ultimate load must be repaired within a defined period. Measures must be taken to prevent water and airstream ingress and to prevent damage propagation before the repair takes place.
(3) Negligible damage may require only a cosmetic damage to restore the surface of the component. A rule of thumb is that delaminations below 20–30 mm in diameter will not reduce the residual strength or grow under compression-dominated fatigue loading when subjected to strains

below 4000 microstrains and can be left unrepaired (Haskin and Baker 1986).

A particular repair technique is selected after considering several factors:

(1) The repair is to be permanent or temporary. In some cases, the damage is such that the part has to be replaced eventually, but it may be necessary to fly the aircraft to a repair facility before this can be done.
(2) The repair is to be made in the field or in a fully equipped repair facility. In the field, all the equipment and repair design procedures may not be available, in which case a simplified procedure becomes necessary. In many field repair situations, storage facilities for composite materials are lacking.
(3) The repair must restore the strength of the part to withstand the design ultimate loads. The repair patch must carry the load across the hole and restore stiffness and strength to the damaged area. At the same time, it must not change the load distribution in the area. Proper design requires a detailed stress analysis, which may not always be possible, particularly for field repairs.
(4) The repair must be watertight to prevent moisture from entering inside the structure, from being absorbed by the core in sandwich structures, or for being absorbed by the skin itself. Moisture strongly affects the mechanical properties of polymer composites and adhesives.
(5) On aerodynamic surfaces, the repair must not affect the airflow. Therefore, the external shape must conform to the profile before damage as much as possible. Where possible, flush repairs should be employed, but if external patches are used, tapering of the patch is required for minimum disturbance of the airflow.
(6) The part is accessible from both sides or only from the outside.

Various protection systems are employed on aircraft: (1) external environmental protection, such as polyurethane paint; (2) lightning strike protection, such as electrical conducting mesh; (3) fuel tank sealant; and (4) surface isolation to prevent galvanic corrosion. The protection systems should be restored after repair.

Impact damage generally weakens the structure locally, and damage may grow under further loading. Most repair procedures attempt to remove the damaged area and provide local reinforcement using some type of patches. Composite repairs can be classified according to the method used to fasten the reinforcing patches to the structure to be repaired. In a battle situation, bolted repairs are used exclusively (Dodd and Smith 1989). Bonded repair techniques are also used extensively. Thermoplastic composites with low-velocity impact consisting of matrix cracks and delaminations can be repaired without patches.

Different types of repair designs used are discussed in this chapter, and their advantages and drawbacks are compared. The procedures for joint design, material selection, and the repair process are also reviewed.

8.2 Mechanically Fastened Patches

As for any joint, mechanical fasteners can be used to secure a reinforcing patch to a composite structure. Mechanically fastened joints are typically used in medium-to-high load transfer application, their main advantage being that they are generally insensitive to surface preparation. Temporary repairs might consist of a bolted aluminum plate, and a definite repair may involve riveted carbon fiber–reinforced patches. When using metallic doublers, a significant design problem is to reach an acceptable compromise between adequate electrical bonding required for lightning protection and protection of the metallic doubler against galvanic corrosion (Tropis 1995). Titanium alloys are used for patches and fasteners because they do not suffer galvanic corrosion when in contact with graphite-epoxy. Aluminum alloys cannot be used with graphite-epoxy unless special coatings is used to prevent corrosion. Often, permanent repairs are made with composite patches tailored to match the parent material.

A typical arrangement is shown in Fig. 8.1. The damaged portion of the skin has been removed and replaced by a metal insert which is bolted to the external metal patch, which in turn will be bolted to the skin. On the top surface a filler material is used to ensure that the repair is flush with the skin surface. A gasket is used to seal the interface between the external patch and the skin. Two sets of bolts are used to transfer loads from the skins to the external patch and also to secure the insert to that external patch.

A simplified bolted repair is shown in Fig. 8.2, where the external metal patch is simply bolted to the skin and the other elements of the joint shown in

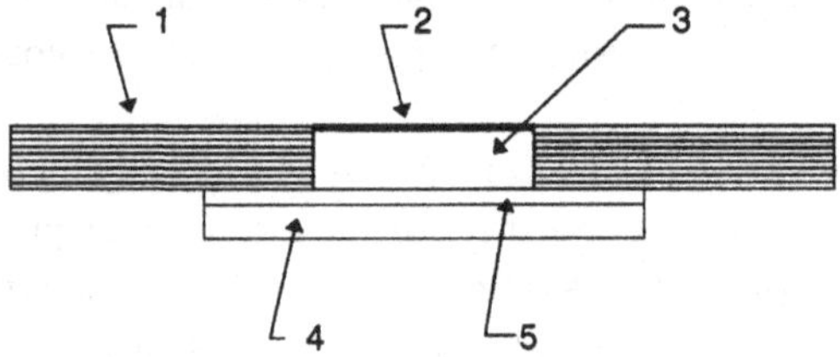

Figure 8.1. Typical bolted repair (1: skin; 2: filler; 3: doubler; 4: patch; 5: gasket material).

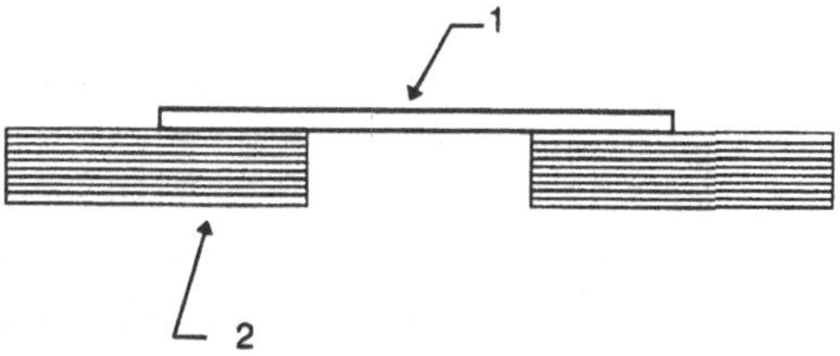

Figure 8.2. Bonded metal patch (1: metal patch; 2: skin).

Fig. 8.1 are omitted. This simplified design is easier to manufacture and will be a prime candidate for temporary field repairs.

Some of the problems encountered with bolted repairs include the following:

(1) Drilling holes for fasteners can introduce damage in the form of splitting and delaminations. This problem can be eliminated either by controlling the thrust force or by supporting the back side during drilling.
(2) Fastener holes introduce stress concentrations. Therefore, proper spacing must be used between the holes themselves and between the holes and the edges of the damaged area.
(3) Large surface areas are needed for clamping. Bolt clamping force has very beneficial effects on joint lives, but through-thickness integrity of composites is relatively weak. It is necessary to ensure that the clamping force is distributed over a sufficient area.

A few simple rules are recommended for patch bolt patterns (Niu 1992):

- Each bolt should be spaced no closer than four diameters to an adjacent bolt or three diameters to an edge of the laminate.
- The edge distance for metallic patches should be two diameters.
- Damage should be surrounded by at least two rows of bolts to protect against biaxial loads and inadvertent oversized bolt holes.

However, this type of joint can be successfully designed to carry high loads. Typically, holes up to 100 mm in monolithic skins with 50 to 100 plies can be repaired using bolted external patches.

8.3 Adhesive Bonded Joints

In the low-to-medium load range, bonded joints can be used between the reinforcing patch and the structure to be repaired. Their main advantage is that no new stress concentrations is introduced. Damage is usually removed, and in that operation, care must be taken to not extend the existing damage or introduce new damage. For cleaning out damaged areas in composite skins, high-speed routers are used. The depth of cut can be controlled and the outline of the removal area can be defined using a template.

With all bonded repairs, the quality of the repair depends on the proper selection of the adhesives used. The mechanical properties of the adhesive have a direct effect on the strength of the bond, but other properties such as handleability, bondability, and fluid resistance must also be considered. The fabrication of composite patches is also critical. Special efforts are required to control the thickness of the patch, to achieve low porosity, and to obtain good hot-wet mechanical properties. The materials used must have good drapability and cure temperatures that do not cause damage to the structure, and it should be possible to cut staged materials in the uncured state.

8.3.1 Peel and Shear Stresses in a Double-Lap Joint

Because of the complexity of a typical bonded composite repair, with anisotropic parts, inserts and patches, and complicated geometries, stress distributions are complex and also difficult to determine accurately. A simpler case is analyzed here to gain insight into the mechanics of a bonded joint. Stresses in a double-lap joint (Fig. 8.3) can be determined by a strength of material approach. The solution to this simpler problem brings some understanding of the mechanics of bonded joints that is useful for the design of bonded repairs.

For the double-lap joint shown in Fig. 8.3, considering the geometry and the loading applied, it might seem that the adhesive is subjected to a uniformly distributed shear stress. However, the following analysis shows that this is true only if the adherents are rigid; in practical cases the situation is different. Because of symmetry, only half of the joint needs to be modeled (Fig. 8.3), and the adherents are modeled as axial rods experiencing uniform displacements u_1 and u_2 and uniform stresses σ_1 and σ_2. The adhesive is assumed to be subjected to shear. From Fig. 8.3, the equations of equilibrium for materials 1 and 2 can be written as

$$\frac{d\sigma_1}{dx} + \frac{\tau}{t_1} = 0, \qquad \frac{d\sigma_2}{dx} - \frac{\tau}{t_2} = 0 \tag{8.1}$$

where τ is the shear stress in the adhesive layer. In terms of the displacements, the shear strain in the adhesive is

$$\gamma = \frac{u_2 - u_1}{t_a}. \tag{8.2}$$

The constitutive equations for the adhesive and the two adherents are

$$\tau = G\gamma, \quad \sigma_1 = E_1 \frac{du_1}{dx}, \quad \sigma_2 = E_2 \frac{du_2}{dx}. \tag{8.3}$$

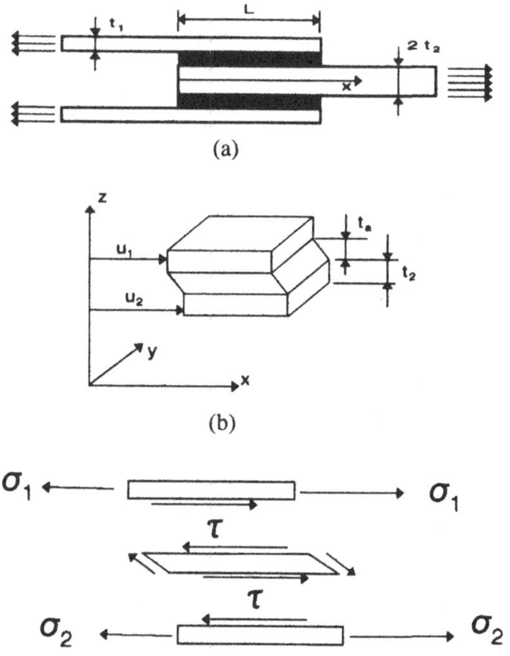

Figure 8.3. Double-lap bonded joint: a) geometry, b) deformation, c) stresses.

These three sets of equations (8.1)–(8.3) can be used to obtain the differential equation

$$\frac{d^2}{dx^2}\left(\frac{\sigma_1}{E_1} - \frac{\sigma_2}{E_1}\right) - \lambda^2\left(\frac{\sigma_1}{E_1} - \frac{\sigma_2}{E_1}\right) = 0 \qquad (8.4)$$

where

$$\lambda^2 = \frac{G}{t_a}\left(\frac{1}{E_1 t_1} + \frac{1}{E_2 t_2}\right). \qquad (8.5)$$

With the boundary conditions

$$\begin{aligned} \sigma_1 &= \sigma_{10}, \quad \sigma_2 = 0 \text{ at } x = 0 \\ \sigma_1 &= 0, \quad \sigma_2 = \sigma_{20} \text{ at } x = L \end{aligned} \qquad (8.6)$$

a solution to (8.4) can be found, and the shear stress distribution in the adhesive can be written as

$$\tau = \frac{G}{t_a \lambda}\left\{\left(\frac{\sigma_{10}}{E_1 \tanh(\lambda L)} + \frac{\sigma_{20}}{E_2 \sinh(\lambda L)}\right)\cosh(\lambda x) - \frac{\sigma_{10}}{E_1}\sinh(\lambda x)\right\}. \qquad (8.7)$$

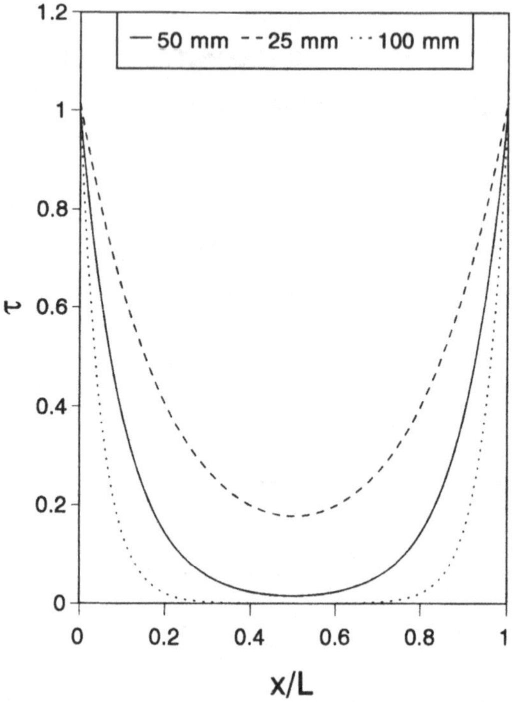

Figure 8.4. Shear stress distribution in double-lap bonded joint.

This equation indicates that the shear stress is not uniformly distributed as might have been expected. Equation (8.5) indicates that the parameter λ becomes small when the stiffness of the adherents becomes large. When λL is small, its hyperbolic cosine remains close to 1 and the hyperbolic sine is negligible, so (8.7) predicts a constant shear stress distribution in the adhesive. However, the following example will show that in practice this situation does not occur and that the shear stress distribution is much more complex.

Considering the special case where $t_1 = t_2 = t$, $E_1 = E_2 = E$, $\sigma_{10} = \sigma_{20} = \sigma_0$, the shear stress can be written in nondimensional form as

$$\bar{\tau} = \tau \frac{t_a \lambda E}{G \sigma_0} = \left(\frac{1}{\tanh(\lambda L)} + \frac{1}{\sinh(\lambda L)} \right) \cosh(\lambda x) - \sinh(\lambda x). \tag{8.8}$$

For the particular case where $E = 180$ MPa, $G = 1700$ MPa, $t_a = 0.2$ mm, and $t = 2.5$ mm, we find the stress distribution shown in Fig. 8.4 for several value of the length L (25, 50, 100 mm). This figure indicates that the shear stress in the joint is definitely not uniformly distributed. Instead, the load is transferred over a length of approximately 15 mm at each end. For longer joints, the central portion does not carry any significant load, and for shorter joints the full load transfer capability is not achieved.

The model used here predicts a strong stress concentration at both ends of the joint with the highest shear stresses occurring at either $x = 0$ or $x = L$. This is in apparent contradiction with the fact that at those two locations the adhesive is stress free. In addition, consider a free body diagram of a portion of adherent 1 near $x = L$ (Fig. 8.3) subjected to the normal stress σ_1 and the shear stress τ. Summing the moments of the forces about point A shows that equilibrium is not possible there with the model used. In fact, transverse normal stresses σ_p called *peeling stresses* are also present near the end (Fig. 8.3). More sophisticated analyses have shown that peeling stresses are significant only near the ends and that in a very small zone near the ends, the shear stress in the adhesive decreases down to zero. However, the value predicted by the present model (8.7) is a very good estimate of the maximum shear stress in the adhesive. Another complication that has not been addressed here is that, under shear, the stress-strain behavior of the adhesive is elastoplastic and that a significant proportion of the load capacity is achieved due to plastic behavior in the adhesive.

The load-carrying capacity of such a joint is limited by the stress concentrations near the ends, but the situation can be improved by tapering or staggering the ends of the adherents. These two modifications have the effect of increasing the load transfer region and resulting in a stronger joint.

The results of this simple analysis of a bonded joint are applicable to the more complex situations found in bonded damage repair. Stress concentrations and peel stresses are found in all bonded joints. Sufficient length should be allowed for proper load transfer, and ways to relieve stress concentrations and increase the load capacity of the joint should be used whenever practical.

8.3.2 Bonded Metal Patch Repair

A single scarf joint produces a nearly uniform stress distribution in the adhesive layer. Peel stresses are low, and this type of joint is very efficient for thick laminates because unlimited thicknesses can be joined and smooth external surfaces can be produced. The drawback is that this type of joint is difficult to produce and can be made only in depot or factory conditions. A large amount of material must be removed when a typical taper ratio of 18/1 or 20/1 is used. The most difficult part of the operation is cutting circular patches to the right size. Patches that are too small leave gaps to be filled with resin, which become a weak point of the component. Large patches will fold up at the edges and produce a weak joint.

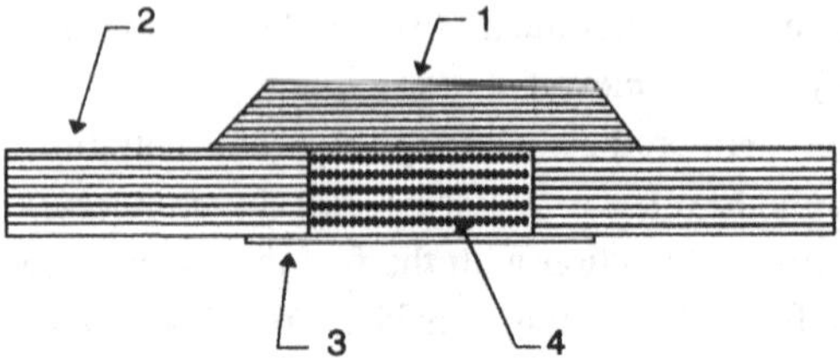

Figure 8.5. Typical bonded composite patch repair (1: external patch; 2: skin; 3: internal sealing plate; 4: insert).

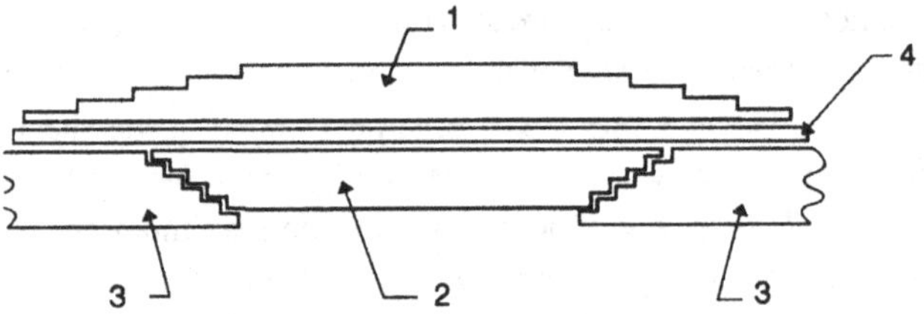

Figure 8.6. Staggered bonded repair (1: external patch; 2: internal patch; 3: skin; 4: adhesive layer).

8.3.3 Bonded Composite Repair

A typical bonded composite repair (Fig. 8.5) consists of a composite insert, an external patch, and an internal sealing plate. The edge of the external patch is tapered in order to lower peel stresses and shear stress concentrations near the edge and provide a larger load transfer region and therefore a larger load capacity. Typical taper ratios are 20/1 for thicknesses up to 16 plies. If only a few plies are used, no taper is necessary. This type of repair is similar to a single-lap joint and presents a problem because of the eccentric load path, which causes severe bending in the patch and peeling stresses in the adhesive. Out-of-plane bending also results in a significant reduction of the stability of the repaired component.

8.3.4 Staggered Bonded Repair

To increase the load capacity of a bonded joint, often a staggered joint is used. The idea is that, since load transfer in a lap joint takes place over a small region near the ends, several load transfer zones can be created with staggered surfaces. This is what is used in the design of the staggered bonded repair in Fig. 8.6, particularly at the interface between the insert and the skin. Peel stresses and shear stress concentrations are also reduced in the bond between the external patch and the skin by increasing the thickness of the patch in steps.

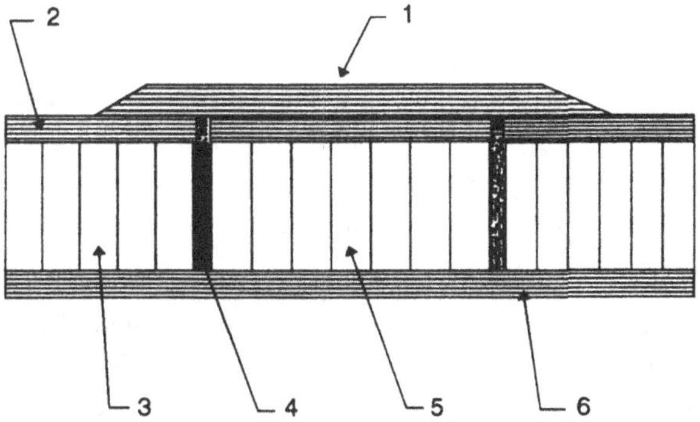

Figure 8.7. Bonded repair on sandwich structure (1: external patch; 2: skin; 3: core; 4: adhesive; 5: core insert; 6: lower skin).

8.3.5 Repair of Composite Sandwich Structures

Low-velocity impact damage on sandwich structures is usually limited to the top skin, the core, and the top skin-core interface. The lower skin is usually undamaged. Repair of impact-damaged sandwich structures typically involve the removal of the damaged portions of the top skin and the core. These areas are then filled in by inserts, and an external patch is then placed over to provide reinforcement (Fig. 8.7). This cure-in-place repair can be performed only at the depot level because of the need for an autoclave. A simpler procedure consists of not removing the damaged honeycomb core but rather stabilizing it using a lightweight filler paste that is cocured with the skin repair patch. A practical approach for field repairs is to stabilize the core with a filler and use a pre-cured patch that requires only low temperature curing for the skin repair.

During elevated temperature curing of repair adhesive, significant pressure can develop within the honeycomb cells, which then blows the skins from the core. This problem can be controlled by drying the repair area prior to curing and by applying pressure during the cure. Using adhesives with cure temperatures below the boiling point of water is another solution to this problem.

8.3.6 Resin Injection

Resin injection is used to repair thermoset matrix composite laminates with delaminations or SMC composites with matrix cracks. Russell and Bowers (1991) examined the resin requirements for successful repair of delaminations that are usually found around fastener holes. The external surfaces of the panel were sealed using a thin layer of sealant adhesive to prevent future leaks due to matrix cracks. The damaged panel was connected to a vacuum pump, and vacuum was applied to the delamination to remove the entrapped air and to slowly draw the epoxy into the hole. Once the hole was filled, pressure was applied to force the epoxy to enter the delamination cavity. Amplitude and time of flight C-scans were used to monitor the progression of the resin front. The flow capabilities of the resin were determined, and a resin system was developed specifically for resin injection repairs. Essential requirements for delamination repair resins include:

(1) the ability to adequately restore the interlaminar strength, fracture toughness, and fatigue crack growth resistance of the rebonded interface;
(2) a cure temperature compatible with the heat resistance of the structure to be repaired;
(3) the ability to cure fully without producing any volatiles, because these volatiles would be unable to escape and would produce voids;

The same approach was also used by Russell and Ferguson (1995) to repair low-velocity impact damage. When used to repair delaminations in the skins of sandwich structures, resin injection is effective in sealing the part and preventing further damage due to moisture ingress and the subsequent corrosion or freeze/thaw delamination.

Liu et al. (1993c) showed that the matrix cracks that develop in a SMC compound during impact can be effectively sealed by resin injection. A low-viscosity, room-temperature curing epoxy is poured into the area with matrix cracks when those cracks can be accessed from an external surface. Otherwise, a small hole is drilled and the resin is injected using a syringe. External patches were also used to compensate for fiber breakage.

8.4 Repair of Thermoplastic Composites

Delaminations and matrix cracking in thermoplastic composites can be repaired easily using what is called a thermoreforming process. Essentially, the damaged specimens are subjected to a sufficiently high temperature and reconsolidate

Table 8.1. *References on repair of thermoplastic composite structures*

Cantwell et al. (1991)
Heimerdinger (1995)
Ong et al. (1989)
Unger et al. (1988)

under pressure. This procedure is usually the same as the original thermoforming procedure. For low-velocity impacts where fiber damage is minimal, thermal re-forming can restore the flexural and compressive strengths to their original levels (Table 8.1). When significant fiber damage is induced by impact, the residual tensile strength of the repaired specimen can be lower than the residual strength before repair because the repair process can introduce severe distortions of the load-bearing fibers. To recover the strength losses due to fiber failures, external thermoplastic composite patches can be added, and both the specimen and the patch can be consolidated at the same time using the original thermoforming procedure.

9

Impact on Sandwich Structures

9.1 Introduction

Composite materials are used extensively in sandwich constructions, and compared with skin-stiffener panels, sandwich construction presents definite advantages: improved stability, weight savings, ease of manufacture, and easy repairs. Using composite materials instead of aluminum for the facesheets results in higher performance and lower weight, even though composite materials are more susceptible to impact damage. Composite sandwich construction has been used extensively for side skin panels, crown skin panels, frames, and longerons in the Boeing 360 helicopter (Llorente 1989). Compared to skin-stringer aluminum construction, the use of sandwich structures lead to an 85% reduction in the number of parts, a 90% reduction in tooling costs, and a 50% reduction in the number of work-hours needed to fabricate the helicopter. In spite of these advantages, the problem of impact on sandwich structures has received only minimal attention.

Contact laws for sandwich structures are significantly different that those for monolithic laminates. With sandwich structures, the indentation is dominated by the behavior of the core material, which becomes crushed as transverse stresses become large. To predict the contact force history and the overall response of the structure to impact by a foreign object, mathematical models should account for the dynamics of the projectile, the dynamics of the structure, and the contact behavior. Failure modes involved in impact damage of sandwich structures and the influence of the several important parameters will be discussed. We will also examine the effect of impact damage on the residual properties of the structure.

9.2 Contact Laws

The indentation of sandwich plates is dominated by the deformation of the core, and it is important to account for the indentation accurately in order to accurately predict the contact force history. As with impacts on monolithic composite structures, the effects of local indentation can be accounted for using a statically determined contact law. The example studied by Lee et al. (1991, 1993) of a foam-core sandwich plate is presented here to illustrate that the contact law for a sandwich structure differs significantly from that for a monolithic laminate. The graphite-epoxy facings have $[0_2, 90_2, 0_2]$ layup, and the polyurethane foam core is 12.7 mm thick. The properties of the graphite-epoxy are

$$\begin{aligned} &E_1 = 139 \text{ GPa}, \quad E_2 = 9.86 \text{ GPa}, \quad G_{12} = 5.24 \text{ GPa}, \\ &\nu_{12} = 0.3, \quad \rho = 1590 \text{ kg/m}^3, \quad \text{ply thickness} = 0.127 \text{ mm} \end{aligned} \tag{9.1}$$

and those of the foam are

$$E_c = 90.0 \text{ MPa}, \quad G_{13c} = G_{23c} = 32.0 \text{ MPa}, \quad \rho_c = 170.6 \text{ kg/m}^3. \tag{9.2}$$

The specimen, fully supported by a rigid platen, was indented by a 12.7-mm-diameter steel sphere. The experimental results for this sandwich plate were fitted by

$$P = k\alpha^n \tag{9.3}$$

and (2.21) and (2.22) with the constants $k = 1.385 \times 10^5$ N/m$^{0.8}$, $n = 0.8$, $q = 1.35$, $\beta = 0.171$, $\alpha_{cr} = 26.24 \times 10^{-5}$ m.

Figure 9.1 shows a comparison between the contact laws of the sandwich plate and those of the monolithic laminate determined from (9.1). For a given force level, the indentation of the sandwich is much larger than that of the solid laminate. The fact that $n = 0.8$ (and not 3/2 as for monolithic laminates) and that $q = 1.35$ (instead of 2.5) further indicates that the contact behavior for a sandwich plate is drastically different than that for a monolithic laminate made of the same material. The difference is attributed to the fact that the core is much softer than the facings in the transverse direction and will therefore experience a much larger deformation.

9.2.1 Compressive Behavior of Core Materials

The mechanical behavior of core materials must be well characterized in order to analyze the indentation of sandwich structures and predict contact laws and the extent of damage. In most applications, the core is made out of foam or honeycomb material, but corrugated cores and balsa cores can also be used effectively. Honeycomb, foam, and wood are what are called *cellular solids*,

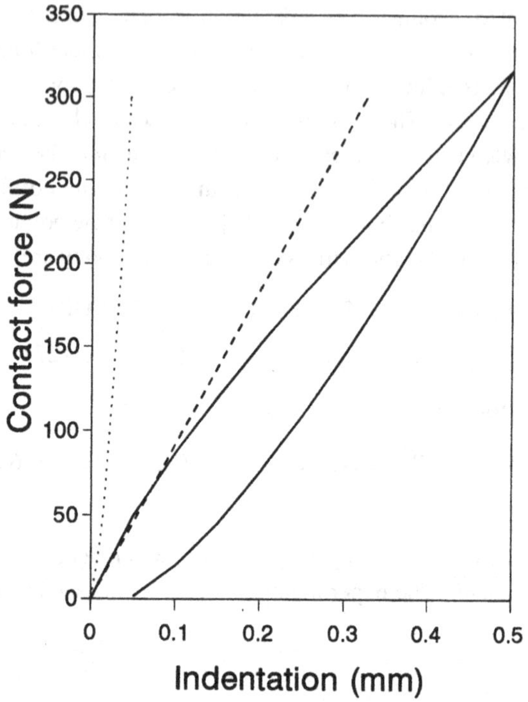

Figure 9.1. Contact law for the sandwich plate. Solid line: experimental results by Lee et al. (1991, 1993); dashed line: plate on foundation model; dotted line: contact law for monolithic laminate.

and the mechanical behavior of cellular solids is discussed at length in the excellent book of Gibson and Ashby (1988).

The stress-strain behavior of foam cores in compression can be idealized as consisting of an initial linear portion, followed by followed by a plateau corresponding to progressive crushing at nearly constant stress level, and then, when densification is completed, by a lock-up phase in which the stress increases rapidly with further deformation. Several investigators characterized the unidirectional compressive behavior of specific foam materials. For example (Throne and Progelhof 1985), the compressive behavior of low-density closed-cell foams can be described by

$$\sigma = \sigma_0 \epsilon^{0.2} + \sigma_1[(1-\epsilon)^{-1.4} - 1]. \tag{9.4}$$

For a LDPE foam with a density of 2 lb/ft^3, $\sigma_0 = 8.0$ psi and $\sigma_1 = 5.0$ psi. Rothschild et al. (1994) showed that it is essential to account for the nonlinear material behavior of cellular sandwich foam core in order to predict the deflections of sandwich beams under four-point bending.

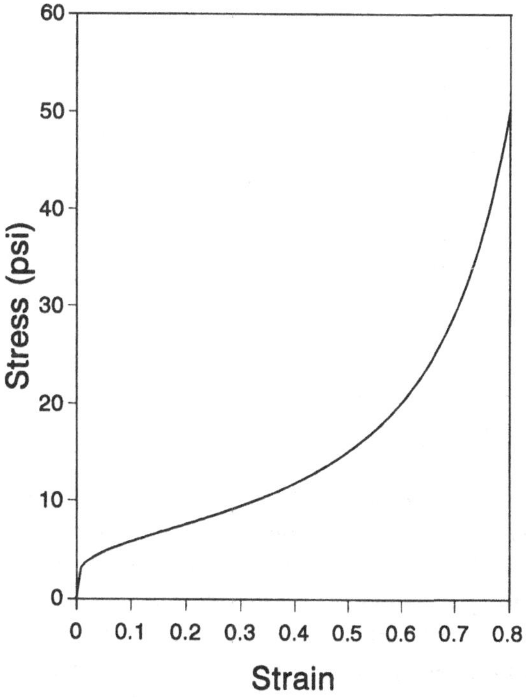

Figure 9.2. Typical stress-strain curve for foam material.

Under the indentor, a complex three-dimensional state of stress develops in the core. With foam cores, indentation sometimes causes core damage only in a small semi-spherical region near the top facing (Hiel et al. 1993). This suggests that the compressive stress in the transverse direction is not uniform through the thickness and is not the only stress component to be considered in determining the onset of core failure. Other failure modes have to be considered. To predict the onset of damage behavior, a three-dimensional failure criterion is needed. In addition to those discussed in Gibson and Ashby (1988), failure criteria for cellular materials under multiaxial loads were proposed by Gibson et al. (1989) and Triantafillou and Gibson (1990). Stress-strain relations were obtained experimentally by Maji et al. (1991) and Schreyer et al. (1994), who developed a three-dimensional plasticity model for foams and honeycombs.

Honeycomb cores are not continua since they are made up of webs arranged in cells which are joined together to form periodic structures. However, for analysis purposes they are modeled as anisotropic continua. In compression, linear elastic behavior is observed until a peak load is reached. As further

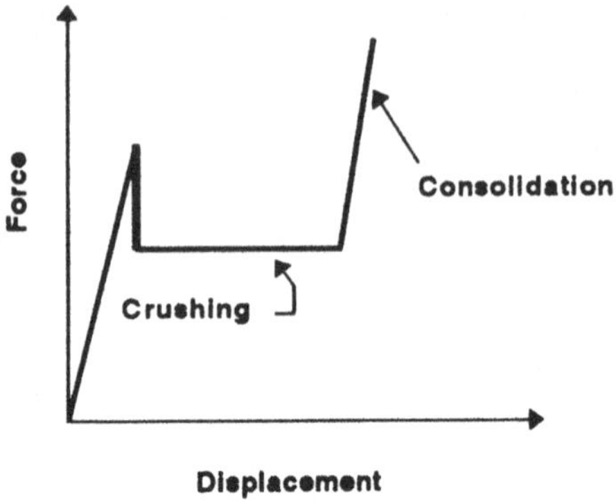

Figure 9.3. Typical stress-strain curves for honeycomb loaded in compression in the transverse direction.

deformation is introduced under stroke control, steady crushing under a nearly constant load is observed. Small oscillations about the mean value correspond to consecutive local buckling. Once the honeycomb is fully compacted, the load rises very rapidly again. This point is called the *densification point*. Unloading after steady crushing has started produces minimal elastic recovery. Goldsmith and Sackman (1991, 1992) showed the high degree of reproducibility in determining the pressure-crush curves for honeycomb. The dynamic crush strength was also found to be slightly higher than the static strength.

The out-of-plane deformation and failure of honeycombs were analyzed in details by Zhang and Ashby (1992). First, relationships are given to determine the elastic properties of the honeycomb subjected to transverse normal and shear loads as a function of the dimensions of the honeycomb cells and walls and the material properties. Two failure modes were identified: buckling and fracture. For low-density honeycombs, the linear elastic regime ends when the cell walls buckle. The initiation of elastic buckling does not cause a complete loss of stiffness and a smooth transition from the linear-elastic regime to the post-buckling regime is observed. For intermediate and high-density honeycombs, fracture of the cell walls is observed, and a sudden drop from a collapse stress to the steady crush strength is observed (Fig. 9.3). Expressions in terms of the cell dimensions and the material properties were obtained for the buckling and fracture strength of honeycombs subjected to transverse normal and shear loadings. Experiments proved the existence of these two failure modes and good agreement with predicted values was obtained. Equivalent constitutive equations for honeycomb structures were derived recently by Shi and Tong (1990) and Kalamkarov (1992) using homogenization theory. Similar results were also presented by Chamis et al. (1988).

Wood has a natural internal structure similar to honeycomb (Gibson and Ashby 1988, Gibson et al. 1995). Balsa wood with its high specific modulus and strength is used as core material in applications where cost considerations are driving the design of the sandwich structure rather than weight savings. Compressive stress-strain curves in the grain direction are similar in shape to those of foam or honeycomb. The modulus and strength are usually much higher than those of foam and honeycomb.

9.2.2 Contact between a Sandwich Beam and a Cylindrical Indentor

Experimentally, the contact behavior is determined by conducting tests in which the back face of the specimen is continuously supported by a rigid platen. Another approach is to keep the specimen supported only along the sides only but to measure the relative displacement between the indentor and the back face of the specimen. In that case, the back face usually experiences only global deformations. Those two experimental methods have been shown to produce similar indentation laws (Williamson and Lagace 1992).

In this section, we study the contact between a sandwich beam and a cylindrical indentor. With the back face supported by a rigid facing, the upper facing can be considered as a beam of rigidity EI supported by a foundation which provides a reaction $r(x)$ per unit length. The equilibrium of the beam is governed by the equation

$$EI\,\frac{d^4w}{dx^4} + r(x) = 0 \tag{9.5}$$

where w is the transverse displacement. When the transverse normal stress remains low, the core behaves elastically and the reaction of the foundation is proportional to the transverse displacement ($r(x) = kw$). In that case, the equilibrium of the top facing is governed by the equation for a beam on linear elastic foundation

$$\frac{d^4w}{dx^4} + \frac{k}{EI}w = 0. \tag{9.6}$$

The stiffness of the foundation k is related to the modulus of the core in the transverse direction E_c, the width of the beam b, and the thickness of the core h_c by

$$k = E_c b/h_c. \tag{9.7}$$

The general solution to (9.6) is

$$w = e^{-\lambda x}(A\sin\lambda x + B\cos\lambda x) + e^{\lambda x}(C\sin\lambda x + D\cos\lambda x) \tag{9.8}$$

where $\lambda^4 = k/(4EI)$. The constants A–D in (9.5) are determined by requiring the solution remains bounded at infinity and the slope at the origin be zero because of symmetry. In terms of α, the displacement at $x = 0$, the deflection of the top facing can be written as

$$w(x) = \alpha e^{\lambda x}(\sin \lambda x + \cos \lambda x). \tag{9.9}$$

The force P required to produce the initial deflection α is twice the value of the shear force at the origin. That is,

$$P = -2V(0) = 2EI\frac{d^3 w}{dx^3}(0) = 8\lambda^3 EI\alpha. \tag{9.10}$$

The contact force increases linearly with the indentation and the initial contact stiffness k_1 is then

$$k_1 = 8\lambda^3 EI = 2^{\frac{3}{2}}\left(\frac{E_c b}{h_c}\right)^{\frac{3}{4}}(EI)^{\frac{1}{4}}. \tag{9.11}$$

Equation (9.11) shows the effect of material and geometric parameters on the initial contact stiffness of the sandwich beam. The radius of curvature at the origin can be calculated from (9.8) and put in the form

$$R = \frac{4\lambda EI}{F}. \tag{9.12}$$

The radius of curvature is inversely proportional to the contact force so that as the force increases, the radius of curvature of the top facing decreases. As long as R remains larger than the radius of the indentor, the top facing will not wrap itself around the indentor and the load can be introduced as a concentrated force.

The foundation behaves elastically as long as the compressive stress in the core does not exceed the value σ_{max}. That is, when the contact force reaches the value

$$P_1 = \frac{2b\sigma_{\text{max}}}{\lambda}. \tag{9.13}$$

As the contact force increases, in a region of length a near the contact zone, the core is assumed to deform under a constant crushing stress σ_{crush}, and in that region, the reaction supplied by the foundation is $r_o = b\sigma_{\text{crush}}$. After substitution into (9.5), the displacement can be written as

$$w = -\frac{r_o}{24EI}x^4 + c_1\frac{x^3}{6} + c_2\frac{x^2}{2} + c_3 x + c_4. \tag{9.14}$$

Requiring that the displacement at $x = 0$ be the indentation α, the slope at the origin be zero because of symmetry, and that the shear force at $x = 0$ be equal to $P/2$, gives

$$w(x) = \alpha - \frac{r_o}{24EI}x^4 + \frac{P}{12EI}x^3 + c_2\frac{x^2}{2}. \tag{9.15}$$

For problems with core damage there is one solution for the damaged region ($x < a$) given by (9.15), and another solution for the undamaged region ($x > a$) which is given by (9.9). In those two expressions, there is a total of seven unknowns: A, B, C, D, α, c_2, and P. The displacement and the slope be bounded at infinity so $C = D = 0$. Four equations are obtained by requiring that the two solutions match at $x = a$. That is, the displacements, slopes, bending moment, and shear forces for the two solutions must be equal at that point. A fifth equation is obtained by requiring that, at $x = a$, the transverse normal strain be equal to the maximum strain before crushing. For a given value of a, these five linear algebraic equations can be solved for the remaining five unknowns. Then, one can plot force-indentation curves by simply determining P and α for various values of a.

9.2.3 Contact between a Sandwich Plate and a Spherical Indentor

The indentation of a sandwich plate by a spherical indentor is studied by modeling the top facing as a plate on elastic foundation and subjected to a concentrated force. The deflection of an infinite isotropic plate on an elastic foundation under a concentrated force P is given by (Timoshenko and Woinowsky-Krieger 1959)

$$w_{\max} = \frac{P}{8(Dk)^{\frac{1}{2}}} \tag{9.16}$$

where D is the bending rigidity of the plate and k is the elastic modulus of the foundation. Many experiments suggest that initially, the load increases linearly with the indentation so that (9.16) is applicable. The local indentation of the sandwich can be modeled by a linear spring with the stiffness given by

$$K = 8(Dk)^{\frac{1}{2}}. \tag{9.17}$$

Equation (9.17) is used to estimate the initial contact stiffness of sandwich plates with quasi-isotropic laminated facings. However, laminated composite facings are not always quasi-isotropic but can be modeled as orthotropic plates if the layup is symmetric and consists of more than six plies (Abrate and Foster 1995). According to the classical plate theory, the equation of motion for a symmetrically laminated plate on elastic foundation is

$$D_{11}\frac{\partial^4 w}{\partial x^4} + 2(D_{12} + 2D_{66})\frac{\partial^4 w}{\partial x^2 \partial y^2} + D_{22}\frac{\partial^4 w}{\partial y^4} + kw = p(x, y) \tag{9.18}$$

where w is the transverse displacement, the D_{ij} are the bending rigidities and p is the distributed loading which can be written as

$$p = P\delta(x-\eta)\delta(y-\xi) \tag{9.19}$$

when a concentrated force P is applied at $x = \eta$ and $y = \xi$. As in the case of beams on elastic foundations, the indentation of the top facing is expected to be localized. Therefore, for indentations away from the plate boundaries, the problem can be studied by considering a rectangular plate with simple supports along the edges. The boundary conditions (3.139), (3.140) are satisfied when the displacements are taken in the form given by (3.141). Substituting (3.141) into the equation of motion (9.18) yields the following expression for the coefficients W_{mn}:

$$W_{mn} = \frac{4P\sin\frac{m\pi\eta}{a}\sin\frac{n\pi\xi}{b}}{ab\left[\frac{\pi^4}{a^4}(D_{11}m^4 + 2(D_{12}+2D_{66})m^2n^2r^2 + D_{22}n^4r^4) + k\right]}. \tag{9.20}$$

This method was used to simulate the indentation test performed by Lee et al. (1991, 1993) on a sandwich plate with foam core. With the properties given by (9.1) and (9.2), the predictions are in good agreement with the experimental results (Fig. 9.1). After failure of the core under the impactor, the reaction supplied by the foundation is no longer proportional to the transverse displacements. Oplinger and Slepetz (1975) used an iterative approach to provide the required corrections in the damaged zone in order to satisfy the nonlinear stress-strain law. Olsson and McManus (1995) developed an analytical solution to this problem dividing the plate into a circular damaged region and a linear outside domain. The facings were assumed to be isotropic.

9.3 Impact Dynamics

To predict the contact force history, the dynamics of both the projectile and the target must be modeled accurately, and the local indentation effects must also be accounted for. The various approaches for studying the impact dynamics described in Chapter 3 can be used to study impacts on sandwich structures.

The static behavior of sandwich beams and plates is discussed in two well-known texts (Plantema 1966, Allen 1969). Suyir and Koller (1986) showed that for large wavelengths, the facings are subjected mainly to compressive or tensile stresses parallel to their middle surface and the core carries transverse shear stresses. This indicates that for large wavelengths, sandwich plates can be modeled using the first-order shear deformation theory (FSDT) and sandwich beams can be modeled using the Timoshenko beam theory.

The FSDT does not account for deformations in the transverse direction. Lee et al. (1991, 1993) developed a new theory based on the assumption that each facing deforms as a first-order shear deformable plate. The difference between the transverse displacements of the two facings accounts for the deformation of the core in that direction. Similarly, Sun and Wu (1991) used shear deformable plate finite elements to model each facing and springs to simulate the transverse normal and shear rigidities of the core in their study of impact dynamics. In the higher-order theory of Meyer-Piening (1989), the sandwich is considered as a three-layer plate with independent linear inplane displacement variations in each layer.

9.4 Impact Damage

9.4.1 Failure Modes

Impacts can induce damage to the facings, the core material, and the core–facing interface. The type of damage usually found in the facings is similar to that observed after impacts on monolithic composites. Damage initiation thresholds and damage size depend on the properties of the core material and the relationship between the properties of the core and those of the facings. Experimental techniques used extensively to detect impact damage in monolithic composites described in Chapter 4 are also employed to detect impact damage in sandwich structures. Table 9.1 lists the materials used in the various experimental studies of impact on composite sandwich structures.

Low-velocity impact damage on sandwich beams and plates with carbon-epoxy facings and honeycomb cores is confined to the top facing, the core–top facing interface, and the core. The lower facing generally is left undamaged. Five different failure modes have been identified: (1) core buckling, (2) delamination in the impacted face sheet, (3) core cracking, (4) matrix cracking, and (5) fiber breakage in the facings. In the upper facing, damage is similar to that observed in laminated composites and consists primarily of delaminations with matrix cracks and some fiber failures. In monolithic laminates, the damage area increases linearly with the kinetic energy of the projectile. With sandwich structures, skin damage also increases almost linearly with impact energy until a maximum value is reached (Table 9.2). At that point, visible damage is noticed and the delamination size remains constant. In honeycomb cores, damage consists of crushing or "buckling" of cell walls in a region surrounding the impact point. Usually honeycomb cores are made out of aluminum or composite materials (e.g., Nomex, glass thermoplastic, or glass-phenolic). With composite honeycombs, the cell walls are usually thicker and do not fail by instability but

Table 9.1. *Materials used in experimental studies of impact on sandwich structures*

Reference	Facings	Core
Adsit and Wazczak (1979)	Graphite-epoxy	Honeycomb
Akay and Hanna (1990)	Graphite-epoxy	Foam
Ambur and Cruz (1995)	Graphite-bismaleimide	Honeycomb
Baron et al. (1995)	Graphite-Peek	Thermoplastic foam
Bernard and Lagace (1987)	Carbon-epoxy	Honeycomb and foam
Caldwell et al. (1990)	Glass-epoxy	Glass thermoplastic, glass-phenolic honeycomb
Caprino and Teti (1994)	Glass reinforced plastic	PVC foam
Charles and Guedra-Degeorges (1991)	Graphite-, aramid-epoxy	Nomex honeycomb
Chen et al. (1991)	Carbon-epoxy	Aluminum honeycomb
Gottesman et al. (1987)	Graphite-epoxy	Aluminum honeycomb
Ishai and Hiel (1992)	Graphite-bismaleimide	Foam
Kim and Jun (1991)	Graphite-epoxy	Nomex honeycomb
Lee et al. (1991, 1993)	Graphite-epoxy	Foam
Llorente and Mar (1989)	Graphite-epoxy	Honeycomb
Llorente et al. (1990)	Thermoset, thermo-plastic matrices	Honeycomb
Manning et al. (1993)	Glass-polyester	Polyurethane foam
Mines et al. (1994)	Glass-, aramid-epoxy Graphite-epoxy	Honeycomb
Minguet (1991)	Graphite-epoxy	Nomex honeycomb
Nemes and Simmonds (1992)	Glass-epoxy	Polyurethane foam
Nettles and Hodge (1990)	Graphite-epoxy	Honeycomb
Nettles et al. (1991)	Carbon-epoxy	Aluminum honeycomb
Oplinger and Slepetz (1975)	Graphite-epoxy	Nomex honeycomb
Palm (1991)	Graphite-epoxy	Honeycomb
Peroni et al. (1989)	Graphite-epoxy,	Honeycomb
Rhodes (1975)	Graphite-epoxy Kevlar-49-epoxy	Honeycomb
Rix and Saczalski (1991)	Graphite-epoxy	Aluminum honeycomb
Saporito (1989)	Graphite-, aramid-epoxy	Nomex honeycomb
Sharma (1981)	Graphite-epoxy	Honeycomb
Shih and Jang (1989)	Various fabric materials	PVC foam
Sun and Wu (1991)	Graphite-epoxy	Foam, aluminum honeycomb
Verpoest et al. (1990)	Glass-epoxy	Foam
Vikstrom et al. (1989)	Glass-polyester	PVC foam
Weems and Llorente (1990)	Graphite-, aramid-epoxy	Nomex honeycomb

rather by fracture of the cell walls. Nettles and Hodge (1990) observed that for glass-phenolic honeycomb cores, at very low energy levels a type of core buckling takes place under the impact point. The phenolic resin had been sufficiently damaged to allow the glass yarns to bend freely. For higher-energy impacts, core cracking was observed when the glass fiber reinforcement was broken.

Table 9.2. *Studies of damage area versus impact energy for impacts on sandwich structures*

Akay and Hanna (1990)
Bernard and Lagace (1987)
Caldwell et al. (1990)
Charles and Guedra-DeGeorges (1991)
Chen, Chen, and Chen (1991)
Gottesman et al. (1987)
Hiel and Ishai (1992)
Palm (1991)

Similar observations were made by other authors (e.g., Palm 1991). With aluminum honeycomb cores, the localized deformations under the indentor cause the cell walls to buckle.

With foam cores, damage takes the form of a depression for high-energy impact and looks more like a crack for low-energy impacts (Bernard and Lagace 1987). Core damage also reaches a maximum value as impact energy increases. The core–facing interface is debonded in a region surrounding the point of impact, and the core experiences permanent deformations. The disbond was found to be nonexistent with Rohacell foam cores, one order of magnitude smaller in size than delaminations with Nomex honeycomb cores, and of the same order of magnitude with higher-density aluminum honeycomb cores (Bernard and Lagace 1987). Also, with the same core, a sandwich with eight-ply facesheets experienced 3 to 4 times more core damage than a sandwich with four-ply facesheets (Palm 1991). However, compression-after-impact tests indicated that cores that provided more facesheet support resulted in higher failure stress levels, even though delamination areas were larger. Therefore, understanding the core–facing interaction is important to explain the damage patterns observed during experiments.

9.4.2 Governing Parameters

Most experimental studies considered a single sandwich configuration and examine the effect of the projectile parameters: mass, shape, and initial kinetic energy on impact damage. Each study results in detailed observations that may appear at odds with conclusions drawn from other studies. Fortunately, two recent studies (Triantafillou and Gibson 1987, Mines et al. 1994) shed some light on the failure mechanisms found during impact of sandwich structures. Triantafillou and Gibson (1987) studied the failure of foam core sandwich beams with aluminum facings subjected to three-point bending and showed that such beams can fail in many ways. The tension and compression faces may fail by yielding or fracture. The compression face may fail by instability induced "wrinkling." Wrinkling involves local buckling of the face into the core over the entire length of the beam. The core can fail in shear, in tension, or in compression. The bond between the face and the core can fail. Since the resin adhesives usually used are brittle, debonding is by brittle fracture. Expressions are derived for calculating the failure load for each of these failure modes. Failure maps are obtained in terms of the core relative density ρ/ρ_s and the ratio of the face thickness t to the span length L. ρ is the density of the foam, and ρ_s is the density of the solid cell wall material. The ρ/ρ_s versus t/L plane is divided in three regions. When both ρ/ρ_s and t/L are small, face wrinkling was the mode of failure. When t/L is large but ρ/ρ_s remains small, core shear was the failure mode. Face yielding was obtained for large values of the relative density. Very good agreement between experimental and predicted values of the failure loads was reported.

Mines et al. (1994) studied the static and impact behavior of sandwich beams and identified four failure mechanisms. Mode I was the most common and started with an upper skin compressive failure, followed by either a stable crushing of the core, which gives a large energy absorbing capability (mode Ia), or core shear failure (mode Ib). For the specimens failing in this manner (mode I), the skin compression strength was lower than the tensile strength. Calculations indicated that upper skin wrinkling did not occur and that the failure mode was a true compressive failure. The roller geometry had negligible effect on beam failure. With woven aramid skins, the upper skin failed in compression but remained intact, no sudden load drop was observed, and final failure resulted from tensile failure of the lower skin (mode II). Mode III is a core shear failure that is attributed to an initial failure at the interface between the upper skin and the core. Beams with glass continuous strand mat (CSM) skins exhibited lower skin tensile failure because the tensile strength of CSM is less than the compressive strength. Mode IV corresponds to a failure of the skin that is loaded in tension, and the load deflection curves show zero residual strength.

Manning et al. (1993) found that compression face wrinkling under flexural loading occurs with composite sandwich panels with low-density foam cores used in surfboards. Llorente et al. (1990) showed that face wrinkling was the mode of failure for both undamaged and impact-damaged sandwich specimens being considered for application in helicopter structures. The critical stress for face wrinkling in compression is estimated using the empirical equation

$$\sigma_{cr} = K_w \langle E_t E_c G_c \rangle^{\frac{1}{3}} \tag{9.21}$$

where $K_w = 0.544$ (Weems and Llorente 1990). Under shear loading the allowable stress is assumed to be equal to the compression wrinkling allowable stress divided by $3^{1/2}$.

Van Voorhees and Green (1991) studied the three-point bending strength of an all-ceramic sandwich system consisting of an alumina open cell ceramic foam core and alumina ceramic faceplates. The applied load versus cross-head displacement curves consisted of an initial linear portion, corresponding to a linear elastic behavior until the core fails by the propagation of a single crack. This leads to a sudden drop in applied load but not to a catastrophic failure. After apparition of this crack, sandwich beams were able to sustain between 40 and 75% of the maximum load. After initial core failure, the applied load gradually decreases, the deflections become large, and the core progressively crushes around the initial crack until tensile failure of the faceplate occurs. Elementary beam theory was used to calculate the bending and shear stresses in the core. The maximum principal stresses and principal directions were calculated using Mohr's circle, and failure is expected to occur when the maximum principal stress exceeds the strength of the ceramic foam core. Excellent agreement is reported between experimental and predicted values of the failures loads and for the orientation of the cracks.

Except for low energy impacts, dents with depths reaching .5 mm to 3 mm are left on the surface (Rix and Saczalski 1991). Dent depth generally increases with the impact energy (Levin 1989). Charles and Guedra-Degeorges (1991) showed that the dent depths increase with impact energy until a maximum value is reached. Rhodes (1975) reported that adding more plies to the facings does not improve the impact resistance of sandwich structures and that increasing the crushing strength of the core improves the impact resistance.

Cantwell et al. (1994) showed that the modulus and strength of end-grain balsa is much larger than that of other core materials typically used in boat construction. Static tests showed that the force required to penetrate a balsa core sandwich is about twice that needed for sandwiches with PVC foam cores. Balsa cores are rate sensitive, and the perforation force increases with loading rate. However, during impact tests, perforation energy is essentially the same for all the core materials used with a 14-mm-diameter impact head. Perforation occurred by shearing the skins and this process absorbed most of the energy. Core properties had little influence on perforation resistance. Tests on sandwich panels with corrugated cores indicated that the large mismatch in Poisson's ratio between the skins and the corrugation was responsible for the separations observed during low-velocity impacts (Jegley 1993).

Fabrication techniques have significant effects on the impact resistance of sandwich panels. With honeycomb core, draping into the cells of the honeycomb before the matrix material hardens causes delaminations in the facings before impact when a one-step process is used to cure the facings and bond the facings to the core. Similarly, with foam cores, localized cell wall collapse and cell coalescence occurs, leading to nonuniform thickness and weaker cores near the skin–core bondline. In both cases, the problem is caused by the high pressure required to cure the facings. Better sandwich structures are obtained by prefabricating the skins and then bonding them to the core since less pressure will be required then. Baron et al. (1995) mentioned that a major drawback of sandwich structures is the adhesive bondline degradation due to moisture intake, which leads to poor impact damage resistance. Using fusion bonding of a thermoplastic foam to thermoplastic composite facesheets, they developed a sandwich construction on which moisture has virtually no effect. Gradin and Howgate (1989) used a thin aramid-reinforced elastomer layer between outer glass fiber laminates and a foamed PVC core to reduce local shear stresses. Noticeable improvements were observed during experiments.

Kosza and Sayir (1994) presented results of a study in which a bullet was fired at a foam core sandwich plate on which a small circular steel plate has been placed in order to distribute the impact load and avoid penetration. A peculiar failure pattern is obtained in this case. Under the steel plate a crack parallel to the facings and located a small distance below the upper skin–core interface is created by longitudinal waves propagating across the thickness direction. After reflection at the free surface, the initial compressive wave is reflected as a tensile wave, and the crack is created where tensile stresses are first generated. Shear stresses are generated outside the area covered by the steel plate, and the failure surface is then oriented towards the lower part of the core. This experiment allows the creation of a simple state of stress under the steel plate, simplified the analysis of the problem, and brought some understanding of the failure process. However, this situation is not typical of most foreign object impacts and oversimplifies the problem.

9.4.3 Improved Sandwich Construction

An approach to improve the impact resistance of sandwich structures by increasing the strength of the core has been suggested. Ishai and Hiel (1992) used interleaved foam cores in sandwich beams. The core was divided into several layers and separated by interlayers consisting of one ply of glass fabric prepreg oriented at $\pm 45^\circ$ to the axis of the beam and embedded between two plies of adhesive film. For traditional sandwich beams, cracks emanating from the impact point and oriented either in the transverse direction or at an oblique angle propagate through the entire thickness of the core. With interleaved core specimens, the damage is limited to the first core layer. Improvements in the residual compressive strength of the impacted skins are attributed to the fact that interleaving stops crack propagation through the core and results in better support of the skin and larger energy absorption during impact.

Weeks and Sun (1994) studied the same type of design where honeycomb cores are divided into several layers separated by laminated layers. The composite layers divide the core into several layers. The height of each honeycomb cell is only a fraction of the total core thickness. Since the crushing strength of honeycomb decreases with cell height, the interleaving process has the effect of increasing the crushing strength of the core. This construction provides better impact resistance and higher residual strength than traditional sandwich construction. Static tests were conducted to determine force-deflection curves, which were then used to estimate the energy required for penetration and estimate the ballistic limit successfully. Therefore no significant rate effect takes place in the local deformation.

9.4.4 High-Velocity Impacts

Higher-energy impacts result in partial or complete penetration of the sandwich structure; this is an area in which little research has been performed. Experiments were performed by Shih and Jang (1989), who considered various combinations of core and facesheet materials. Contact force histories were reported for many cases and consistently showed two peaks corresponding to perforation of the top and bottom facesheets. The valley separating these two peaks indicate that the core material presents little resistance to penetration. A limited number of test results presented by Akay and Hanna (1990) confirmed this finding. Penetration resistance is governed by the overall rigidity of the target and the resistance of the facings to perforation.

9.5 Residual Properties

Low-velocity impact damage on sandwich plates introduces significant reductions in tensile, compressive, shear, and bending strengths (Table 9.3). Typically, the residual strength is not affected until the impact energy exceeds a threshold value, drops sharply, and then levels off at a value as low as 40 or 50% of the initial strength.

Hiel et al. (1993) reported results on impact of sandwich structures with syntactic foam and skins made with a hybrid glass fiber-reinforced plastic/carbon fiber-reinforced plastic. The outer layer of GFRP acts as a sacrificial protective coating and serves to visualize the impact-damaged area because of the discoloration that takes place. The imprint created corresponds to delaminations

Table 9.3. *Experimental studies of residual strength of sandwich structures after impact*

Tension
Akay and Hanna (1990)
Caprino and Teti (1995)
Llorente and Mar (1989)
Rhodes (1975)
Sharma (1981)
Compression
Akay and Hanna (1990)
Caldwell et al. (1990)
Cantwell et al. (1994)
Ishai and Hiel (1992)
Llorente and Mar (1989)
Minguet (1991)
Nettles and Lance (1991,1993)
Rhodes (1975)
Saporito (1989)
Weeks and Sun (1994)
Weems and Llorente (1990)
Wong and Abbott (1990)
Shear
Caldwell et al. (1990)
Chen, Chen, and Chen (1991)
Nettles and Hodge (1990)
Weems and Llorente (1990)
Bending
Cantwell et al. (1994)
Charles and Guedra-DeGeorges (1991)
Chen, Chen, and Chen (1991)
Hiel and Ishai (1992)
Kassapoglou et al. (1988)
Llorente et al. (1990)
Palm (1991)
Wong and Abbott (1990)
Fatigue
Wong and Abbott (1990)

and is slightly elliptical in shape, D being the length of the major axis of the ellipse. Tests showed that specimens in which a circular hole of diameter D had been drilled exhibited the same residual strength and the same failure modes as the specimens with impact damage when D varied between 10 and 20 mm. This implies that knowing the damage size, the residual strength of the facing can be predicted using available methods for predicting the residual strength of open-hole composite samples such as those presented by Whitney and Nuismer

(1975), Nuismer and Labor (1979). Nettles and Lance (1991, 1993) also investigated the effect of using different materials for the outer skin layers on impact damage and residual properties. Results indicate that only slight improvements were achieved.

Caprino and Teti (1994) studied the effect of impacts on sandwich panels with glass-polyester facings and PVC foam cores. After impact the facings were removed from the core and tested to determine the residual tensile strength. Reductions in residual tensile strength are found to depend on the size of fiber damage. The residual tensile strength σ_c can be expressed as

$$\frac{\sigma_c}{\sigma_o} = \left(\frac{U_o}{U}\right)^m \tag{9.22}$$

where σ_o is the unnotched strength, U_o the energy below which no strength reduction takes place and m is an exponent determined either experimentally or through fracture mechanics considerations. In this case, m was found to be 0.31 from a best fit through the data in good agreement with the fracture mechanics prediction of 0.29.

Peroni et al. (1989) studied the effect of low-velocity impact damage on the natural frequencies and mode shapes of a sandwich panel. Good correlation is obtained between experimental results and predictions from a finite element analysis, but it must be noted that the fundamental natural frequency is not very sensitive to this type of damage. This explains why low-velocity impact damage is not included in mathematical models for analyzing the impact dynamics. Kassapoglou et al. (1988) used a one-term Rayleigh-Ritz approximation for predicting the buckling strength of an elliptical composite plate. This solution was used to estimate the compressive strength of sandwich composite plates with low-velocity impact damage. Buckling of the delaminated area often does not immediately cause final failure, and delamination buckling is sometimes difficult to detect experimentally. However, a good correlation is obtained between experimental values of the final failure load and the simple solution for delamination buckling. Minguet (1991) presented a model of a sandwich plate with low-velocity impact damage in the form of core damage and initial indentation and subjected to inplane compression loading. The nonlinear material behavior of the core in the damaged region is accounted for, and the model predicts an unstable propagation of the initial dent across the width of the panel as observed during experiments.

References

Abrate, S. 1991. "Impact on laminated composite materials." *Applied Mechanics Reviews* 44(4):155–190.

Abrate, S. 1993. "On minimizing thermal deflections of laminated plates." *Proc. of the 10th ASME Biennial Conference on Reliability, Stress Analysis and Failure Prevention*, Sept. 19–22, 1993, Albuquerque, NM. *ASME Publ. DE-Vol. 55*:1–8.

Abrate, S. 1994a. "Vibration of composite plates with internal supports." *Int. J. Mechanical Sciences* 36(11):1027–1043.

Abrate, S. 1994b. "Impact on laminated composites: Recent advances." *Applied Mechanics Reviews* 47(11):517–544.

Abrate, S. 1995a. "Stability of rectangular laminated plates with multiple internal supports." *Int. J. Solids and Structures* 32(10):1331–1347.

Abrate, S. 1995b. "Free vibration of point supported rectangular composite plates." *Composites Science and Technology* 53:325–332.

Abrate, S. 1995c. "Stability and optimal design of laminated plates with internal support." *Int. J. Solids and Structures* 32(10):1331–1347.

Abrate, S. 1995d. "Free vibration of point supported triangular plates." *Computers and Structures* 58(2):327–336.

Abrate, S. 1995e. "Design of multispan composite plates to maximize the fundamental natural frequency." *Composites* 26(10):691–697.

Abrate, S. 1996a. "Maximizing the fundamental natural frequency of triangular composite plates." *J. Vibration and Acoustics* 118:141–146.

Abrate, S. 1996b. "Foreign object impact on composite sandwich beams." *Proc. 41st Int. SAMPE Symposium*, Anaheim, CA, March 24–28:766–777.

Abrate, S. 1997. "Localized impact on sandwich structures with laminated facings." *Applied Mechanics Reviews* 50(2):69–82.

Abrate, S., and Foster, E. 1994. "Vibrations of composite plates with intermediate line supports." *J. Sound and Vibration* 179(5):793–815.

Abrate, S., Spoerre, J., and Schoeppner, G. 1997. Discussion. *ASCE J. of Engineering Mechanics* 123(6):648–649.

Abramovich, H., and Elishakoff, I. 1990. "Influence of shear deformation and rotary inertia on vibration frequencies via Love's equation." *J. Sound and Vibration* 137:516–522.

Adan, M., Sheinman, I., and Altus, E. 1994. "Buckling of multiply delaminated beams." *J. Composite and Materials* 28(1):77–90.

Adsit, N.R., and Wazczak, J.P. 1979. "Effect of near visual damage on the properties of graphite/epoxy." *ASTM STP 674*:101–117.

Akay, M., and Hanna, R. 1990. "A comparison of honeycomb-core and foam-core carbon fibre/epoxy sandwich panels." *Composites* 21(4):325–331.
Allen, H.G. 1969. *Analysis and Design of Structured Sandwich Panels*. Pergamon Press, Oxford.
Altus, E., and Ishai, O. 1992. "Delamination buckling criterion for composite laminates: A macro approach." *Engineering Fracture Mechanics* 41(5):737–751.
Ambur, D.R., Prasad, C.B., and Waters, W.A. 1995. "A dropped weight apparatus for low speed impact testing of composite structures." *Experimental Mechanics* 35(1):77–82.
Arndt, S.M., and Coltman, J.W. 1990. "Design trade-offs for ceramic/composite armor materials." *Proc. 22nd Int. SAMPE Technical Conf.*, Nov. 6–8:278–292.
Avery, J.G., and Porter, T.R. 1975. "Comparison of the ballistic response of metals and composites for military applications." *ASTM STP 568*:3–29.
Avery, J.G., Bradley, S.J., and King, K.M. 1981. "Fracture control in ballistic-damaged graphite/epoxy wing structure." *ASTM STP 743*:338–359.
Avery, W.B. 1989. "A semi-discrete approach to modeling post-impact compression strength of composite laminates." *Proc. of the 21st Int. SAMPE Technical Conf.*, Sept. 25–29:141–151.
Avery, W.B., and Grande, D.H. 1990. "Influence of materials and lay-up parameters on impact damage mechanisms." *Proc. 22nd Int. SAMPE Technical Conf.*, Nov. 6–8:470–483.
Avva, V.S. 1983. "Effect of specimen size on the buckling behavior of laminated composites subjected to low velocity impact." *ASTM STP 808*:140–154.
Avva, V.S., Vala, J.R., and Jeyaseelan, M. 1986. "Effect of impact and fatigue loads on the strength of graphite/epoxy composites." *ASTM STP 893*:196–206.
Awerbuch, J., and Hahn, H.T. 1976. "Hard object impact damage of metal matrix composites." *J. Comp. Mater.* 10:231–257.
Bachrach, W.E. 1988. "The development of a finite element model for impact damage of a composite cylinder using a mixed variational principle." In W.J. Ammann, W.K. Liu, J.A. Studer, and T. Zimmermann (eds.), *Impact: Effects of Fast Transient Loadings*, A.A. Balkema, Rotterdam:251–272.
Bachrach, W.E., and Hansen, R.S. 1989. "Mixed finite element method for composite cylinder subjected to impact." *AIAA J.* 27(5):632–638.
Baird, J., Heslehurst, R., Clark, B., and Williamson, H. 1993. "Holographic determination of the structural integrity of aircraft structures." *Proc. Int. Conf. on Advanced Composite Materials 1993*, Wollongong, Australia:379–382.
Balis-Crema, L., Castellani, A., and Peroni, I. 1985. "Modal tests on composite material structures: Application in damage detection." *Proc. 3rd IMAC II*:708–713.
Baron, W.G., Smith, W.G., and Czarnecki, G.J. 1995. "Damage tolerance of composite sandwich structure." *Proc. of 36th AIAA/ASME/ASCE/AHS/ASC Structures, Structural Dynamics and Materials Conference*, April 10–13, 1995, New Orleans, LA:1413–1418.
Basehore, M.L. 1987. "Low velocity impact damage in composites." *Annual Forum Proceedings American Helicopter Soc. 43rd* 2:577–585.
Bhaskar, K., and Varadan, T.K. "Interlaminar stresses in composite cylindrical shells under transient loads." *J. Sound and Vibration* 168(3):469–477.
Bishop, S.M. 1985. "The mechanical performance and impact behavior of carbon fibre reinforced PEEK." *Composite Structures* 3:295–318.
Blalock, T.N., and Yost, W.T. 1986. "Detection of fiber damage in a graphite epoxy composite using current injection and magnetic field mapping." *Review of*

Progress in Quantitative Nondestructive Evaluation, Vol. 5B, D.O. Thompson, and D.E. Chimenti (eds.), Plenum Press, NY:1207–1213.

Bless, S.J., and Hartman, D.R. 1989. "Ballistic penetration of S-2 glass laminates." *Proc. 21st SAMPE Tech. Conf.*, Atlantic City, NJ.

Bless, S.J., Benyami, M., and Hartman, D. 1990. "Penetration through glass-reinforced phenolic." *Proc. 22nd Int. SAMPE Technical Conf.*, Nov. 6–8:293–303.

Blodgett, E.D., and Miller, J.G. 1986. "Correlation of ultrasonic polar backscatter with the deply technique for assessment of impact damage in composite laminates." *Review of Progress in Quantitative Nondestructive Evaluation*, Bol. 5B, D.O. Thompson, and D.E. Chimenti (eds.), Plenum Press, NY:1227–1238.

Bolduc, M., and Roy, C. 1993. "Evaluation of impact damage in composite materials using acoustic emission." In W.W. Stinchcomb, and N.E. Ashbaugh (eds.), *ASTM STP 1156*:127–138.

Boll, D.J., Bascom, W.D., Weidner, J.C., and Murri, W.J. 1986. "A microscopic study of impact damage of epoxy carbon fibre composites." *J. Mater. Sci.* 21:2667–2677.

Bottega, W.J., and Maewal, A. 1983. "Delamination buckling and growth in laminates." *J. Applied Mechanics* 50:185–189.

Bouadi, H., Marple, L.R., and Marshall, A.P. 1992. "Normal and oblique impact on thick-section composite laminates." *ASME Winter Annual Meeting*, Anaheim, CA. ASME Publ. AD v 30:157–169.

Bowles, K.J. 1988. "The correlation of low velocity impact resistance of graphite fiber reinforced composites with matrix properties." *ASTM STP 972*:124–142.

Breivik, N.L., Gurdal, Z., and Griffin, O.H. 1992. "Compression of laminated composite beams with initial damage." *Proc. of the 7th Technical Conf. of the Am. Soc. for Composites*, Pennsylvania State University, Oct. 13–15:972–981.

Brosey, W.D., Whittaker, J.W., Bell, Z.W., Cantrell, J.L., and Henneke, E.G. 1989. "Nondestructive evaluation of impact-damaged filament wound spherical composite structures." *ASME NDE 5*:147–155.

Bucinell, R.B., Nuismer, R.J., and Koury, J.L. 1991. "Response of composite plates to quasi-static impact events." In T.K. O'Brien (ed.), *ASTM STP 1110*:528–549.

Butcher, B.R. 1979. "The impact resistance of unidirectional CFRP under tensile stress." *Fibre Science and Technol.*:295–326.

Butcher, B.R., and Fernback, P.J. 1981. "Impact resistance of unidirectional CFRP under tensile stress: further experimental variables." *Fibre Sci. Technol.*:41–58.

Buynak, C.F., and Moran, T.J. 1986. "Characterization of impact damage in composites." *Review of Progress in Quantitative Nondestructive Evaluation* 6B:1203–1211.

Buynak, C.F., Moran, T.J., and Donaldson, S. 1988. "Characterization of impact damage in composites." *SAMPE J.* 24(2):35–39.

Byun, C., and Kapania, R.K. 1992. "Nonlinear impact response of thin imperfect laminated plates using a reduction method." *Composites Engineering* 2(5–7):391–410.

Cairns, D.S. 1991. "Simple elasto-plastic contact laws for composites." *J. Reinforced Plastics and Composites* 10(4):423–433.

Cairns, D.S., and Lagace, P.A. 1987. "Thick composite plates subjected to lateral loading." *J. Applied Mechanics* 14:611–616.

Cairns, D.S., and Lagace, P.A. 1989. "Transient response of graphite/epoxy and kevlar/epoxy laminates subjected to impact." *AIAA J.* 27(11):1590–1596.

Cairns, D.S., and Lagace, P.A. 1990. "Residual tensile strength of graphite/epoxy and kevlar/epoxy laminates with impact damage." *ASTM STP 1059*:48–63.

Cairns, D.S., and Lagace, P.A. 1992. "A consistent engineering methodology for the treatment of impact in composite materials." *J. Reinforced Plastics and Composites* 11(4):395–412.

Cairns, D.S., Minguet, P.J., and Abdallah, M.G. 1994. "Theoretical and experimental response of composite laminates with delaminations loaded in compression." *Composite Structures* 27:431–437.

Caldwell, M.S., Borris, P.W., and Falabella, R. 1990. "Impact damage tolerance testing of bonded sandwich panels." *Proc. 22nd Int. SAMPE Technical Conf.*, Nov. 6–8:509–512.

Campanelli, R.W., and Engblom, J.J. 1995. "The effect of delaminations in graphite PEEK composite plates on modal dynamic characteristics." *Composite Structures* 31:195–202.

Caneva, C., Olivieri, S., Santulli, C., and Bonifazi, G. 1993. "Impact damage evaluation on advanced stitched composites by means of acoustic emission and image analysis." *Composites Structures* 25:121–128.

Cantwell, W.J. 1981. "The influence of the fibre stacking sequence on the high velocity impact response of CFRP." *J. Mater. Sci. Letters* 7(7):756–758.

Cantwell, W.J. 1988a. "The influence of target geometry on the high velocity response of CFRP." *Composite Structures* 10(3):247–265.

Cantwell, W.J. 1988b. "The influence of the fibre stacking sequence on the high velocity impact response of CFRP." *J. Mater. Sci. Letters* 7(7):756–758.

Cantwell, W.J., and Morton, J. 1984. "Low velocity impact damage in carbon fibre reinforced plastic laminates." *Proc. of V. Int. Congress on Experimental Mech.*, Montreal, Canada, June 10–15:314–319.

Cantwell, W.J., and Morton, J. 1985a. "Ballistic perforation of CFRP." *Proc. Conf. on Impact of Polymeric Materials*, Guildford, Surrey, 17/1–17/6.

Cantwell, W.J., and Morton, J. 1985b. "Detection of impact damage in CFRP laminates." *Composite Structures* 3:241–257.

Cantwell, W.J., and Morton, J. 1989a. "Geometrical effect in the low velocity impact response of CFRP." *Composite Structures* 12(1):39–60.

Cantwell, W.J., and Morton, J. 1989b. "Comparison of the low and high velocity impact response of CFRP." *Composites* 20(6):545–551.

Cantwell, W.J., and Morton, J. 1989c. "The influence of varying projectile mass on the impact response of CFRP." *Composite Structures* 13:101–104.

Cantwell, W.J., and Morton, J. 1990. "Impact perforation of carbon fibre reinforced plastic." *Composite Science and Technology* 38:119–141.

Cantwell, W.J., and Morton, J. 1991. "The impact resistance of composite materials – a review." *Composites* 22(5):347–362.

Cantwell, W., Curtis, P., and Morton, J. 1983. "Post impact fatigue performance of carbon fibre laminates with non-woven and mixed woven layers." *Composites* 14(3):301–305.

Cantwell, W.J., Curtis, P.T., and Morton, J. 1984. "A study of the impact resistance and subsequent O-compression fatigue performance of non-woven and mixed woven composites." In J. Morton (ed.), *Structural Impact and Crashworthiness 2*. Elsevier Applied Sci. Publ., London.

Cantwell, W., Curtis, P., and Morton, J. 1986. "An assessment of the impact performance of CFRP reinforced with high-strain carbon fibres." *Comp. Sci. Technol.* 25(2):133–148.

Cantwell, W.J., Davies, P., and Kausch, H.H. 1991. "Repair of impact-damaged carbon fiber PEEK composites." *SAMPE J.* 27(6):30–35.

Cantwell, W.J., Dirat, C., and Davies, P. 1994. "A comparative study of the mechanical properties of sandwich materials for nautical construction." *SAMPE J.* 30(4):45–51.

Caprino, G. 1983. "On the prediction of residual strength for notched laminates." *J. Mater. Sci.* 18:2269–2273.

Caprino, G. 1984. "Residual strength prediction of impacted composite laminates." *J. Comp. Mater.* 18:508–518.

Caprino, G., and Teti, R. 1994. "Impact and post-impact behavior of foam core sandwich structures." *Composite Structures* 29(1):47–55.

Caprino, G., and Teti, R. 1995. "Residual strength evaluation of impacted GRP laminates with acoustic emission monitoring." *Composites Sci. Technology* 53(1):13–19.

Caprino, G., Crivelli Visconti, I., and Langella, F. 1983. "Sulla resistanza residua di compositi in vetro-resina dopo impatto a bassa velocita." *Acts of 1 Convegno ASMI*, Milano, October 26–27.

Caprino, G., Crivelli-Visconti, I., and Di Ilio, A. 1984. "Elastic behavior of composite structures under low velocity impact." *Composites* 15(3):231–234.

Cawley, P., and Adams, R.D. 1979a. "A vibration technique for nondestructive testing of fibre composite structures." *J. Composite Materials* 13:161–175.

Cawley, P., and Adams, R.D. 1979b. "The location of defects in structures from measurements of natural frequencies." *J. Strain Analysis* 14(2):49–57.

Chai, H., Babcock, C.D., and Knauss, G. 1981. "One-dimensional modeling of failure in laminated plates by delamination buckling." *Int J. Solids Struct.* 17(11):1069–1083.

Chai, H., Knauss, W.G., and Babcock, C.D. 1983. "Observation of damage growth in compressively loaded laminates." *Experimental Mechanics* 23(3):329–337.

Challenger, K.D. 1986. "The damage tolerance of carbon fiber composites. A workshop summary." *Composite Structures* 6:295–318.

Chang, F.K., and Choi, H. 1990. "Damage of laminated composites due to low velocity impact." *31st AIAA/ASME/ASCE/AHS/ASC Struct., Struct. Dyn. Materials Conf.*, Long Beach, CA, April 2–4:930–940.

Chang, F.K., Choi, H.Y., and Jeng, S.T. 1989. "Characterization of impact damage in laminated composites." *Proc. 34th Int. SAMPE Symp.*, Reno, Nevada, May 8–11:702–713.

Chang, F.K., Choi, H.Y., and Jeng, S.T. 1990. "Characterization of impact damage in laminated composites." *SAMPE J.* 26(1):18–25.

Chang, F.K., Choi, H.Y., and Wang, H.S. 1990. "Damage of laminated composites due to low velocity impact." *31st AIAA/ASME/ASCE/AHS/ASC Structures, Structural Dynamics and Materials Conf.*, Long Beach, CA, April 2–4:930–940.

Charles, J.P., and Guedra-Degeorges, D. 1991. "Impact damage tolerance of helicopter sandwich structures." *Proc. 23rd Int. SAMPE Conf.*, Kiamesha Lake, NY, Oct. 21–24, 1991:51–61.

Chaudhuri, J., Choe, G.H., and Vinson, J.R. 1993. "Impact characterization of graphite fiber reinforced thermoplastic laminates." *J. Reinforced Plastics and Composites* 12:677–685.

Chen, C.F., and Frederick, D. 1993. "Contact of isotropic square plates with rigid spherical indentors." *Int. J. Solids Structures* 30(5):637–650.

Chen, C.H., Chen, M.Y., and Chen, J.P. 1991. "Residual shear strength and compressive strength of composite sandwich structure after low velocity impact." *Int. SAMPE Symp. and Exhibition* 36(Part 1):932–943.

Chen, H.P. 1991. "Shear deformation theory for compressive delamination buckling and growth." *AIAA J.* 29:813–819.

Chen, H.P. 1993. "Transverse shear effects on buckling and postbuckling of laminated and delaminated plates." *AIAA J.* 31:163–169.

Chen, H.P. 1994. "Free vibration of prebuckled and postbuckled plates with delamination." *Composites Science and Technology* 51:451–462.

Chen, H.P., Tracy, J.J., and Nonato, R. 1995. "Vibration analysis of delaminated composite laminates in prebuckled states based on a new constrained model." *J. Composite Materials* 29(2):229–256.

Chen, J.K., and Sun, C.T. 1983. "Impact response of buckled composite laminates." *Recent Adv. in Eng. Mech. and their Impact on Civil Eng. Practice*, Proc. 4th Eng. Mech. Div. Specialty Conf., Purdue Univ., West Lafayette, IN, May 23–25:515–519.

Chen, J.K., and Sun, C.T. 1985a. "Analysis of impact response of buckled composite laminates." *Composite Structures* 3:97–118.

Chen, J.K., and Sun, C.T. 1985b. "Dynamic large deflection response of composite laminate subjected to impact." *Composite Structures* 4(1):59–73.

Chen, V.L., Wu, H.Y.T., Yeh, H.Y. 1993. "A parametric study of residual strength and stiffener for impact damaged composites." *Composite Structures* 25:267–275.

Choi, H.Y., and Chang, F.K. 1992. "A model for predicting damage in graphite/epoxy laminated composites resulting from low-velocity point impact." *J. Composite Materials* 26(14):2134–2169.

Choi, H.Y., Wang, H.S., and Chang, F.K. 1990. "Effect of laminate configuration on low-velocity impact damage of laminated composites." *Proc. 22nd Int. SAMPE, Technical Conf.* Nov. 6–8:484–493.

Choi, H.Y., Downs, R.J., and Chang, F.K. 1991a. "A new approach toward understanding damage mechanisms and mechanics of laminated composites due to low-velocity impact: Part I – Experiments." *J. Composite Materials* 25:992–1011.

Choi, H.Y., Wu, H.T., and Chang, F.K. 1991b. "A new approach toward understanding damage mechanisms and mechanics of laminated composites due to low-velocity impact: Part II – Analysis." *J. Composite Materials* 25:1012–1038.

Choi, H.Y., Wang, H.S., and Chang, F.K. 1992. "Effect of laminate configuration and impactor's mass on the initial impact damage of graphite-epoxy composite plates due to line loading impact." *J. Composite Materials* 26(6):804–827.

Choi, I.H., and Hong, C.S. 1994. "Low-velocity impact response of composite laminates considering higher-order shear deformation and large deflection." *Mech. Composite Mat. Struct.* 1(2):157–170.

Chou, S., and Wu, C.J. 1992. "Study of the physical properties of epoxy resin composites reinforced with knitted glass fiber fabrics." *J. Reinforced Plastics and Composites* 11(11):1239–1250.

Chou, S., and Wu, C.J. 1994. "Impact study of epoxy resin composites reinforced with glass fiber fabrics." *J. Polymer Research* 1(3):255–264.

Chou, S., Chen, H.C., and Chen, H.E. 1992. "Effect of weave structure on mechanical fracture behavior of three-dimensional carbon fiber fabric reinforced epoxy resin composites." *Composites Science and Technology* 45:23–35.

Christoforou, A.P. 1993. "On the contact of a spherical indenter and a thin composite laminate." *Composite Structures* 26:77–82.

Christoforou, A.P., and Swanson, S.R. 1990. "Analysis of simply-supported orthotropic cylindrical shells subject to lateral impact loads." *J. Applied Mechanics* 57:376–382.

Christoforou, A.P., and Swanson, S.R. 1991. "Analysis of impact response in composite plates." *Int. J. Solids Structures* 27(2):161–170.

Chuanchao, Z., and Kaida, Z. 1991. "The investigation of the behavior of laminates subject to low velocity impact." *Proc. of the 8th Int. Conf. on Composite Mater. (ICCM/8)*, Honolulu, HI, July 15–19, 32.I:1–9.

Clark, G. 1989. "Modeling of impact damage in composite laminates." *Composites* 20(3):209–214.

Clark, G., and Saunders, D.S. 1991. "Morphology of impact damage growth by fatigue in carbon fibre composite laminates." *Materials Forum* 15(4):333–342.

Clerico, M., Ruvinetti, G., Cipri, F., and Pelosi, M. 1989. "Analysis of impact damage and residual static strength in improved CFRP." *Int. J. Mat. Product. Tech.* 4(1):61–70.

Cohen, A., Yalvac, S., and Wetters, D.G. 1992. "Impact fatigue behavior of a syntactic composite material." *37th Int. SAMPE Symposium and Exhibition*, Covina, CA, March 9–12, 37:641–653.

Cortes, R., Navarro, C., Martinez, M.A., Rodriguez, J., and Sanchez-Galvez, V. 1992a. "Numerical modeling of normal impact on ceramic composite armours." *Int. J. Impact Eng.* 12(4):639–651.

Cortes, R., Navarro, C., Martinez, M.A., Rodriguez, J., and Sanchez-Galvez, V. 1992b. "Use of mathematical models of ceramic faced plates subjected to impact loading." *Proc. 2nd Int. Conf. on Structures Under Shock and Impact*, Portsmouth, England:357–368.

Cowper, G.R. 1966. "The shear coefficient in Timoshenko's beam theory." *J. Applied Mechanics* 33:335–340.

Craig, D.M., and Chapman, C.E. 1991. "NDI of impact damaged composite panels." *British J. of Non-Destructive Testing* 33(2):64–68.

Cristescu, N., Malvern, L.E., and Sierakowski, R.L. 1975. "Failure mechanisms in composite plates impacted by blunt-ended penetrators." *ASTM STP* 568:159–172.

Crivelli Visconti, I., Caprino, G., Di Ilio, A., and Carrino, L. 1983. "Impact tests on CFRP: a static-dynamic analogy." *Acts of 5th Int. SAMPE European Chapter*, Bordeaux.

Crook, A.W. 1952. "A study of some impacts between metal bodies by a piezoelectric method." *Proc. Royal Soc.*, London, Series A, 212:377–390.

Cui, W.C., and Wisnom, M.R. 1992. "Contact finite element analysis of three and four-point short beam bending of unidirectional composites." *Composites Science and Technology* 45:323–334.

Curson, A.D., Moore, D.R., and Leach, D.C. 1990. "Impact failure mechanisms in carbon fiber/PEEK composites." *J. Thermoplastic Composites* 3:24–31.

Curtis, P.T., and Bishop, S.M. 1984. "An assessment of the potential of woven carbon fibre-reinforced plastics for high performance applications." *Composites* 15(4):259–264.

Curtis, P.T., Lawrie, D., and Young, J.B. 1984. "The effect of impact on pre-fatigued fibre composites. In J. Morton (ed.), *Structural Impact and Crashworthiness*, London: Elsevier Appl. Sci. Publ.:494–509.

Czarnecki, G. 1992a. "A preliminary investigation of dual mode fracture sustained by graphite/epoxy laminates impacted by high-velocity spherical metallic projectiles." MS Thesis University of Dayton.

Czarnecki, G. 1992b. "Dual mode fracture of graphite/epoxy laminates penetrated by spherical projectiles." *Proc. of 18th Congress of the Int. Council of the Aeronautical Sciences*, Beijing, China, Sept. 20–25, 2:1–7.

Dan-Jumbo, E., Leewood, A.R., and Sun, C.T. 1989. "Impact damage characteristics of bismaleimides and thermoplastic composite laminates." *ASTM STP 1012*:356–372.

Davidson, B.D. 1989. "On modeling the residual strength of impact damaged, compression loaded laminates." *21st Int. SAMPE Technical Conf.*, Sept. 25–28:109–119.

Davidson, B.D. 1991. "Delamination buckling: Theory and experiment." *J. Composite Materials* 25:1351–1378.

Davies, G.A.O., and Zhang, X. 1995. "Impact damage prediction in carbon composite structures." *Int. J. Impact Engineering* 16(1):149–170.

Davies, G.A.O., Zhang, X., Zhou, G., and Watson, S. 1994. "Numerical Modeling of Impact Damage." *Composites* 25(5):342–350.

De Goeje, M.P., and Wapenaar, K.E.D. 1992. "Non-destructive inspection of carbon-fibre-reinforced plastics using eddy current methods." *Composites* 23(3):147–157.

Delfosse, D., Pageau, G., Bennett, A., and Pousatrip, A. 1993. "Instrumented impact testing at high velocities." *J. Composite Tech. Res.* 15(1):38–45.

Dempsey, R.L., and Horton, R.E. 1990. "Damage tolerance evaluation of several elevated temperature graphite composite materials." *SAMPE Int. Symp.* 35:1292.

Demuts, E. 1990. "Damage tolerance of composites." *Proc. of the American Soc. for Composites*, 4th Technical Conf., Virginia Polytechnic Institute and State University, Blacksburg, VA, Oct. 3–5:425–433.

Demuts, E. 1993. "Low velocity impact in a graphite/PEEK." *34th AIAA Struct. Struct. Dyn. and Materials Conf.*, La Jolla, CA, 1993, Part 2:901–908.

Demuts, E., and Sharpe, P. 1987. "Toughen advanced composite structures." *28th AIAA Structures, Structural Dynamics and Materials Conf.*, Monterey, CA, Part 1:385–393.

Demuts, E., Whitehead, R.S., and Deo, R.B. 1985. "Assessment of damage tolerance in composites." *Composite Structures* 4:45–58.

Dobyns, A.L. 1980. "Analysis of simply supported plates subjected to static and dynamic loads." *AIAA J.* 19(5):642–650.

Dobyns, A.L., and Porter, T.R. 1981. "A study of the structural integrity of graphite composite structure subjected to a low velocity impact." *Polym. Eng. Science* 21(8):493–498.

Dodd, S.M., and Smith, H. 1989. "Expert systems for design of battle damage repairs." *Proc. of the 21st Int. SAMPE Technical Conf.*, Sept. 25–29:239–248.

Dorey, G. 1987. "Impact damage in composites: development, consequences, and prevention." *Proc. of the 6th Int. Conf. on Comp. Mater. combined with the 2nd European Conf. on Comp. Mater.*, London, England:3.1–3.26.

Dorey, G., Sidey, G.R., and Hutchings, J. 1978. "Impact properties of carbon fiber/kevlar 49 fibre hybrid composites." *Composites* 9:25–32.

Dorey, G., Bishop, S., and Curtis, P. 1985. "On the impact performance of carbon fibre laminates with epoxy and PEEK matrices." *Comp. Sci. Technol.* 23(3):221–237.

Dost, E.F., Ilcewicz, L.B., and Avery, W.B. 1991. "Effects of stacking sequence on impact damage resistance and residual strength for quasi-isotropic laminates." In T.K. O'Brien (ed.), *ASTM STP 1110*:476–500.

Dow, M.B., and Smith, D.L. 1989. "Damage-tolerant composite materials produced by stitching carbon fabrics." *21st Int. SAMPE Tech. Conf.*, Sept., 995–605.

Dransfield, K., Baillie, C., and Mai, Y.W. 1994. "Improving the Delamination resistance of CFRP by stitching: A review." *Composite Science and Technology* 50:305–317.

Elber, W. 1985. "Effect of matrix and fiber properties on impact resistance." *Tough Composite Materials: Recent Developments*, Noyes Publ, Park Ridge, NJ:89–110.

El-Zein, M.S., and Reifsnider, K.L. 1990a. "The strength prediction of composite laminates containing a circular hole." *J. Composites Technology and Research*, 12(1):24–30.

El-Zein, M.S., and Reifsnider, K.L. 1990b. "On the prediction of tensile strength after impact of composite laminates." *J. Composites Technology and Research* 12(3):147–154.

Epstein, J.S., Deason, V.A., and Abdallah, M.G. 1992. "Impact wave propagation in a thick composite plate using Moire interferometry." *Optics and Lasers in Engineering* 17(1):35–46.

Essenburg, F. 1962. "On surface constraints in plate problems." *J. Appl. Mech.* 29(2):340–344.

Evans, K.E., and Alderson, K.M. 1992. "Transverse impact of filament-wound pipes. Part II: Residual properties and correlation with impact damage." *Composite Structures* 20:47–52.

Evans, R.E., and Masters, J.E. 1987. "A new generation of epoxy composites for primary structural applications: materials and mechanics." *ASTM STP 937*:413–436.

Finn, S.R., He, Y.F., and Springer, G.S. 1992. "Compressive strength of damaged and repaired composite plates." *J. Composite Materials* 26(12):1796–1825.

Found, M.S., and Howard, I.C. 1995. "Single and multiple impact behavior of a CFRP laminate." *Composite Structures* 32:159–163.

Freeman, S.M. 1982. "Characterization of lamina and interlaminar damage in graphite/epoxy composites by the deply technique." *ASTM STP 787*:50–62.

Friedman, Z., and Kosmatka, J.B. 1993. "An improved two-node Timoshenko beam finite element." *Computers and Structures* 47(3):473–481.

Frock, B.G., Moran, T.J., Shimmin, K.D., and Martin, R.W. 1988. "Imaging of impact damage in composite materials." *Rev. Prog. Quant. Nondestr. Eval.* 7B:1093–1099.

Gambone, L.R., Marusic, M., Kung, D., and Janke, G. 1992. "Acoustic emission monitoring and destructive testing of glass fibre reinforced plastic bucket truck booms." *ASTM STP 1139*:29–49.

Gandhe, G.V., and Griffin, O.H. Jr. 1989. "Post-impact characterization of interleaved composite materials." *SAMPE Quarterly* 20(4):55–58.

Gardiner, D.S., and Pearson, L.H. 1985. "Acoustic emission monitoring of composite damage occurring under static and impact loading." *Exp. Tech.* 9(11):22–28.

Garg, A.C. 1988. "Delamination: A damage mode in composite structures." *Eng. Fract. Mech.* 29(5):557–584.

Garg, A.C. 1993. "Prediction of residual strength of impact damaged aerospace composite structure." *Proc. of Int. Conf. on Advanced Composite Materials*, Wollongong, Australia, 1993:587–590.

Ghaffari, S., Tan, T.M., and Awerbuch, J. 1990. "An experimental and analytical investigation on the oblique impact of graphite/epoxy laminates." *Proc. 22nd Int. SAMPE Technical Conf.*, Nov. 6–8:494–508.

Ghasemi Nejhad, M.N., and Parvizi-Majidi, A. 1990. "Impact behavior and damage tolerance of woven carbon fibre-reinforced thermoplastic composites." *Composites* 21(2):155–168.

Gibson, L.J., and Ashby, M.F. 1988. *Cellular Solids, Structure and Properties.* Pergamon Press, Oxford.

Gibson, L.J., Ashby, M.F., Zhang, J., and Triantafillou, T.C. 1989. "Failure surfaces of cellular materials under multiaxial loads. I. Modeling." *Int. J. Mechanical Sciences* 31(9):635–663.

Gibson, L.J., Ashby, M.F., Karam, G.N., Wegst, U., and Shercliff, H.R. 1995. "The mechanical properties of natural materials. II. Microstructures for mechanical efficiency." *Proc. Royal Society of London*, Series A, 450:141–162.

Gillepsie, J.W., and Carlsson, L.A. 1991. "Buckling and growth of delamination in thermoset and thermoplastic composites." *J. Eng. Materials and Technology* 113:93–98.

Gladwell, G.M.L. 1980. *Contact Problems in the Classical Theory of Elasticity.* Sijthoff and Noordhoff, Alphen aan den Rijn, Netherlands.

Goldsmith, W. 1960. *Impact: The Theory and Physical Behavior of Colliding Solids.* Edward Arnold Publishers, Ltd. London.

Goldsmith, W., and Sackman, J.L. 1991. "Energy absorption by sandwich plates: a topic in crashworthiness." In T. Khalil et al. (eds.), *Proc. of Symp. on Crashworthiness and Occupant Protection in Transportation Systems*, D-Vol. 126/BED-Vol. 19, NY:1–30,

Goldsmith, W., and Sackman, J.L. 1992. "An experimental study of energy absorption in impact on sandwich plates." *Int. J. Impact Engineering* 12(2):241–262.

Goldsmith, W., Dharan, C.H.K., and Chang, H. 1995. "Quasi-static and ballistic perforation of carbon fiber laminates." *Int. J. Solids Structures* 32(1):89–103.

Gong, J.C., and Sankar, B.V. 1991. "Impact properties of three-dimensional braided graphite/epoxy composites." *J. Composite Materials* 25(6):715–731.

Gong, S.W., Toh, S.L., and Shim, V.P.W. 1994. "The elastic response of laminated orthotropic cylindrical shells to low velocity impacts." *Composites Engineering* 4(2):247–266.

Gong, S.W., Shim, V.P.M., and Toh, S.L. 1995. "Impact response of laminated shells with orthogonal curvatures." *Composites Engineering* 5(3):257–275.

Gong, S.W., Shim, V.P.M., Toh, S.L. 1996. "Central and Noncentral normal impact on orthotropic composite cylindrical shells." *AIAA Journal* 34(8).

Gottesman, T., Bass, M., and Samuel, A. 1987. "Criticality of impact damage in composite sandwich structures." *Proc. of the 6th Int. Conf. on Comp. Mater. Combined with the 2nd European Conf. on Comp. Mater.*, London, England, July 20–24:327–335.

Gottesman, T., Grishovich, S., Drukker, E., Sela, N., Log, J. 1994. "Residual strength of impacted composites: analysis and tests." J. *Composites Technology and Research* 16(3):244–255.

Grady, J.E., and Sun, C.T. 1986. "Dynamic delamination crack propagation in a graphite/epoxy laminate." *ASTM STP 907*:5–31.

Grady, J.E., and Meyn, E. 1989. "Vibration testing of impact damaged composite laminates." *Proc. 30th AIAA/ASME/ASCE/AHS/ASC, Struct., Struct. Dyn. Mat. Conf.*, Mobile, AL, April 3–5, 1989.

Grady, J.E., Chamis, C.C., and Aiello, R.A. 1989. "Dynamic delamination buckling in composite laminates under impact loading: computational simulation." *ASTM STP 1012*:137–149.

Graves, M.J., and Koontz, J.S. 1988. "Initiation and extent of impact damage in graphite/epoxy and graphite/peek composites." *Proc. of the 29th AIAA Struct. Struct. Dyn. Mat. Conf.*, Williamsburg, VA, April 18–20:967–975.

Green, E.R. 1991a. "The effect of different impact time histories on the response of a fibre composite plate." *ASME Publication NDE* 10:9–21, Proc. of WAM, Atlanta, GA, Dec. 1–6.

Green, E.R. 1991b. "Transient impact response of a fiber composite laminate." *Acta Mechanica* 86(1–4):153–165.

Green, E.R. 1993. "Response of a fiber composite laminate to a time varying surface

line load." *J. Applied Mechanics* 60:217–221.
Green, W.A., Rogerson, G.A., and Milosaviljevic, D.I. 1992. "Transient waves in six-ply and eight-ply fibre composite plates." *Composites Science and Technology* 44:151–158.
Greszczuk, L.B. 1982. "Damage in composite materials due to low velocity impact." *Impact Dynamics*. In Z.A. Zukas et al. (eds.), J. Wiley, NY:55–94.
Griffin, C.F. 1987. "Damage tolerance of toughened resin graphite composites." *ASTM STP 937*:23–33.
Griffin, C.F., and Becht, G.J. 1991. "Fatigue behavior of impact damage BMI and thermoplastic graphite composites." *Int. SAMPE Symp. and Exhibition* 36, Part 2:2197–2209.
Guedra-Degeorges, D., Maison, S., and Renault, M. 1991. "Numerical simulation of the behavior after impact of a carbon epoxy laminated plate." *Proc. of the 8th Int. Conf. on Composite Mater. (ICCM/8)*, Honolulu, HI, July 15–19, 32.A.1–12.
Gu, Z.L., and Sun, C.T. 1987. "Prediction of impact damage regions in SMC composites." *Composite Structures* 7(3):179–190.
Guynn, E.G., and O'Brien, T.K. 1985. "The influence of lay-up and thickness on composite impact damage and compression strength." *AIAA/ASME/SAE 26th Struct, Struct. Dyn. and Mater. Conf.*, Orlando, FL:187–196.
Hamstad, M.A., Whittaker, J.W., and Brosey, W.D. 1992. "Correlation of residual strength with acoustic emission from impact-damaged composite structures." *J. Composite Materials* 26(15):2307–2328.
Hassiotis, S., and Jeong, G.D. 1995. "Identification of stiffness reductions using natural frequencies." *J. Engineering Mechanics* 121(10):1106–1113.
Heimerdinger, M.W. 1995. "Repair technology for thermoplastic aircraft structures." *AGARD Conf. Proceedings 550*, Composite repair of military Aircraft Structures: 15.1–15.12.
Hetherington, J.G., and Rajagopalan, B.P. 1991. "Investigation into the energy absorbed during ballistic perforation of composite armours." *Int. J. Impact Engineering* 11(1):33–40.
Hetherington, J.G., and Lemieux, P.F. 1994. "The effect of obliquity on the ballistic performance of two component composite armours." *Int. J. Impact Eng.* 15(2):131–137.
Hiel, C., and Ishai, O. 1992. "Design of highly damage-tolerant sandwich panels." *37th Int SAMPE Symposium and Exhibition*, Covina, CA, March 9–12, 37:1228–1242.
Hiel, C., and Ishai, O. 1992. "Effect of impact damage and open hole on compressive strength of hybrid composite skin laminates." *ASTM Symp. on Compression Response of Composite structures*, Nov. 16–17.
Hiel, C., Dittman, D., and Ishai, O. 1993. "Composite sandwich construction with syntactic foam core: A practical assessment of post-impact damage and residual strength." *Composites* 24(5):447–450.
Hillman, D.J., and Hillman, R.L. 1985. "Thermographic inspection of carbon epoxy structures." *ASTM STP 876*:481–493.
Hirschbuehler, K.R. 1985. "An improved 270°F performance interleaf system having extremely high impact resistance." *SAMPE Quarterly* 17(1):46–49.
Hirschbuehler, K.R. 1987. "A comparison of several mechanical tests used to evaluate the toughness of composites." *ASTM STP 937*:61–73.
Hitchen, S.A., and Kemp, R.M.J. 1995. "The effect of stacking sequence on impact damage in a carbon fibre/epoxy composite." *Composites* 26(3):207–214.
Ho, V. 1989. "A BMI system with high impact strength." *Proc. 34th Int. SAMPE Symp.*, Reno, Nevada, May 8–11:514–528.

Hofer, B. 1987. "Fibre optic damage detection in composite structures." *Composites* 18(4):309–316.
Hong, S., and Liu, D. 1989. "On the relationship between impact energy and delamination area." *Experimental Mechanics* 29(2):115–120.
Horban, B., and Palazotto, A. 1987. "Experimental buckling of cylindrical composite panels with eccentrically located circular delaminations." *AIAA J. Spacecraft and Rockets* 24(4):349–352.
Hsieh, C.Y., Mount, A., Jang, B.Z., and Zee, R.H. 1990. "Response of polymer composites to high and low velocity impact." *Proc. 22nd Int. SAMPE Technical Conf.*, Nov. 6–8:14–27.
Hull, D., and Shi, Y.B. 1993. "Damage mechanism characterization in composite damage tolerance investigations." *Composite Structures* 23:99–120.
Hunter, S.C. 1957. "Energy absorbed by elastic waves during impact." *J. Mech. Phys. Solids* 5:162–171.
Husman, G.E., Whitney, J.M., and Halpin, J.C. 1975. "Residual strength characterization of laminated composite subjected to impact loading." *ASTM STP 568*:92–113.
Ilcewicz, L.B., Dost, E.F., and Coggeshall, R.L. 1989. "A model for compression after impact strength evaluation." *Proc. of the 21st International SAMPE Technical Conf.*, Sept. 25–29:130–140.
Ishai, O., and Shragai, A. 1990. "Effect of impact loading on damage and residual compressive strength of CFRP laminated beams." *Composite Structures* 14:319–337.
Ishai, O., and Hiel, C.C. 1992. "Damage tolerance of composite sandwich panels with interleaved foam core." *J. Composite Technology and Research* 14(3):155–168.
Ishikawa, H., Koimai, T., Natsumura, T., and Funatogawa, O. 1993. "Experimental and analytical studies on impact damage in CFRP laminate." *Proc. 25th Int. SAMPE Technical Conf.*, Philadelphia, PA, Oct. 26–28, 1993:444–455.
Ishikawa, H., Koimai, T., Natsumura, T., and Funatogawa, O. 1993. "Experimental and analytical studies on impact damage in CFRP laminate." *Proc. 25th Int. SAMPE Technical Conf.*, Philadelphia, PA, Oct. 26–28, 1993:444–455.
Jackson, W.C., and Poe, C.C. 1993. "Use of impact force as a scale parameter for the impact response of composite laminates." *J. Composite Technology and Research* 15(4):282–289.
Jang, B., Kowbel, W., and Jang, B.Z. 1992. "Impact behavior and impact fatigue testing of polymer composites." *Composite Sci. Technology* 44(2):107–118.
Jegley, D. 1993. "Impact-damaged graphite-thermoplastic trapezoidal-corrugation sandwich and semi-sandwich panels." *J. Composite Materials* 27(5):526–538.
Jenq, S.T., and Goldsmith, W. 1988. "Effect of target bending in normal impact of a flat-ended cylindrical projectile near the ballistic limit." *Int. J. Solids and Structures* 24:1243–1266.
Jenq, S.T., Wang, S.B., and Wu, J.D. 1991. "Effect of normal and oblique impact damages on the strength degradation of composite laminates." *Proc. of the 8th Int. Conf. on Composite Mater. (ICCM/8)*, Honolulu, HI, July 15–19:32.F.1–9.
Jenq, S.T., Wang, S.B., and Sheu, L.T. 1992a. "Model for predicting the residual strength of GFRP laminates subject to ballistic impact." *J. Reinforced Plastics and Composites* 11(10):1127–1141.
Jenq, S.T., Yang, S.M., and Wu, J.D. 1992b. "An experimental study of the strength degradation of GFRP composite laminates subject to ballistic normal impact." *J. Chinese Soc. of Mechanical Engineers* 13(1):1–9.
Jenq, S.T., Jing, H.S., and Chung, C. 1994. "Predicting the ballistic limit for plain

woven glass/epoxy composite laminate." *Int. J. Impact Engineering* 15(4):451–464.
Johnson, K.L. 1985. *Contact Mechanics*. Cambridge University Press, Cambridge.
Johnson, S.R., and Sun, C.T. 1988. "Impact damage in polyimide composite laminates." *33rd SAMPE Int. Symp.*, March 7–10:1200–1209.
Johnson, W. 1972. *Impact Strength of Material*. Edward Arnold, London.
Jones, R., Broughton, W., Mousley, R.F., and Potter, R.T. 1985. "Compression failures of damaged graphite epoxy laminates." *Composite Structures* 3:167–186.
Jones, R., Paul, J., Tay, T.E., and Williams, J.F. 1988. "Assessment of the effect of impact damage in composites: some problems and answers." *Composite Structures* 10(1):51–73.
Joshi, S.P., and Sun, C.T. 1985. "Impact induced fracture in a laminated composite." *J. Comp. Mat.* 19:51–66.
Joshi, S.P., and Sun, C.T. 1987. "Impact-induced fracture in a quasi-isotropic laminate." *J. Composites Technology and Research* 9(2):40–46.
Kaczmarek, H., and Maison, S. 1994. "Comparative ultrasonic analysis of damage in CFRP under static indentation and low-velocity impact." *Composites Science and Technology* 51:11–26.
Kageyama, K., and Kimpara, I. 1991. "Delamination failures in polymer composites." *Materials Science and Engineering A: Structural Materials: Properties, Microstructures and Processing*, A143(1–2):167–174.
Kalamkarov, A.L. 1992. *Composite and Reinforced Elements of Construction*. J. Wiley, Chichester.
Kalker, J.J. 1990. *Three-Dimensional Elastic Bodies in Rolling Contact*. Kluwer Academic Publishers, Dordrecht, The Netherlands.
Kaneko, T. 1975. "On Timoshenko's correction for shear in vibrating beams." *J. Phys. D: Appl. Phys.* 8:1927–1936.
Kant, T., and Mallikarjuna. 1991. "Non-linear dynamics of laminated plates with a higher-order theory and C° finite elements." *Int. J. Nonlinear Mechanics* 26(3–4):335–343.
Kapania, R.K., and Wolfe, D.R. 1987. "Delamination buckling and growth in axially loaded beam plate." *Proc. AIAA/ASME/AHS/ASCE 28th Structures, Structural Dynamics, and Materials Conference*, Monterey, CA, Part 1:766–775.
Kapania, R.K., and Wolfe, D.R. 1989. "Buckling of axially loaded beam-plate with multiple delaminations." *J. Pressure Vessel Technology* 111:151–158.
Karas, K. 1939. "Platten unter seitchem stoss." *Ingenieur Archiv*. 10:237–250.
Kardomateas, G.A. 1989. "End fixity effects on the buckling and post buckling of delaminated composites." *Composite Science and Technology* 34:113–128.
Kardomateas, G.A., and Schmueser, D.W. 1987. "Effect of transverse shearing forces on buckling and postbuckling of delaminated composites under compressive loads." *Proc. AIAA/ASME/AHS/ASCE 28th Structures, Structural Dynamics, and Materials Conference*, Monterey, CA, Part 1:757–765.
Kassapoglou, C. 1988. "Buckling, post-buckling and failure of elliptical delaminations in laminates under compression." *Composite Structures* 9:139–159.
Kassapoglou, C., and Abbot, R. 1988. "A correlation parameter for predicting the compressive strength of composite sandwich panels after low speed impact." *Proc. 29th AIAA Structures Struct. Dyn. Mater. Conf.*:642–650.
Kassapoglou, C., Jonas, P.J., and Abbott, R. 1988. "Compressive strength of composite sandwich panels after impact damage: an experimental and analytical study." *J. Composites Technology and Research* 10(2):65–73.
Keer, L.M., and Silva, M.A.G. 1970. "Bending of a cantilever brought gradually into

contact with a cylindrical supporting surface." *Int. J. Mech. Sci.* 12:751–760.
Keer, L.M., and Ballarini, R. 1983a. "Smooth indentation of an initially stressed orthotropic beam." *AIAA J.* 21(7):1035–1042.
Keer, L.M., and Miller, G.R. 1983b. "Smooth indentation of a finite layer." *ASCE J. Eng. Mech.* 109(3):706–717.
Keer, L.M., and Miller, G.R. 1983c. "Contact between an elastically supported circular plate and a rigid indenter." *Int. J. Eng. Sci.* 21(6):681–690.
Keer, L.M., and Lee, J.C. 1985. "Dynamic impact of an elastically supported beam: large area contact."*Int. J. Eng. Sci.* 23(10):987–997.
Keer, L.M., and Schonberg, W.P. 1986a. "Smooth indentation of an isotropic cantilever beam." *Int. J. Solids Struct.* 22:87–103.
Keer, L.M., and Schonberg, W.P. 1986b. "Smooth indentation of a transversely isotropic cantilever beam." *Int. J. Solids Struct.* 22(9):1033–1053.
Kenner, V.H., Staab, G.H., and Jing, H.S. 1985. "Quantification of the tapping technique for the detection of edge defects in laminated plates." *ASTM 876*:465–480.
Kim, B.S., and Moon, F.C. 1979. "Impact induced stress waves in an anisotropic plate." *AIAA J.* 17:1126–1133.
Kim, C.G., and Jun, E.J. 1991. "Impact characteristics of composite laminated sandwich structures." *Proc. of the 8th Int. Conf. on Composite Mater. (ICCM/8)*, Honolulu, HI, July 15–19:32.G.1–8.
Kim, C.G., and Jun, E.J. 1992. "Impact resistance of composite laminated sandwich plates." *J. Composite Materials* 26(15):2247–2261.
Kim, J.K., Mackay, D.B., and Mai, Y.W. 1993. "Drop-weight impact damage tolerance of CFRP with rubber-modified epoxy matrix." *Composites* 24(6):485–494.
Kimpara, I., Kageyama, K., Susuki, T., and Ohsawa, I. 1991. "Fatigue and impact strength of aramid/glass hybrid laminates for marine use." *Proc. of the 8th Int. Conf. on Composite Mater. (ICCM/8)*, Honolulu, HI, July 15–19:5.B.1–10.
Koller, M., and Busenhart, M. 1986. "Elastic impact of spheres on thin shallow spherical shells." *Int. J. Impact Engineering* 4(1):11–21.
Kosza, P., and Sayir, M.B. 1994. "Failure patterns in the core of sandwich structures under impact loading." *Int. J. Impact Engineering* 15(4):501–517.
Krafchack, T.M., Petra, J.M., Davidson, B.D., and Chen, G.S. 1993. "Effect of impact damage on the compression fatigue behavior of composite tubes." *34th AIAA Struct. Dyn. and Materials Conf.*, La Jolla, CA, Part 2:859–866.
Krishna Murty, A.V., and Reddy, J.N. 1993. "Compressive failure of laminates and delamination buckling: A review." *The Shock and Vibration Digest* 25(3).
Kumar, P., and Narayanan, M.D. 1990. "Energy dissipation of projectile impacted panels of glass fabric reinforced composite." *Composite Structures* 15:75–90.
Kumar, P., and Rai. 1991. "Impact damage on single interface GFRP laminates. An experimental study." *Composite Structures* 18(1):1–10.
Kutlu, Z., and Chang, F.K. 1992. "Modeling compression failure of laminated composites containing multiple through the width delaminations." *J. Composite Materials* 26(3):350–389.
Lagace, P.A., and Wolf, E. 1993. "Impact damage resistance of several laminated material systems." *34th AIAA Struct. Struct. Dyn. and Materials Conf.*, La Jolla, CA, Part 4:1863–1872.
Lagace, P.A., Ryan, K.F., and Graves, M.J. 1994. "Effect of damage on the impact response of composite laminates." *AIAA J.* 32(6):1328–1330.
Lagace, P.A., Williamson, J.E., Tsang, P.H.W., Wolf, E., and Thomas, S. 1992. "The use of force as an (impact) damage resistance parameter." *Proc. of the 7th*

Technical Conf. of the Am. Soc. for Composites, Pennsylvania State University, Oct. 13–15:991–1000.
Lal, K.M. 1982. "Prediction of residual tensile strength of transversely impacted composite laminates." In J.M. Housner, and A.K. Noor (eds.), *Research in Structural and Solid Mechanics*, NASA CP 2245:97–112.
Lal, K.M. 1983a. "Low velocity transverse impact behavior of 8-ply, graphite-epoxy laminates." *J. Reinf. Plastics Composites* 2:216–225.
Lal, K.M. 1983b. "Residual strength assessment of low velocity impact damage of graphite-epoxy laminates." *J. Reinforced Plastics and Composites* 2:226–238.
Lal, K.M. 1984. "Evaluation of residual strength of composite laminates damaged by low velocity impacts." *Proc. of the 6th Int. Conf. on Fracture (ICF6)*, Pergamon Press:2933–2943.
Lam, K.Y., Liew, K.M., and Chow, S.T. 1992. "Use of two-dimensional orthogonal polynomials for vibration analysis of circular and elliptical plates." *J. Sound and Vibration* 154(2):261–269.
Lane, S.S., Moore, R.H., Groger, H.P., Gandhe, G.V., and Griffin, O.H. 1991. "Eddy current inspection of graphite/epoxy laminates." *J. Reinforced Plastics and Composites* 10(2):158–166.
Lauder, A.J., Amateau, M.F., and Queeney, R.A. 1993. "Fatigue resistance of impact damaged specimens vs. machined hole specimens." *Composites* 24(4):443–445.
Leach, D.C., and Moore, D.R. 1985. "Toughness of aromatic polymer composites reinforced with carbon fibres." *Composites Science and Technology* 23:131–161.
Leach, D.C., Curtis, D.C., and Tamblin, D.R. 1987. "Delamination behavior of carbon fiber/poly (etheretherketone) (PEEK) composites." *ASTM STP 937*:358–380.
Lee, L.J., Pai, C.K., and Shiau, L.C. 1989. "Contact behaviors of laminated composite thin shells and a rigid ball." *Proc. of Third European Conference on Composite Materials*, Bordeaux, France, March 20–23:315–320.
Lee, L.J., Huang, K.Y., and Fann, Y.J. 1991. "Dynamic responses of composite sandwich plates subjected to low velocity impact." *Proc. of the 8th Int, Conf. on Composite Materials (ICCM/8)*, Honolulu, HI, July 15–19:32.D.1–10.
Lee, L.J., Huang, K.Y., and Fann, Y.J. 1993. "Dynamic responses of composite sandwich plate impacted by a rigid ball." *J. Composite Materials* 27(13):1238–1256.
Lee, S.W.R., and Sun, C.T. 1991. "Modeling penetration process of composite laminates subjected to a blunt-ended punch." *Int. SAMPE Tech. Conf. 23*, Kiamesha Lake, NY, Oct. 21–24:624–638.
Lee, S.W.R., and Sun, C.T. 1993a. "Dynamic penetration of graphite/epoxy laminates impacted by a blunt-ended projectile." *Composite Science and Technology* 49:369–380.
Lee, S.W.R., and Sun, C.T. 1993b. "A quasi-static penetration model for composite laminates." *J. Composite Materials* 27(3):251–271.
Lekhnitskii, S.G. 1981. *Theory of Elasticity of an Anisotropic Body*. Mir Publishers, Moscow.
Lesser, A.J., and Filippov, A.G. 1991. "Kinetics of damage mechanisms in laminated composites." *Int. SAMPE Symp. and Exhibition* 36(Part 1):886–900.
Levin, K. 1986. "Effect of low-velocity impact on compressive strength of quasi-isotropic laminate." *Proc. of the American Society for Composites*, First Technical Conference, Dayton, OH:313–326.
Levin, K. 1989. "Damage tolerance of carbon fibre reinforced plastic sandwich panels." *Proc. of Third European Conference on Composite Materials*, Bordeaux, France, March 20–23, 1989:509–514.

Levinson, M. 1981a. "A new rectangular beam theory." *J. Sound and Vibration* 74(1):81–87.

Levinson, M. 1981b. "Further results on a new beam theory." *J. Sound and Vibration* 77(3):440–444.

Li, S., Reid, S.R., and Soden, P.D. 1992. "Geometrically non-linear analysis of thin filament wound laminated tubes under lateral indentation." *7th UK ABAQUS User Group Conference*, University of Cambridge, England, Sept. 17–18, 1992:246–257.

Li, S., Soden, P.D., Reid, S.R., and Hinton, M.J. 1993. "Indentation of laminated filament wound composite tubes." *Composites* 24(5):407–421.

Liao, C.L., and Tsai, J.S., 1994. "Dynamic analysis of laminated composite plates subjected to transverse impact using a partial mixed 3-D finite element." *Computers and Structures* 53(1):53–61.

Lih, S.S., and Mal, A.J. 1995. "On the accuracy of approximate plate theories for wave field calculations in composite laminates." *Wave Motion* 21:17–34.

Lin, H.J., and Lee, Y.J. 1990. "On the inelastic impact of composite laminated plate and shell structures." *Composite Structures* 14(2):89–111.

Lin, L.C., and Bhatnagar, A. 1991. "Ballistic energy absorption of composites. II. " *Int. SAMPE Tech. Conf. 23*, Kiamesha Lake, NY, Oct 21–24:669–683.

Liu, D. 1987. "Delamination in stitched and nonstitched composite plates subjected to low velocity impact." *Proc. of the American Society for Composites, 2nd Technical Conference*, Newark, DE, Sept. 23–25, Publ. by Technomic Publishing: 147–155.

Liu, D. 1988. "Impact induced delamination: A view of bending stiffness mismatching." *J. Comp. Mater.* 22:674–692.

Liu, S. 1993. "Indentation damage initiation and growth of thick-section and toughened composite materials." In C.W. Bert, V. Birman, and D. Hui (eds.), *ASME Publ. AD-Vol. 37*, AMD-Vol. 179:103–116.

Liu, D., and Malvern, L.E. 1987. "Matrix cracking in impacted glass/epoxy plates." *J. Comp. Mat.* 21:594–609.

Liu, S., and Chang, F.K. 1994. "Matrix cracking effect on delamination growth in composite laminates induced by a spherical indenter." *J. Composite Materials* 28(10):940–977.

Liu, S., Kutlu, Z. and Chang, F.K. 1993a. "Matrix cracking and delamination propagation in laminated composites subjected to transversely concentrated loading." *J. Composite Materials* 27(5):436–470.

Liu, S., Kutlu, Z., and Chang, F.K. 1993b. "Matrix cracking-induced stable and unstable delamination propagation in graphite/epoxy laminated composites due to a transversely concentrated load." In W.W. Stinchcomb, and N.E. Ashbaugh (eds.), *ASTM STP 1156*:86–101.

Liu, D., Lee, C.Y., and Lu, X. 1993c. "Repairability of impact-induced damage in SMC composites." *J. Composite Materials* 27(13):1257–1271.

Liu, D., Lillycrop, L.S., Malvern, L.E., and Sun, C.T. 1987. "The evaluation of delamination: An edge replication study." *Experimental Techniques* 11(5):20–25.

Llorente, S. 1989. "Honeycomb sandwich primary structure applications on the Boeing Model 360 Helicopter." *Proc. 34th Int. SAMPE Symp. and Exhibition*, Reno, Nevada, May 8–11, 1989, Book 1:824–838.

Llorente, S.G., and Mar, J.V.V. 1989. "The residual strength of filament wound graphite/epoxy sandwich laminates due to impact damage and environmental conditioning." Paper 89-275-CP, *Proc. 30th AIAA/ASME/ASCE/AHS/ASC Struct., Struct. Mater. Conf.*, Mobile, AL, April 3–5.

Llorente, S., Weems, D., and Fay, R. 1990. "Evaluation of advanced sandwich structures designed for improved durability and damage tolerance." *Annual Forum Proc. Helicopter Soc.* 2:825–831.

Love, A.E.H. 1929. "The stress produced in a semi-infinite solid by pressure on part of the boundary." *Phil. Trans. Roy. Soc. Lond. Series A*, 228:377–420.

Ma, C.C.M., Huang, Y.H., and Chang, M.J. 1991a. "Effect of fiber orientation and laminate thickness on the mechanical properties of PEEK/C.F. and PPS/C.F. after impact loading." *Proc. of the 8th Int. Conf. on Composite Mater. (ICCM/8)*, Honolulu, HI, July 15–19:32.H.1–9.

Ma, C.C.M., Huang, Y.H., Chang, M.J., and Ong, C.L. 1991b. "Effect of impact damage on the morphology of carbon fiber reinforced polyether ether ketone and polyphenylene sulfide composites." *Int. SAMPE Symp. and Exhibition* 36(Part 1): 872–885.

Ma, C.C.M., Tai, N.H., Wu, G.Y., Lin, S.H., Lin, J.M., Ong, C.L., Chang, Y.C., and Sheu, M.F., 1996. "Tension-compression fatigue behavior of carbon fiber reinforced epoxy quasi-isotropic laminated composites subjected to low energy impact." *Proc. 41st Int. SAMPE Symposium*, Anaheim, CA, March 24–28, 1996:462–473.

Madaras, E.I., Poe, C.C., Illg, W., and Heyman, J.S. 1986. "Estimating residual strength in filament wound casings from nondestructive evaluation of impact damage." *Proc. Review of Progress in Quantitative Nondestructive Evaluation* 6B:1221–1230.

Malvern, L.E., Sun, C.T., and Liu, D. 1987. "Damage in composite laminates from central impacts at subperforation speeds." *Recent Trends in Aeroelasticity, Structures and Structural Dynamics*, Proc. of Symp. in Memory of Prof. Bisplinghoff, P. Hajela (ed.), Univ. of Florida Press, Gainesville, FL:298–312.

Malvern, L.E., Sun, C.T., and Liu, D. 1989. "Delamination damage in central impacts at subperforation speeds on laminated kevlar/epoxy plates." *ASTM STP 1012*:387–405.

Manders, P.W., and Harris, W.C. 1986. "A parametric study of composite performance in compression-after-impact testing." *SAMPE J.* 22:47–51.

Manning, J.A., Crosky, A.G., and Bandyopadhyay, S. 1993. "Flexural and Impact properties of sandwich panels used in surfboard construction." *Proc. Int. Conf. on Advanced Composite Materials 1993*, Wollongong, Australia:123–127.

Marshall, A.P., and Bouadi, H. 1993. "Low velocity impact damage on thick-section graphite/epoxy laminated plates." *J. Reinforced Plastics and Composites* 12:1281–1294.

Marshall, R.D., Sandorff, P.E., and Lauraitis, K.N. 1988. "Buckling of a damaged sublaminate in an impacted laminate." *J. Composite Technology and Research* 10(3):107–113.

Matemilola, S.S., and Stronge, W.J. 1995. "Impact induced dynamic deformations and stresses in CFRP composite laminates." *Composites Engineering* 5(2):211–222.

Matsuhashi, H., Graves, M.J., Dugundji, J., and Lagace, P. 1993. "Effect of membrane stiffening in transient impact analysis of composite laminated plates." *34th AIAA Struct. Struct. Dyn. and Materials Conf.*, La Jolla, CA, Part 5:2668–2678.

Medick, M.A. 1961. "On classical plate theory and wave propagation." *J. Applied Mechanics 28*:223–228.

Meffert, B., and Milewski, G. 1988. "Aramid liners as armor augmentation." *Proc. of Annual Reliability and Maintainability Symposium*, IEEE:68–74.

Meyer, P.I. 1988. "Low-velocity hard-object impact of filament-wound kevlar/epoxy composite." *Composite Science and Technology* 33:279–293.

Meyer-Piening, H.R. 1989. "Remarks on higher order sandwich stress and deflection analyses." *Proc. of 1st Int. Conf. on Sandwich Constructions*, Stockholm, Sweden, June 1989:107–127.

Mines, R.A.W., Worral, C.M., and Gibson, A.G. 1994. "The static and impact behavior of polymer composite sandwich beams." *Composites* 25(2):95–110.

Minguet, P.J. 1991. "A model for predicting the behavior of impact damaged minimum gage sandwich panels under compression." *Proc. of 32nd AIAA/ASME/ASCE/AHS/ASC Struct. Struct. Dyn., Mater. Conf.*, Baltimore, MD, April 8–10:1112–1122.

Minguet, P. 1993. "Damage tolerance evaluation of new manufacturing techniques for composite helicopter drive shafts." *34th AIAA Struct. Struct. Dyn. and Materials Conf.*, La Jolla, CA, Part 2:867–876.

Mittal, R.K., and Khalili, M.R. 1994. "Analysis of impact of a moving body on an orthotropic elastic plate." *AIAA J.* 32(4):850–856.

Moon, D., and Shively, J.H. 1990. "Towards a unified method of causing impact damage in thick laminated composite." *Proc. 35th Int. SAMPE Symposium and Exhibition*, Anaheim, CA, April 2–5, 35:1466–1478.

Moon, D.G., and Kennedy, J.M. 1994. "Predicting post-impact damage growth and fatigue failures in stitched composites." *Proc. of ASC 9th Technical Conf.*, U. of Delaware, Sept. 20–22:991–998.

Morton, J. 1988. "Scaling of impact-loaded carbon-fiber composites." *AIAA J.* 26(8):989–994.

Morton, J., and Godwin, E.W. 1989. "Impact response of tough carbon fibre composites." *Composite Structures* 13:1–19.

Mousley, R.F. 1984. "Inplane compression of damaged laminates." In J. Morton (ed.), *Structural Impact and Crashworthiness*. Elsevier Appl. Sci. Publ., London: 494–509.

Mujumdar, P.M., and Suryanarayan, S. 1988. "Flexural vibrations of beams with delaminations." *J. Sound and Vibration* 125:441–461.

Murri, W.J., Sermon, B.W., and Pearson, L.H. 1985. "Ultrasonic backscatter studies of impact damage in graphite/epoxy composite laminate materials." *Proceedings of the 15th Symposium on Nondestructive Evaluation*, publ. by Southwest Research Inst., San Antonio, TX:219–233.

Naganarayana, B.P., and Atluri, S.N. 1995. "Energy release rate evaluation for delamination growth prediction in a multi-plate model of a laminate composite." *Computational Mechanics* 15(5):443–459.

Navarro, C., Martinez, M.A., Cortes, R., and Sanchez-Galvez, V. 1993. "Some observations on the normal impact on ceramic faced armours backed by composite plates." *Int. J. Impact Eng.* 13(1):145–156.

Nemes, J.A., and Simmonds, K.E. 1992. "Low velocity impact response of foam-core sandwich composites." *J. Composite Materials* 26(4):500–519.

Nettles, A.T., and Hodge, A.J. 1990. "Impact testing of glass/phenolic honeycomb panels with graphite/epoxy facesheets." *Proc. 35th Int. SAMPE Symposium and Exhibition*, Anaheim, CA, April 2–5, 35:1430–1440.

Nettles, A.T., and Hodge, A.J. 1991. "Compression-after-impact testing of thin composite materials." *Proc. 23rd Int. SAMPE Conf.*, Kiamesha Lake, NY, Oct. 21–24:177–183.

Nettles, A.T., and Lance, D.G. 1993. "On the enhancement of impact damage tolerance of composite laminates." *Composites Engineering* 3(5):383–394.

Nettles, A.T., Lance, D.G., and Hodge, A.J. 1991. "Damage tolerance comparison of 7075 T6 aluminum alloy and IM7/977-2 carbon/epoxy." *Int. SAMPE Symp. and*

Exhibition 36(Part 1):924–931.
Nuismer, R.J., and Whitney, J.M. 1975. "Uniaxial failure of composite laminates containing stress concentrations." *ASTM STP 593*:117–142.
Nuismer, R.J., and Labor, J.D. 1979. "Application of the average stress criterion: Part II – compression." *J. Composite Materials* 13:49–60.
Ochiai, S.I., Lew, K.Q., and Green, J.E. 1982. "Instrumented impact testing of structural fiber-reinforced plastic sheet materials and the simultaneous AE measurements." *J. Acoustics Emission* 1(3):191–192.
O'Kane, B., and Benham, P. 1986. "Damage threshold for low velocity impact on aircraft structural composites." *Aeron. J.* 90(899):368–372.
Olesen, K.A., Falabella, R., and Buyny, R.A. 1992. "Mechanical behavior of three generations of bismaleimide composite prepregs." *37th Int. SAMPE Symposium and Exhibition*, Covina, CA, March 9–12, 37:705–715.
Olsson, R. 1992. "Impact response of orthotropic composite plates predicted from a one-parameter differential equation." *AIAA J.* 30(6):1587–1596.
Olsson, R., and McManus, M.L. 1995. "Simplified theory for contact indentation of sandwich panels." *Proc. of 36th AIAA/ASME/ASCE/AHS/ASC Structures, Structural Dynamics and Materials Conference*, April 10–13, 1995, New Orleans, LA:1812–1820.
Ong, C.L., Sheu, M.F., and Liou, Y.Y. 1989. "The repair of thermoplastic composites after impact." *Proc. 34th Int. SAMPE Symp.*, Reno, Nevada, May 8–11, 1989:458–469.
Ong, C.L., Sheu, M.F., Liou, Y.Y., and Hsiao, T.J. 1991. "Study of the fatigue characteristics of the composite after impact." *Int. SAMPE Symp. and Exhibition* 36(Part 1):912–923.
Oplinger, D.W., and Slepetz, J.M. 1975. "Impact damage tolerance of graphite/epoxy sandwich panels." *ASTM STP 568*:30–48.
Palazotto, A., Perry, R., and Sandhu, R. 1991. "Impact response of graphite-epoxy cylindrical panels." *Proc. 32th AIAA/ASME/ASCE/AHS/ASC Structures, Structural Dynamics, and Materials Conf.*, Baltimore, MD, April 8–10:1130–1136.
Palm, T.E. 1991. "Impact resistance and residual compression strength of composite sandwich panels." *Proc. 8th Int. Conf. on Composite Mater. (ICCM/8)*, Honolulu, HI, July 15–19:3.G.1–13.
Pakdemirli, M., and Nayfeh, A.H. 1994. "Nonlinear vibrations of a beam-spring-mass system." *J. Vibration and Acoustics* 116:433–439.
Peck, S.O., and Springer, G.S. 1991. "The behavior of delaminations in composite plates: Analytical and experimental results." *J. Composite Materials* 25:907–929.
Peijs, A.A.J.M., Venderbosch, R.W., and Lemstra, P.J. 1990. "Hybrid-composites based on polyethylene and carbon fibres. Part 3: Impact resistant structural composites through damage management." *Composites* 21(6):522–530.
Peijs, A.A.J.M., Venderbosch, R.W., and de Kok, J.M.M. 1993. "Hybridization as a concept for damage management in advanced composites." *Proc. Int. Conf. on Advanced Composite Materials 1993*, Wollongong, Australia:755–762
Pelstring, R.M. 1989. "Stiching to improve damage tolerance of composites." *Proc. 34th Int. SAMPE Symp.*, Reno, Nevada, May 8–11:1519–1528.
Peroni, I., Paolozzi, A., and Bramante, A. 1989. "Diagnosis of debonding damage in a sandwich panel by the impulse technique." *Proc. 1st Int. Machinery Monitoring and Diagnostics Conf. and Exhibit*, Las Vegas, NV:796–802.
Phillips, D.C., Park, N., and Lee, R.J. 1990. "The impact behavior of high performance ceramic matrix fibre composites." *Composite Science and Technology*

37(1–3):249–265.
Pintado, P., Vogler, T.J., and Morton, J. 1991a. "Impact damage development in thick composite laminates." *Composites Engineering* 1(4):195–210.
Pintado, P., Vogler, T.J., and Morton, J. 1991b. "Impact damage tolerance of thick graphite-epoxy composite material systems." *Proc. of the 8th Int. Conf. on Composite Mater. (ICCM/8)*, Honolulu, HI, July 15–19:32.F.1–12.
Plantema, F.J. 1966. *Sandwich Construction*. Wiley, Chichester, 1966.
Poe, C.C. 1990. "Summary of a study to determine low-velocity impact damage and residual tension strength for a thick graphite/epoxy motor case." *Proc. 17th Congress ICAS*, Stockholm, Sweden, Sept. 9–14, 1:994–1004.
Poe, C.C. 1991a. "Relevance of impactor shape to nonvisible damage and residual tensile strength of a thick graphite/epoxy laminate." In T.K. O'Brien (ed.), *ASTM STP 1110*:501–527.
Poe, C.C. 1991b. "Simulated impact damage in thick graphite/epoxy laminate using spherical indenters." *J. Reinf. Plastics and Composites* 10(3):293–307.
Poe, C.C. 1992. "Impact damage and residual tension strength of a thick graphite/epoxy rocket motor case." *J. Spacecraft and Rockets* 29(3):394–404.
Poe, C.C., and Illg, W. 1986. "Hidden impact damage in thick composites." *Review of Progress in Quantitative Nondestructive Evaluation*, Bol. 5B, D.O. Thompson, and D.E. Chimenti (eds.), Plenum Press, NY:1215–1226.
Potet, P., Jeannin, P., and Bathias, C. 1987. "The use of digital image processing in vibro-thermographic detection of impact damage in composite materials." *Materials Evaluation* 45:466–470.
Potet, P., Bathias, C., and Degrigny, B. 1988. "Quantitative characterization of impact damage in composite materials: A comparison of computerized vibrothermography and x-ray tomography." *Materials Evaluation* 46:1050–1051.
Prandy, J., Boyd, J., Recker, H., and Altstadt, V. 1991. "Effect of absorbed energy on the CAI performance for composite materials." *Int. SAMPE Symp. and Exhibition* 36(Part 1):901–911.
Preuss, T.E., and Clark, G. 1988. "Use of time-of-flight C-scanning for assessment of impact damage in composites." *Composites* 19(2):145–148.
Prevorsek, D.C., and Chin, H.B. 1992. "Origins of damage tolerance in polyethylene fibers and composites." *Int. SAMPE Electronics Conf.* 24:307–318.
Prevorsek, D.C., Kwon, Y.D., Harpell, G.A., and Li, H.L. 1989. "Spectra composite armor: Dynamics of absorbing the kinetic energy of ballistic projectiles." *Proc. 34th Int. SAMPE Symp.*, Reno, Nevada, May 8–11:1780–1791.
Prevorsek, D.C., Kwon, Y.D., and Chin, H.B. 1994. "Analysis of the temperature rise in the projectile and extended chain polyethylene fiber composite armor during ballistic impact and penetration." *Polymer Engineering and Science* 34(2):141–152.
Pritchard, J.C., and Hogg, P.J. 1990. "The role of impact damage in post-impact compression testing." *Composites* 21(6):503–511.
Qian, Y., and Swanson, S.R. 1989. "Experimental measurement of impact response in carbon/epoxy plates." AIAA Paper 89-1276-CP, *Proc. 30th AIAA/ASME/ASCE/AHS/ASC, Struct., Struct. Dyn. Mater. Conf.*, Mobile, AL.
Qian, Y., and Swanson, S.R. 1990. "A comparison of solution techniques for impact response of composite plates." *Composite Structures* 14:117–192.
Ramkumar, R.L. 1983. "Effect of low-velocity impact damage on the fatigue behavior of graphite-epoxy laminates." *ASTM STP 813*:116–135.
Ramkumar, R.L., and Thakar, Y.R. 1987. "Dynamic response of curved laminated plates subjected to low velocity impact." *J. Eng. Mater. and Technol.* 109:67–71.

Ramkumar, R.L., Kulkarni, S.V., and Pipes, R.B. 1979. "Free vibration frequencies of a delaminated beam." *Proc. 34th Annual Tech. Conf.*, Reinforced Composites Institute, Soc. Plastics Industry Inc., Section 22-E:1–5.

Razi, H., and Kobayashi, A.S. 1993. "Delamination in cross-ply laminated composite subjected to low-velocity impact." *AIAA J.* 31(8):1498–1502.

Recker, H.G., Allspach, T., Alstadt, V., Folda, T., Heckman, W., Itteman, P., Tesch, H., and Weber, T. 1989. "Highly damage tolerant carbon fiber epoxy composites for primary aircraft applications." *SAMPE Quarterly* 21(1):46–51.

Reddy, J.N. 1984. "Energy and variational methods in applied mechanics." John Wiley and Sons, New York.

Rhodes, M.D. 1975. "Impact fracture of composite sandwich structures." *Proc. ASME/AIAA/SAE, 16th Struct., Struct. Dyn. and Materials Conf.*, Denver, CO.

Rhodes, M.D., Williams, J.G., and Starnes, J.H. 1981. "Low velocity impact damage in graphite-fiber reinforced epoxy laminates." *Polymer Composites* 2(1):36–44.

Rix, C., and Saczalski, T. 1991. "Damage tolerance of composite sandwich panels." *Proc. 8th Int. Conf. on Composite Mater. (ICCM/8)*, Honolulu, HI, July 15–19:3.I.1–10.

Rogerson, G.A. 1992. "Penetration of impact waves in a six-ply fibre composite laminate." *J. Sound and Vibration* 158(1):105–120.

Romeo, G., and Gaetani, G. 1990. "Effect of low velocity impact damage on the buckling behavior of composite panels." *Proc. 17th Congress ICAS*, Stockholm, Sweden, Sept. 9–14, 1:994–1004.

Rotem, A. 1988. "Residual flexural strength of FRP composite specimens subjected to transverse impact loading." *SAMPE J.* 24(2):19–25.

Rothschild, Y., Echtermeyer, A.T., and Arnesen, A. 1994. "Modeling of the non-linear material behavior of cellular sandwich foam core." *Composites* 25(2):111–117.

Russell, A.J., and Bowers, C.P. 1991. "Resin requirements for successful repair of delaminations." *Int. SAMPE Symp. and Exhibit.* 36(Part 2):2279–2290.

Salpekar, S.A. 1993. "Analysis of delamination in cross-ply laminates initiating from impact induced matrix cracking." *J. Composites Technology and Research* 15(2):88–94.

Sankar, B.V. 1987. "An approximate Green's function for beams and application to contact problems." *J. Appl. Mech.* 54:735.

Sankar, B.V. 1989a. "An integral equation for the problem of smooth indentation of orthotropic beams." *Int. J. Solids Structures* 25(3):327–337.

Sankar, B.V. 1989b. "Smooth indentation of orthotropic beams." *Composites Sci. Technol.* 34:95–111.

Sankar, B.V., and Sun, C.T. 1983. "Indentation of a beam by a rigid cylinder." *Int. J. Solids Struct.* 19(4):293–303.

Sankar, B.V., and Sun, C.T. 1986. "Low-velocity impact damage in graphite epoxy laminates subjected to tensile initial stresses." *AIAA J.* 24(3):470–472.

Saporito, J. 1989. "Sandwich structures on aerospatiale helicopters." *Proc. 34th Int SAMPE Symp.*, Reno, Nevada, May 8–11:2506–2513.

Saunders, D.S., and Van Blaricum, T.J. 1988. "Effect of load duration on the fatigue behavior of graphite/epoxy laminates containing delaminations." *Composites* 19(3):217–228.

Savrun, E., Tan, C.Y., and Robinson, J. 1991. "Ballistic performance of ceramic faced composite armors." *Int. SAMPE Tech. Conf. 23*, Kiamesha Lake, NY, Oct. 21–24:661–667.

Sayir, M.B., and Koller, M.G. 1986. "Dynamic behavior of sandwich plates." *ZAMP* 37:78–103.

Schoeppner, G.A. 1993. "Low velocity impact response of tension preloaded

composite laminates." *10th DoD/NASA/FAA Conference on Fibrous Composites in Structural Design*, Nov. 1–4, Hilton Head Island, S.C.
Schreyer, H.L., Zuo, Q.H., and Maji, A.K. 1994. "Anisotropic plasticity model for foams and honeycombs." *J. Eng. Mech.* 120(9):1913–1930.
Segal, C.L. 1991. "High performance organic fibers, fabrics, and composites for soft and hard armor applications." *Int. SAMPE Tech. Conf. 23*, Kiamesha Lake, NY, Oct. 21–24:651–657.
Seifert, G., and Palazotto, A. 1986. "The effect of a centrally located midplane delamination on the instability of composite panels." *Experimental Mechanics* 26(4):330–336.
Sharma, A.V. 1981. "Low velocity impact tests on fibrous composite sandwich structures." *ASTM STP* 734:54–70.
Shaw, D., and Tsai, H.Y. 1989. "Analysis of delamination in compressively loaded laminates." *Composite Science and Technology* 34:1–17.
Shen, M.H.H., and Grady, J.E. 1992. "Free vibrations of delaminated beams." *AIAA J.* 30:1361–1370.
Shi, G., and Tong, P. 1995. "Derivation of equivalent constitutive equations of honeycomb structures by a two scale method." *Computational Mechanics* 15(5):395–407.
Shih, W.K., and Jang, B.Z. 1989. "Instrumented impact testing of composite sandwich panels." *J. Reinf. Plastics and Composites* 8(1–3):270–298.
Shim, V.P.W., Toh, S.L., and Gong, S.W. 1996. "The elastic impact response of glass/epoxy laminated ogival shells." *Int. J. Impact Engineering* 18(6):633–655.
Shivakumar, K.N., and Whitcomb, J.D. 1985. "Buckling of a sublaminate in a quasi-isotropic laminate." *J. Composite Materials* 19:2–18.
Shivakumar, K.N., Elber, W., and Illg, W. 1985a. "Prediction of impact force and duration due to low-velocity impact on circular composite laminates." *J. Appl. Mech.* 52:674–680.
Shivakumar, K., Elber, W., and Illg, W. 1985b. "Prediction of low-velocity impact damage in thin circular laminates." *AIAA J.* 23(3):442–449.
Sierakowski, R.L. 1991. "Damage tolerance of composites." *Mechanics Computing in the 1990's and Beyond*, ASCE Eng. Mech. Spec. Conf.:1010–1014.
Sierakowski, R.L., Malvern, L.E., and Ross, C.A. 1976. "Dynamic failure modes in impacted composite plates." In T.T. Chaio (ed.), *Failure Modes in Composite III*, AIME:73–88.
Simitses, G.J., Sallam, S., and Yin, W.L. 1985. "Effect of delamination of axially loaded homogeneous laminated plates." *AIAA J.* 23(9):1437–1444.
Singh, B., and Chakraverty, S. 1994. "Use of characteristic orthogonal polynomials in two dimensions for transverse vibration of elliptic and circular plates with variable thickness." *J. Sound and Vibration* 173(3):289–299.
Sjoblom, P., and Hwang, B. 1989. "Compression-after impact: The $5,000 data point." *Proc. 34th Int. SAMPE Symp.*, Reno, Nevada, May 8–11:1411–1421.
Smith, B.T., and Yamaki, Y.R. 1990. "Assessment of impact damage in carbon-carbon plates with correlation to residual compression strength after impact." *Proc. of the IEEE 1990 Ultrasonics Symposium*, Honolulu, HI, Dec. 4–7, 2:949–952.
Spamer, C.T., and Brink, N.D. 1988. "Investigation of the compression strength after impact properties of carbon/PPS and carbon/APC-2 thermoplastic materials." *33rd SAMPE Int. Symp.* 33:284–295.
Srinivasan, K., Jackson, W.C., and Hinkley, J.A. 1991. "Response of composite materials to low velocity impact." *Int. SAMPE Symp. and Exhibition* 36(Part 2): 850–862.
Srinivasan, K., Jackson, W.C., Smith, B.T., and Hinkley, J.A. 1992. "Characterization

of damage modes in impacted thermoset and thermoplastic composites." *J. Reinf. Plastics and Composites* 11(10):111–1126.
Strait, L.H., Karasek, M.L., and Amateau, M.F. 1992. "Effects of stacking sequence on the impact resistance of carbon fiber reinforced thermoplastic toughened epoxy laminates." *J. Composite Materials* 26(12):1725–1740.
Su, K.B. 1989. "Delamination resistance of stitched thermoplastic matrix composite laminates." *Advances in Thermoplastic Matrix Composite Materials*, ASTM STP 1044:279–300.
Suemasu, H., Herth, S., and Maier, M. 1994. "Indentation of spherical head indentors on transversely isotropic composite plates." *J. Composite Materials* 28(17):1723–1739.
Sun, C.T. 1977. "An analytical method for evaluation of impact damage energy of laminated composites." *ASTM STP 617*:427–440.
Sun, C.T., and Huang, S.N. 1975. "Transverse impact problems by higher order finite elements." *Comput. Struct.* 5:297–303.
Sun, C.T., and Tan, T.M. 1984. "Wave propagation in a graphite/epoxy laminate." *J. Astronautical Sciences* 32(3):269–284.
Sun, C.T., and Chen, J.K. 1985. "On the impact of initially stressed composite laminates." *J. Composite Materials* 19(11):490–504.
Sun, C.T., and Sankar, B.V. 1985. "Smooth indentation of an initially stressed orthotropic beam." *Int. J. Solids Struct.* 21:161.
Sun, C.T., and Jih, C.J. 1987. "On strain energy release rates for interfacial cracks in bi-material media." *Engineering Fracture Mechanics* 28(1):13–20.
Sun, C.T., and Grady, J.E. 1988. "Dynamic delamination fracture toughness of a graphite/epoxy laminate under impact." *Comp. Sci. Technol.* 31:55–72.
Sun, C.T., and Rechack, S. 1988. "Effect of adhesive layers on impact damage in composite laminates." *ASTM STP 972*:97–123.
Sun, C.T., and Liou, W.J. 1989. "Investigation of laminated composite plates under impact dynamic loading using a three-dimensional hybrid stress finite element method." *Computers and Structures* 33(3):879–884.
Sun, C.T., and Manoharan, M.G. 1989. "Growth of delamination cracks due to bending in a $[90_5,0_5,90_5]$ laminate." *Composite Science and Technology* 34:365–377.
Sun, C.T., and Jih, C.J. 1990. "Mechanics of delamination in composite laminates subjected to low velocity impact." In A.K. Mal, and Y.D.S. Rajapakse (eds.), *Impact Response and Elastodynamics of Composites*, ASME Publication AMD-116:1–10.
Sun, C.T., and Wu, C.L. 1991. "Low velocity impact of composite sandwich panels." *Proc. 32nd AIAA/ASME/ASCE/AHS/ASC Struct., Struct. Dyn., Mater. Conf.*, Baltimore, MD, April 8–10:1123–1129.
Sun, C.T., and Jih, C.J. 1992. "A quasi-static treatment of delamination crack propagation in laminates subjected to low velocity impact." *Proc. of the 7th Technical Conf. of the Am. Soc. for Composites*, Pennsylvania State University, Oct. 13–15:949–961.
Sun, C.T., and Potti, S.V. 1993. "High velocity impact and penetration of composite laminates." *Proc. of ICCM-9, Ninth Int. Conf. on Composite Materials*, Madrid, Spain, July 12–16, 4:157–165.
Sun, C.T., and Potti, S.V. 1995. "Modeling dynamic penetration of thick section composite laminates." *Proc. of 36th AIAA/ASME/ASCE/AHS/ASC Structures, Structural Dynamics and Materials Conference*, April 10–13, 1995, New Orleans, LA:383–393.
Sun, C.T., Dicken, A., and Wu, H.F. 1993. "Characterization of impact damage in

ARALL laminates." *Composite Science and Technology* 49:139–144.
Swanson, S.R. 1991a. "Impact of composite cylinders." *Proc. of the 8th Int. Conf. on Composite Mater. (ICCM/8)*, Honolulu, HI, July 15–19:32.C.1–10.
Swanson, S.R. 1991b. "Impact response of fiber composite structures." *Appl. Mechanics Reviews* 44(11), Part 2:S256–S263.
Swanson, S.R. 1992. "On impact damage in fiber composite structures." *ASME Winter Annual Meeting*, Anaheim, CA, ASME Publ. AMD 142:135–144.
Swanson, S.R. 1993a. "Mechanics of transverse impact in fiber composite plates and cylinders." *J. Reinforced Plastics and Composites* 12:256–413.
Swanson, S.R. 1993b. "Interpretation of scaling of damage and failure in fiber composite laminates." *Proc. of the ASC 8th Technical Conference*, Cleveland, OH, Oct. 19–21:245–254.
Swanson, S.R., and Rezaee, H.G. 1990. "Strength loss in composites from lateral contact loads." *Composite Science and Technology* 38:43–54.
Swanson, S.R., Smith, N.L., and Qian, Y. 1991. "Analytical and experimental strain response in impact of composite cylinders." *Composite Structures* 18(2):95–108.
Swanson, S.R., Cairns, D.S., Guyll, M.E., and Johnson, D. 1993. "Compression fatigue response for carbon fiber with conventional and toughened epoxy matrices with damage." *J. Engineering Materials and Technology* 115:116–121.
Sykes, G.F., and Stoakley, D.M. 1980. "Impact penetration studies of graphite/epoxy laminates." *Proc. 12th Nat. SAMPE Tech Conf.*, Oct. 7–9:482–493.
Takamatsu, K., Kimura, J., and Tsuda, N. 1986. "Impact resistance of advanced composite structures." In K. Kawata, S. Umekawa, and A. Kobayashi (eds.), *Composites'86: Recent Advances in Japan and the United States*, Proc. Japan-U.S. CCM-III, Tokyo:77–84.
Takeda, N., Sierakowski, R.L., and Malvern, L.E. 1981a. "Studies of impacted glass fiber reinforced laminates." *SAMPE Quarterly* 12(2):9–17.
Takeda, N., Sierakowski, R.L., and Malvern, L.E. 1982a. "Microscopic observations of cross sections of impacted composite laminates." *Composite Technology Review* 4:40–44.
Tan, T.M., and Sun, C.T. 1985. "Use of static indentation laws in the impact analysis of laminated composite plates." *J. Applied Mechanics* 52:6–12.
Teh, K.T., and Morton, J. 1993. "Impact damage development and residual compression performance of advanced composite material systems." *34th AIAA Struct. Struct. Dyn. and Materials Conf.*, La Jolla, CA, Part 2:877–886.
Throne, J.L., and Progelhof, R.C. 1985. "Impact stress-strain behavior of low-density closed-cell foam." *Proc. of 43rd Annual Technical Conf. Soc. Plastics Engineers*, Washington D.C., April 29–May 2:442–446.
Tian, Z., and Swanson, S.R. 1992. "Residual tensile strength prediction on a ply-by-ply basis for laminates containing impact damage." *J. Composite Materials* 26(8):1193–1206.
Timoshenko, S.P. 1913. "Zur frage nach der wirkung eines Stosses auf einen Balken." *Z.A.M.P.* 62:198–209.
Timoshenko, S., and Woinowsky-Krieger, S. 1959. *Theory of plates and Shells*. McGraw-Hill, New York.
Timoshenko, S.P., and Goodier, J.N. 1970. *Theory of Elasticity*. McGraw-Hill, New York.
Toh, S.L., Gong, S.W., and Shim, V.P.W. 1995. "Transient stresses generated by low velocity impact on orthotropic laminated cylindrical shells." *Composite Structures* 31:213–228.
Tracy, J.J., and Pardoen, G.C. 1989. "The effect of delamination on the fundamental

and higher order buckling loads of laminated beams." *J. Composites Technology and Research* 11(3):87–93.
Tracy, J.J., Dimas, D.J., and Pardoen, G.C. 1985. "The effect of impact damage on the dynamic properties of laminated composite plates." *Proc. 5th Int. Conf. Composite Materials, ICCM-V*, San Diego, CA, July 29–Aug. 1:111–125.
Triantafillou, T.C., and Gibson, L.J. 1987. "Failure mode maps for foam core sandwich beams." *Materials Science and Engineering* 95:37–53.
Triantafillou, T.C., and Gibson, L.J. 1990. "Constitutive modeling of elastic-platic open-cell foams." *J. Engineering Mechanics* 116(12):2772–2778.
Tsai, X., and Tang, J. 1991. "Impact behavior of laminated glass fiber composites by weight dropping testing method." *Int. SAMPE Symp. and Exhibition* 36, Part 1: 1118–1127.
Tsai, Y.M. 1971. "Dynamic contact stresses produced by the impact of an axisymmetrical projectile on an elastic half-space." *Int. J. Solids Struct.* 7:543–558.
Tsang, P.H., and Dugundji, J. 1992. "Damage resistance of graphite-epoxy sandwich panels under low speed impacts." *J. of the American Helicopter Society* 37(1):75–81.
Unger, W., Ko, H., Altus, E., and Hansen, J.S. 1988. "Healing of Fibre-Reinforced Thermoplastic Structures." *Canadian Aeronautics and Space J.* 34(4):233–238.
Van Voorhees, E.J., and Green, D.J. 1991. "Failure behavior of cellular core ceramic sandwich composites." *J. Am. Ceramics. Soc.* 74(11):2747–2752.
Vasudev, A., and Mehlman, M.J. 1987. "A comparative study of the ballistic performance of glass reinforced plastic materials." *SAMPE Quart.* 18(4):43–48.
Verpoest, I., Marien, J., Devos, J., and Wevers, M. 1987. "Absorbed energy, damage and residual strength after impact of glass fiber epoxy composites." *Proc. of the 6th Int. Conf. on Comp. Mater. Combined with the 2nd European Conf. on Comp. Mater.*, London, England:3.485–3.479.
Verpoest, I., Marien, J., Devos, J., and Wevers, M. 1987. "Absorbed energy, damage and residual strength after impact of glass fiber epoxy composites." *Proc. of the 6th Int. Conf. on Comp. Mater. Combined with the 2nd European Conf. on Comp. Mater.*, London, England:3.485–3.479.
Verpoest, I., Wevers, M., Ivens, J., and De Meester, P. 1990. "3D fabrics for compression and impact resistant composite sandwich structures." *35th Int. SAMPE Symposium and Exhibition*, Anaheim, CA, April 2–5.
Vikstrom, M., Backlund, J., and Olsson, K.A. 1989. "Nondestructive testing of sandwich constructions using thermography." *Composite Structures* 13(1):49–65.
Vizzini, A.J., and Lagace, P.A. 1987. "An elastic foundation model to predict the growth of delaminations." *Proc. AIAA/ASME/AHS/ASCE 28th Structures, Structural Dynamics and Materials Conference*, Monterey, CA, Part 1:776–782.
Vinson, J.R., and Sierakowski, R.L. 1987. *The Behavior of Structures Composed of Composite Materials*. Kluwer Academic Publishers, Dordrecht.
Wang, C.Y., and Yew, C.H. 1990. "Impact damage in composite laminates." *Computers and Structures* 37(6):967–982.
Wang, H., and Khanh, T.V. 1991. "Low velocity impact damage of carbon/PEEK cross-ply laminates." *Proc. of the 8th Int. Conf. on Composite Mater. (ICCM/8)*, Honolulu, HI, July 15–19:32.E.1–10.
Wang, C.J., Jang, B.Z., Panus, J., and Valaire, B.T. 1991. "Impact behavior of hybrid-fiber and hybrid-matrix composites." *J. Reinforced Plastics and Composites* 10(4):356–378.
Wang, J.T.S., Liu, Y.Y., and Gibby, J.A. 1982. "Vibrations of split beams." *J. Sound*

and Vibration 84:491–502.

Wardle, B., and Lagace, P. 1996. "Importance of instability in the impact response of composite shells." *Proc. of* 37th *AIAA/ASME/ASCE/AHS/ASC Struct., Struct. Dyn., and Materials Conf.*, April 15–17, 1996, Salt Lake City, UT, Part III:1363–1373.

Weeks, C.A., and Sun, C.T. 1993. "Design and characterization of multi-core composite laminates." *Proc. 38th Int. SAMPE Symposium*, Anaheim, CA, May 10–13, 1993:1736–1750.

Weeks, C.A., and Sun, C.T. 1994. "Multi-core composite laminates." *J. of Advanced Materials* 25(3):28–37.

Weems, D., and Llorente, S. 1990. "Evaluation of a simplified approach to inplane shear testing for damage tolerance evaluation." *Proc. 46th Annual Forum American Helicopter Soc.*, Vol. 1:713–720.

Whitcomb, J.D. 1981. "Instability related delamination growth." *J. Composite Materials* 15:403–419.

Whitney, J.M. 1985. "Elasticity analysis of orthotropic beams under concentrated loads." *Composites Sci. Technol.* 22:167–184.

Whitney, J.M., and Pagano, N.J. 1970. "Shear deformation in heterogeneous anisotropic plates." *J. Appl. Mech.* 92:1031–1036.

Wilder, B., and Palazotto, A. 1988. "A study of damage in curved composite panels." *Eng. Const. Oper. Space*, Proc. Space 88, Publ. by ASCE, New York, NY:518–528.

Williams, J.G., and Rhodes, M.D. 1982. "Effects of resin on impact damage tolerance of graphite/epoxy laminates." *ASTM STP 787*:450–480.

Williams, J.F., Stouffer, D.C., Ilic, S., and Jones, R. 1986. "An analysis of delamination behavior." *J. Composite Structures* 5:203–216.

Williamson, J.E., and Lagace, P.A. 1993. "Response mechanisms in the impact of graphite-epoxy honeycomb sandwich panels." *Proc. of the ASC 8th Technical Conf.*, Cleveland, OH, October 19–21:287–297.

Wong, R., and Abbot, R. 1990. "Durability and damage tolerance of graphite-epoxy honeycomb structures." *Proc. of 35th Int. SAMPE Symp.*, April:366–380.

Wu, E., and Shyu, K. 1993. "Response of composite laminates to contact loads and relationship to low-velocity impact." *J. Composite Materials* 27(15):1443–1464.

Wu, E., and Liau, J. 1994. "Impact of unstitched and stitched laminates bi-line loading." *J. Composite Materials* 28(17):1641–1658.

Wu, E., and Yen, C.-S. 1994. "The contact behavior between laminated composite plates and rigid spheres." *J. Applied Mechanics* 61:60–66.

Wu, E., Chao, J.C., and Yen, C.S. 1993. "Smooth contact of orthotropic laminates by rigid cylinders." *AIAA J.* 31(10):1916–1921.

Wu, E., Sheen, H.J., Chen, Y.-C., and Chang, L.-C. 1994a. "Penetration force measurement of thin plates by laser doppler anemometry." *Experimental Mechanics* 34(2):93–99.

Wu, E., Yeh, J.C., and Yen, C.S. 1994b. "Impact on composite laminated plates: An inverse method." *Int. J. Impact Engineering* 15(4):417–434.

Wu, H.T., and Springer, G.S. 1986. "Impact damage of composites." *1st Tech. Conf. of the Am. Soc. for Composites*, Dayton, OH:346–351.

Wu, H.T., and Chang, F.K. 1989. "Transient dynamic analysis of laminated composite plates subjected to transverse impact." *Computers Structures* 31(3):453–466.

Wu, H.Y., and Springer, G.S. 1988a. "Measurements of matrix cracking and delamination caused by impact on composite plates." *J. Comp. Materials* 22:518–532.

Wu, H.Y., and Springer, G.S. 1988b. "Impact induced stresses, strains, and delaminations in composite plates." *J. Comp. Materials* 22:533–560.

Wu, H.Y., and Springer, G.S. 1988c. "Predicting impact induced delaminations in composite plates." *33rd SAMPE Int. Symp.*:352–356.

Xiong, Y., Poon, C., Straznicky, P.V., and Vietinghoff, H. 1995. "A prediction method for the compressive strength of impact damaged composite laminates." *Composite Structures* 390:357–367.

Yang, S.H., and Sun, C.T. 1982. "Indentation law for composite laminates." *ASTM STP 787*:425–449.

Yao, J.C. 1966. "An analytical and experimental study of cylindrical shells under localized impact loads." *The Aeronautical Quarterly* 17(Part 2):72–82.

Yeh, M.K., and Tan, C.M. 1994. "Buckling of elliptically delaminated composite plates." *J. Composite Materials* 28(1):36–52.

Yigit, A.S., and Christoforou, A.P. 1995. "Impact between a rigid sphere and a thin composite laminate supported by a rigid substrate." *Composite Structures* 30(2):169–177.

Yin, W.L. 1985. "Axisymmetric buckling and growth of a circular delamination in a compressed laminate." *Int. J. Solids and Structures* 21(5):503–514.

Yin, W.L. 1986. "Cylindrical buckling of laminated and delaminated plates." *Proc. AIAA/ASME/AHS/ASCE 27th Structures, Structural Dynamics, and Materials Conference*, San Antonio, TX, Part 1:159–164.

Yin, W.L. 1987. "Energy balance and the speed of crack growth in a buckled strip delamination." *Proc. AIAA/ASME/AHS/ASCE 28th Structures, Structural Dynamics, and Materials Conference*, Monterey, CA, Part 1:784–756.

Yin, W.L., and Fei, Z. 1984. "Buckling load of a circular plate with a concentric delamination." *Mechanics Research Communications* 11(5):337–344.

Yin, W.L., and Wang, J.T.S. 1984. "The energy release rate in the growth of a one-dimensional delamination." *J. Applied Mechanics* 51:939–941.

Yin, W.L., and Fei, Z. 1985. "Delamination buckling and growth in a clamped circular plate." *Proc. AIAA/ASME/AHS/ASCE 26th Structures, Structural Dynamics, and Materials Conference*, Orlando, Florida, Part 1:274–282.

Yin, W.L., and Jane, K.C. 1988. "Vibration of a delaminated beam-plate relative to buckled states." *Proc. AIAA/ASME/ASCE/AHS 29th Structures Structural Dynamics and Materials Conference*, Williamsburg, VA, April 18–20, Part II:860–870.

Yin, W.L., Sallam, S.N., and Simitses, G.J. 1986. "Ultimate axial load capacity of a delaminated beam-plate." *AIAA J.* 24(1):123–128.

Ying, L. 1983. "Role of fiber/matrix interphase in carbon fiber epoxy composite impact toughness." *SAMPE Quarterly* 14(3):26.

Yoshida, H., Ogasa, T., and Uemura, M. 1990. "Local stress distribution in the vicinity of loading points in flexural test of orthotropic beams." *Composite Materials Technology 1990*, ASME Publ. PD, 32:213–217.

Yoshida, H., Urabe, K., Ogasa, T., and Watabe, M. 1990. "Nondestructive inspections for internal defects in CFRP by using SLAM techniques." *Proc. 22nd Int. SAMPE Technical Conf.*, Nov. 6–8.

Zee, R.H., and Hsieh, C.Y. 1993. "Energy loss partitioning during ballistic impact of polymer composites." *Polymer Composites* 14(3):265–271.

Zee, R.H., Wang, C.J., Mount, W.A., Jang, B.Z., and Hsieh, C.Y. 1991. "Ballistic response of polymer composites." *Polymer Composites* 12(3):196–202.

Zener, C. 1941. "The intrinsic inelasticity of large plates." *Physical Review* 59:669–673.

Zhang, J., and Ashby, M.F. 1992. "The out-of-plane properties of honeycombs." *Int. J. Mechanical Sciences* 34(6):475–489.

Zheng, S., and Sun, C.T. 1995. "Double plate finite element model for the impact-induced delamination problem." *Composite Sci. Tech.* 53(1):111–118.

Zhou, G. 1995. "Prediction of impact damage thresholds of glass fibre reinforced laminates." *Composite Structures* 31:185–193.

Zhou, G., and Davies, G.A.O. 1995. "Impact response of thick glass fibre reinforced polyester laminates." *Int. J. Impact Engineering* 16(3):357–374.

Zhu, G., Goldsmith, W., and Dharan, C.K.H. 1992a. "Penetration of laminated Kevlar by projectiles: Analytical model." *Int. Solids Structures* 29(4):421–436.

Zhu, G., Goldsmith, W., and Dharan, C.K.H. 1992b. "Penetration of laminated Kevlar by projectiles: experimental investigation." *Int. Solids Structures* 29(4):399–420.

Index